# Experimental Pulse NMR
## A Nuts and Bolts Approach

# Experimental Pulse NMR
## A Nuts and Bolts Approach

## Eiichi Fukushima
*University of California*
*Los Alamos National Laboratory*

## Stephen B.W. Roeder
*Department of Chemistry and Department of Physics*
*San Diego State University*

1981

**Addison-Wesley Publishing Company, Inc.**
Advanced Book Program
Reading, Massachusetts

London·Amsterdam·Don Mills, Ontario·Sydney·Tokyo

BCDEFGHIJ-MA-8987654

CONTENTS

This book is about pulse nuclear magnetic resonance (NMR), with its techniques, the information to be obtained, and practical advice on performing experiments. The emphasis is on the motivation and physical ideas underlying NMR experiments and the actual techniques, including the hardware used. The level is generally suitable for those to whom pulse NMR is a new technique, be they students in chemistry or physics on the one hand and research workers in biology, geology, or agriculture, on the other. The book can be used for a senior or first year graduate course where it could supplement the standard NMR texts.

The contents of this book are not overly scholarly, mathematical, nor complete but, at the same time, includes subjects which are seldom discussed in NMR books. We also attempt to provide some physical insight into subjects whose canonical treatments may be difficult to understand. At the same time, an effort was made to avoid discussions which can easily be found elsewhere. There are two recent books on experimental NMR by Shaw and by Martin, Delpuech, and Martin, both referenced in Appendix A, which are largely complementary to ours.

Unlike many other books, ours does not build the subject systematically from first principles. Rather, it discusses concepts at levels most useful in interfacing to an actual experiment or to the specialized literature. In this way this book is more of a guidebook and a handbook. When someone is in unfamiliar terrain, he consults a guidebook to gain information about the immediate countryside without having to learn the geography of the entire country. Similarly, a handbook is referred to for brief information on individual items. This book is a collection of essays to facilitate its use in these modes. We have cited references to help you look up additional information rather than for giving credit where credit is due and they are grouped at the end of each sub-section.

You can see that our book covers a lot of ground. In practical terms, this means that the book is not one to be read in an orderly way from beginning to end, except possibly by reviewers. Instead, different parts of the book should appeal to different groups of people. A unique feature of this book is the discussion of hardware in Chapter V. In it, we cover how the hardware works and how they may be modi-fied as well as what to look for in commercial instruments. We are trying to continue in the tradition of many earlier workers to impart the message that what happens inside the spectrometer "box" is not as mysterious as it might seem.

In addition to the practical necessity for understanding the experimental technique, such an understanding will give added insight into the experimental results. Thus, the knowl-edge of the experimental methods gives insight into the "why" as well as the "how".

Above all, NMR experiments are great fun, at least to many of us. So many different kinds of information can be gathered by performing the experiments in different ways. We hope that our effort will introduce you to something new and enjoyable and we welcome any feedback.

One of us (EF) has benefited immeasurably from working and talking with Dr. Atholl A. V. Gibson of Texas A&M University. Several of the insights and designs in this book are his, not always so identified. He has also read more than one-half of the manuscript, as has Professor Irving J. Lowe of Pittsburgh. Dr. Robert E. London of Los Alamos has also been very helpful, especially in the area of high resolution solution NMR. There are many other people who contributed ideas to the book as well as tolerated all manner of questions on their ideas, published or otherwise. To all these kind people, we are very grateful for their interest and valuable suggestions. The above contribution notwithstanding, all the errors in the book are the sole responsibility of the authors and we would like to be informed of them at the readers' convenience.

One of us (EF) wrote a part of the manuscript at the University of Florida while on leave from Los Alamos and acknowledge the kind hospitality provided by the university and the solid state physics group of the late Professor Thomas A. Scott.

Eiichi Fukushima
Stephen B.W. Roeder

# Experimental Pulse NMR

A Nuts and Bolts Approach

CHAPTER I

## BASICS OF PULSE AND FOURIER TRANSFORM NMR

## I.A.  INTRODUCTION

Nuclear magnetic resonance (NMR) can mean different things to different people. To someone interested in studying slow molecular diffusion, it might mean measuring relaxation times of one kind or another as a function of temperature and possibly pressure. To someone interested in magnetism of metals, it might mean sweeping through a large frequency range (like tens of MHz, where MHz = megahertz = million cycles per second) and monitoring the signal height (obtained by spin echoes). To the majority of those coming into contact with NMR, it usually means running some kind of a high resolution spectrum and in the past decade this has come to mean obtaining the spectrum by Fourier transformation of a time varying signal called the free induction decay (FID). All these different kinds of NMR experiments are related and

Eiichi Fukushima and Stephen B. W. Roeder, Experimental Pulse NMR: A Nuts and Bolts Approach

we propose to explain some of the considerations necessary in either understanding or in carrying out such experiments.

The Fourier transformation relates representations in two complementary domains. Specifically, the transform as used in NMR relates the spectrum in the frequency domain, as we know it from cw (continuous wave) NMR, to the signal obtained in the time domain called the free induction decay (FID). Such a relation was first demonstrated by Lowe and Norberg (1957). In order to make this correspondence more plausible, consider the following example. If we want to find the frequency response of a high fidelity amplifier, i.e., what notes (frequencies) it amplifies and by how much, one way of doing this is to feed in a pure tone of constant volume (amplitude) and measure the output volume as the tone is changed. In that way, we might learn that the amplifier does not pass very low tones (say, below 100 Hz or cycles per second) nor very high tones (say, above 20 kHz) but it passes everything else in between with quite constant amplification (gain). The plot of the response vs. frequency is called a spectrum and is entirely analogous to the NMR spectrum.

This is a good method for accomplishing our goal and we would be perfectly happy with it as long as we did not have to measure the frequency response of a large number of amplifiers. If we were the impatient sort, however, we could think of several faster methods. One would be to get several tone generators (frequency sources) which put out different tones and use them with tuned detectors which can separately but simultaneously detect those tones. You could perform the required measurement all at once but it is simply not practical to assemble the required number of separate tone generators and detectors if any kind of frequency resolution were required.

This example, with its impracticality, contains the seed for the best solution to the problem. We somehow want to put all the frequencies into the amplifier at once and detect all the frequencies which come out. We state here without proof that a short square pulse contains a continuous distribution of frequencies up to frequencies of the order of the reciprocal of the pulse length in order to shape the sharp corners of the pulse. Thus, for such a pulse, the amplifier sees many frequencies coming into it and will amplify them according to its characteristics. For example, if the amplifier does not respond to the high frequencies needed to shape the sharp corners, the corners will be rounded off in the pulse coming out of the amplifier. If some method were available to decompose the output pulse into its frequency components so they can be plotted out as a spectrum and this is compared with the spectrum of the input pulse, we will have accomplished our goal. The Fourier transform performs the desired decomposition.

To recapitulate, the frequency response of the amplifier can be obtained by measuring the gain at each frequency or alternatively by measuring the response of the output of the amplifier to a rectangular input pulse and a subsequent Fourier transform. (Clearly, there are other pulse shapes which contain many different frequency components, too, but the square pulse is easy to form and its frequency distribution is well known.) The efficiency of the latter method is due to the multiplexing effect, i.e., the fact that we are not taking the data at one frequency at a time and not due to the Fourier transform itself, and this fact was first demonstrated in NMR by Ernst and Anderson (1966). All the spectral information desired is already contained in the transient response to a pulse and the transform merely allows us to decompose the signal into its spectral components.

The above considerations are common to many phenomena and it turns out that an NMR spectrum of a sample is analogous to the frequency response of the amplifier. The difference is that in NMR, we are dealing with the nuclei responding to radio frequencies (rf) in the MHz range. The inefficient solution in the above example corresponds to a cw NMR experiment where each part of the spectrum is measured sequentially whereas the final solution corresponds to Fourier transform NMR.

Having said that the Fourier transform process enables us to transform information in the time domain into equivalent information in the frequency domain and vice versa, we will now consider the NMR experiment in the time domain. NMR is possible because nuclei of many atoms possess magnetic moments and angular momenta. (Electrons possess moments about 1000 times larger than the largest nuclear moment.) Now a magnetic moment interacts with a static magnetic field in such a way that the field tries to force the moment to line up along it just like a compass needle lines up with the earth's field. The significance of the angular momentum, which is proportional to the moment, is that it makes the nuclei precess around the magnetic field when it experiences the torque due to the field acting on the moment. The result is that the nuclei precess about the field rather than oscillating in a plane like a compass needle. This precession is exactly analogous to a top, with its angular momentum along its spinning axis, precessing about the earth's gravitational field. The precession frequency of the moment is proportional to and uniquely determined by the gyromagnetic ratio $\gamma$ and the strength of the magnetic field $H_0$. It is given by the Larmor relation $\omega_0 = \gamma H_0$ where $\omega_0$ is called the Larmor speed

which is $2\pi$ times the Larmor frequency $\nu_0$. $\gamma$ is the proportionality constant between the moment $\mu$ and the angular momentum I$\hbar$, so that $\mu = \gamma I \hbar$. Therefore, in a given magnetic
field, the precession frequency is different for every distinct
nucleus because each has a uniquely defined $\gamma$. For typical
nuclei used in NMR, the Larmor frequency falls in the range
of a few to a few hundred MHz in magnetic fields of commonly
available magnets.

Now consider the nuclear moment precessing about a
static magnetic field. Because there is no friction, it will
precess forever while maintaining a fixed angle with respect
to the magnetic field if left alone. Quantum mechanics requires that the orientation of a magnetic moment with respect
to the field be quantized. The number of allowed orientation
is 2I+1 where I is the nuclear spin quantum number. For
example, protons, carbon-13, and fluorine-19, being spin-½
nuclei, are restricted to two orientations. The two allowed
orientations for these nuclei are not collinear with the magnetic field but rather at some angle so that the nuclear moment
can be thought of as precessing about the field on the surface
of a cone with its axis either parallel or anti-parallel to the
field. This quantum mechanical representation is not essential
to our presentation and we will assume for the sake of these
discussions that the permitted orientations are parallel and
anti-parallel to the field.

Classically, the energy of a magnetic moment $\overline{\mu}$ in a mag-
netic field $\overline{H}_0$ is $-\overline{\mu} \cdot \overline{H}_0$ so that a positive moment will always
want to line up parallel with the field. Such a moment already
parallel to the field, i.e., in a lower energy state, can undergo a transition to the higher energy state by receiving an
appropriate amount of energy. By Planck's law, the quantum
of work required to reverse the orientation of a moment is $\hbar\omega$

where $\hbar = h/2\pi$ is Planck's constant divided by $2\pi$ and $\omega$ is the angular velocity of the applied rf field. We will develop a formalism in the next section whereby the frequency of the rf irradiation needed to cause such a transition is equal to the Larmor velocity $\omega_0$. If the frequency of the radiation applied to induce transitions is too high or too low, the nucleus will not undergo the transition; it is this fact of requiring exactly the right frequency to do something which makes NMR very useful and provides the word "resonance" in its name. The fact that the frequency required for causing transitions is precisely the Larmor frequency of precession is no coincidence. We will see that the applied rf radiation provides a rotating magnetic field component and when it is rotating at the same angular velocity and in the same sense as the nuclei, it applies torque to the nucleus causing its angle with the static field to change.

Let us now consider a macroscopic NMR sample, such as 1 gm of water containing roughly $10^{23}$ protons in a static magnetic field. All protons have identical positive moments and they would all line up with the field, i.e., be in the lower energy state, if it were not for the thermal motion of the molecules and atoms counteracting the effect of the field. The population ratio of the two states depends on the sample's absolute temperature T and the strength of the magnetic field H by the Boltzmann factor $\exp(-\mu H/kT)$ with k being the Boltzmann constant. For NMR experiments at room temperature, this factor differs from unity by only $10^{-5}$ or $10^{-6}$ so that most of the protons are randomly distributed (parallel and anti-parallel to the field) and their effects cancel. Even so, $10^{17}$ to $10^{18}$ protons do line up and provide a macroscopic magnetization $\overline{M}$ whose precession can be detected under certain conditions. In this book, we will mainly deal with the behavior of such a macroscopic magnetization vector both as we forcibly rotate it by applying an external rf magnetic field and as it precesses and relaxes, that is, changes direction and magnitude towards an equilibrium state, in the absence of an external rf field in the laboratory magnetic field.

We have now established that at equilibrium there exists a macroscopic magnetization vector parallel to the applied field. Although it results from $10^{23}$ nuclei whose individual orientations with respect to the static magnetic field are quantized, it may assume any orientation because it is a sum of so many contributions. The macroscopic magnetization obeys the Larmor relation and precesses about any static magnetic field with which it is not collinear at a rate proportional to the field strength and the individual nuclear moment. The precession occurs with a constant cone angle in the absence of external work or friction. If somehow the magnetization is rotated away from the field, it can relax back to thermal equilibrium, i.e., towards the field, only by giving up quanta of energy to the surroundings, like the kinetic energy of a molecule, in multiples of $h\nu=\hbar\omega$. The surroundings which can absorb this energy from the nuclei are collectively called "the lattice," a term due to some of the earliest workers in the field who were solid state physicists.

If we apply a magnetic field rotating at the Larmor frequency in the plane perpendicular to the static field, we can cause the individual nuclear moments to flip with the result that the entire macroscopic magnetization may be turned towards or away from the static field. In order to understand this and other more complicated operations, we now discuss the rotating frame representation of pulse NMR. The major problem in trying to visualize the motion of the magnetization in a static magnetic field is the fact that all motions towards and away from the static field direction are superimposed on the rapid Larmor precession itself. The effect of the static field can be cancelled by working in a coordinate system rotating about the static field at the applied rf.

Let us designate the laboratory Cartesian axes as x, y, z and the rotating frame axes as x', y', z'. We might visualize the laboratory frame as the three axes, x, y, z, chalked in the corner of the laboratory with the z direction up and the x and y directions on the floor. If we have a phonograph turntable in the laboratory, consider the spindle to be the z' axis and two mutually perpendicular lines on the turntable be the x' and y' axes. As the turntable rotates, the z and z' axes continue to point in the same direction, but the x' and y' axes rotate with respect to the x and y axes. If there is a magnetization rotating at the Larmor frequency about the z axis as seen in the laboratory frame, and we now climb on the turntable rotating at the Larmor frequency, the mag-

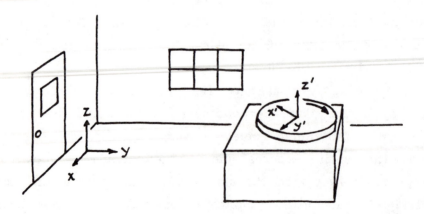

netization will appear to be stationary. The equations, describing the transformation from the laboratory frame to a frame rotating at the Larmor frequency, cause the static field to disappear from the equations of motion of the magnetization. This may seem surprising, but we can think of the static field now as imparting the Larmor precession to the rotating frame within which no static field effects exist. Of course, a rotating magnetic field on resonance will appear stationary in the rotating frame. For these reasons, the transformation to the rotating frame simplifies the understanding of NMR experiments.

Let us make this concrete by seeing how we would do a simple pulse NMR experiment. We place a sample and a coil in the field of a laboratory magnet in such a way that the rf magnetic field generated in the coil is perpendicular to the static field.

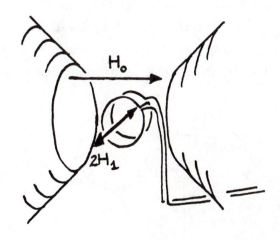

After the magnet has polarized the magnetic moments, there will be a magnetization pointing along the laboratory field. In the laboratory frame, it stays in this orientation indefinitely because a field collinear with the magnetization exerts no torque. In the rotating frame, it is stationary because there are no fields at all. When we turn on the transmitter, the sinusoidal rf current in the transmitter coil generates a linear-ly polarized magnetic field $2H_1\cos\omega t$.

This linear motion can be decomposed into two counter rotating vectors of magnitude $H_1$ as shown in the sketch.

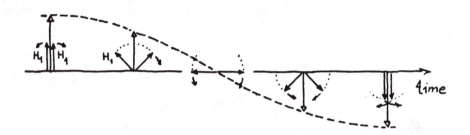

In a frame rotating at $\omega$, one of these rotating fields will have the correct frequency and sense of precession to be station-ary. (The other component can be neglected because it ro-tates at $2\omega$ in the rotating frame and will have little interaction with the magnetization.) By the Larmor theorem, the magne-tization $\overline{M}$ will rotate in the rotating frame about this $H_1$ field

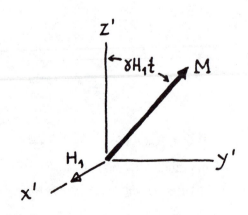

at angular speed $\overline{\omega}_1 = \gamma \overline{H}_1$ and since $\overline{H}_1$ is perpendicular to the laboratory field $\overline{H}_0$, $\overline{M}$ will change its orientation with respect to the static field because the rotating field will do work on the magnetization. Therefore, an appropriate combination of the intensity and the duration of the rotating field can rotate the magnetization by any desired amount. The usual rotating field intensity is about 1000 times smaller than the laboratory field intensity so the precession about the rotating field takes place at tens and hundreds of kHz instead of the tens and hundreds of MHz of the laboratory frame Larmor frequency. Therefore, the rotating field $\overline{H}_1$ required to turn a proton magnetization by 90° (called a 90 degree or a $\pi/2$ pulse) usually lasts between 1 and 50 μs in a modern pulse NMR experiment. Such a magnetization turned 90° from $\overline{H}_0$ is stationary in the rotating frame after $\overline{H}_1$ has been turned off but is precessing in a plane perpendicular to the static magnetic field in the laboratory frame at the Larmor frequency.

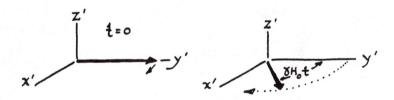

After the rf field $\overline{H}_1$ is turned off, the magnetization will precess freely. In an actual experiment, the precessing magnetization decays owing to various mechanisms which usually are the objects of an NMR study. In a coil with its axis perpendicular to the laboratory field, this decaying magnetization will induce an rf current at the Larmor frequency in complete analogy with an electrical generator. (This coil could be the same one that was used to generate the $H_1$ field or it could be a separate receiver coil.) The signal induced in the coil is a free precession signal and, owing to its decay, is called a free induction decay (FID).

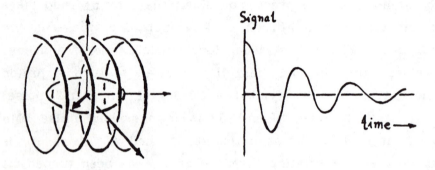

Unfortunately, NMR is a reasonably insensitive experiment. It ordinarily requires a sample containing about $10^{21}$ to $10^{23}$ spins which is six orders of magnitude worse than EPR (electron paramagnetic resonance). In considering a "new" nucleus for NMR experiments, one of the crucial parameters to consider is the sensitivity, because if it is a factor of 100 or more worse than the most sensitive nucleus (protons), we may have doubts about doing the experiment at all, at least at room temperature. Of course, it is possible to design and build amplifiers to amplify any small signal to any desired level provided that the information content is not masked by random disturbance called noise. So the real problem is not so much the small signal in NMR but the small signal-to-noise ratio (S/N). The S/N depends on the design of the electronics as well as the kind and the number of nuclear spins and these considerations will be discussed in Chapters V and VI.

The sources of noise are many. Normal laboratory surroundings generate a variety of noise. For example, fluorescent lights generate noise which can be picked up in the spectrometer. Also, other electronic equipment can generate noise, even in normal operation and especially upon turn-on. An ironic but common source is the minicomputer which is used in order to improve S/N.

Depending on the noise source, the frequency spectrum of the noise can be quite different. Some sources emit specific frequency components, for example 60 or 120 Hz, while others emit noise with frequencies distributed quite uniformly throughout some range. The minicomputer, cited as an example above, emits noise which is quite specific, in fact, to the operation it is performing at the particular time. In any case, the noise components at frequencies far from that of interest, namely the Larmor frequency, can be removed by suitable filtering as discussed in VI.C. The noise components with frequencies coincident with the Larmor frequency, however, cannot be distinguished from the desired signal on the basis of frequency. In this case, the only recourse is to "signal average" which means to repeat the experiment and sum the results so that the signal, which is not random, will accumulate faster than the random noise which, due to its random character, will accumulate more slowly. This process is also known as "CATing," which comes from the acronym for computer for average transients, "co-adding" which literally means adding together, and several other descriptions no more appropriate than "signal averaging."

The accumulation of the non-random signal is clearly proportional to the number of experiments. On the other hand, the noise accumulates at a rate proportional to the square root of the number of experiments, a fact we are stating without proof here. Therefore, S/N increases with the number of experiments n as $n/\sqrt{n} = \sqrt{n}$. Since the number of experiments n is proportional to the time t required to take data, S/N increases at $\sqrt{t}$. This means that the duration of the experiment must be quadrupled in order to double the S/N so that long time signal averaging becomes futile after a certain point. Generally, it is a losing proposition to signal average for much more than a day since there may be other experiments which can be performed in that time or the expected growth in the S/N may not materialize because of instrumental instabilities over a long period. Further discussion of noise and how to cope with it will appear in Chapter VI.

## REFERENCES

R. R. Ernst and W. A. Anderson, "Application of Fourier transform spectroscopy to magnetic resonance," Rev. Sci. Instrum. 37, 93-102 (1966).

I. J. Lowe and R. E. Norberg, "Free-induction decays in solids", Phys. Rev. 107, 46-61 (1957).

## I.B.  THE ROTATING FRAME

After the introductory remarks of the last section, we now dive into the rotating frame concept and consider how various magnetic fields and the magnetization behave in that frame.

The Larmor theorem, mentioned in the last section, states that a magnetic moment $\overline{M}$ in a magnetic field $\overline{H}_0$ obeys the equation $d\overline{M}/dt = \overline{M} \times \gamma \overline{H}_0$ where $\gamma$ is the gyromagnetic ratio defined as the moment divided by the angular momentum. This vector equation states that the rate of change of the magnetization vector for a given nucleus having a specific $\gamma$, is proportional to both the magnetization and the field, and the direction of the change is at right angles to both the magnetization and the field. You can try drawing diagrams or use your fingers to convince yourself that the result is a precession of $\overline{M}$ about $\overline{H}_0$ at an angular velocity $\gamma \overline{H}_0$ and without any change in the length of $\overline{M}$. The angular velocity $\gamma \overline{H}_0$ corresponds to the Larmor frequency $\gamma H_0/2\pi$ and, in the case of a nucleus with a particular $\gamma$ in a laboratory field $\overline{H}_0$, is the "NMR frequency".

We now state without proof that the time derivative $\delta \overline{M}/\delta t$ of a vector $\overline{M}$ as viewed in the rotating frame (analogous to a turntable) with angular velocity $\overline{\Omega}$ is $\delta \overline{M}/\delta t = d\overline{M}/dt - (\overline{\Omega} \times \overline{M})$ where the derivative $d\overline{M}/dt$ is that defined in the stationary reference frame. Using this fact and the Larmor theorem, we see that the time derivative of $\overline{M}$ viewed from the reference frame rotating at $\overline{\Omega}$ is the same as that in the stationary reference frame except for a substitution of a new field $\overline{H}_e$, i.e., $\delta \overline{M}/\delta t = \overline{M} \times \gamma(\overline{H}_0 + \overline{\Omega}/\gamma) = \overline{M} \times \gamma \overline{H}_e$ with the effective field $\overline{H}_e = \overline{H}_0 + \overline{\Omega}/\gamma$. The term $\overline{\Omega}/\gamma$ represents a fictitious magnetic field component in the rotating frame to account for the altered and, in fact, simplified behavior of the magnetization in that frame. If the rotating frame angular velocity $\overline{\Omega}$ is chosen to be equal to the Larmor velocity $-\gamma \overline{H}_0$, the effective field for that particular nucleus vanishes and the magnetization is stationary in the rotating frame. This simply means that the magnetization is precessing about the laboratory field at frequency $-\gamma H/2\pi$ which we already knew.

Suppose that we now introduce a magnetic field $H_1(t)$ rotating at angular velocity $\Omega$ near the Larmor frequency with the rotation axis collinear with $\overline{H_0}$ in the stationary frame. We do this by creating a linearly oscillating magnetic field which is, in fact, a sum of two oppositely rotating magnetic fields. The rotating frame is "locked" to one of these rotating fields in which case the other rotating field rotates at twice the frequency in the rotating frame and this is too high a frequency to affect anything. Let us arbitrarily choose the phase relationship between the rotating field and the rotating frame so that this field lies fixed along the rotating x' axis. Then the total effective field in the rotating frame is $\overline{H}_e = (H_0+\Omega/\gamma)\overline{k} + H_1\overline{i}$ where $\overline{i}$ and $\overline{k}$ are unit vectors along the rotating x' and z' axes and $H_1$ is the magnitude of $H_1(t)$. Let us rewrite the Larmor speed $-\gamma H_0$ as $\omega$ in which case the effective field becomes $\overline{H}_e = [(\Omega-\omega)/\gamma]\overline{k} + H_1\overline{i}$. This relationship can be sketched as shown and gives us a very physical picture of the nuclear resonance phenomenon. In particular, when the static field H has the value $H_0$, the Larmor speed $\omega$ of the magnetization will equal $\Omega$, and $\overline{H}_e=H_1\overline{i}$. Therefore, any magnetization which starts out along z' precesses in the y'-z'

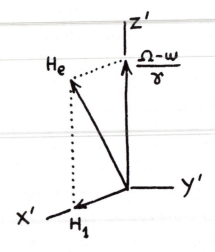

plane. Since the angle between the magnetization and H is changing, the rotating field $H_1(t)$ is doing work on the spins. This is the magnetic resonance condition.

The Larmor relation we use, $\omega_0=\gamma H_0$, is a scalar relation which comes from the vector equation $\overline{\omega}_0 = \gamma \overline{H}_0$. In this book, we will usually consider only the scalar relation so that the field along z' in the rotating frame will be written as $H_0 - \Omega/\gamma$.

Although the usefulness of the rotating frame concept will become more obvious in the following sections where we discuss the various NMR experiments, we will give one instructive example here. Consider an experiment called the adiabatic fast passage which is an instructive example even though it is not a pulse experiment. We start with a sample in a static magnetic field and with a rotating magnetic field $H_1(t)$. If the static field is different enough from $\Omega/\gamma$ so that the magnetization is way off resonance, i.e., $|(\Omega-\omega)/\gamma| \gg H_1$, then $\overline{H}_e$ is nearly collinear with the z' axis in the rotating frame and the magnetization $\overline{M}$ will precess around it at angular velocity $\gamma \overline{H}_e$. Now as the

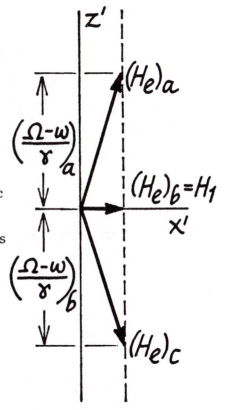

static field intensity is changed so the system is closer to resonance, $|(\Omega-\omega)/\gamma|$ decreases and the effective field $\overline{H}_e$ rotates away from the z' axis towards the x' axis (and gets shorter) and carries $\overline{M}$ along with it. At resonance when $\Omega-\omega=0$, $\overline{H}_e=\overline{H}_1$ and the effective field is perpendicular to the z' axis. As the field continues to change past the resonance condition, $\overline{H}_e$ continues to rotate and gets longer again so that when the system is again very far from resonance but in the opposite direction, the orientation of $\overline{M}$ has been reversed.

In an adiabatic fast passage experiment, the sweep rate is chosen so that no appreciable relaxation takes place during the sweep (the "fast" condition), and the nuclear precession around $\overline{H}_e$ is always rapid compared with the rotation of $\overline{H}_e$ (the "adiabatic" condition). The first condition means that $\overline{M}$ remains constant in length during the sweep while the second condition ensures that whatever initial relationship the magtization had with respect to $\overline{H}_e$ remains throughout the sweep. In particular, if the nuclei were in thermal equilibrium with

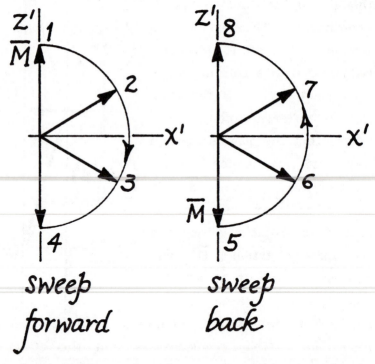

the lattice to begin with so that the macroscopic magnetization pointed along the static field (which coincides with $\overline{H}_e$ far from resonance), it will follow $\overline{H}_e$ during the passage and will end up 180 degrees rotated by the passage. In the laboratory frame, this motion is superimposed on the much more rapid precession about the static field. A detector coil in a plane perpendicular to the static field, for example, along the x' axis, will pick up a signal at the Larmor frequency (or the NMR frequency) whose amplitude increases from zero to a maximum and decreases back to zero as the resonance is traversed.

In the absence of relaxation, when the process is reversed the signal is retraced and since it was symmetric about the resonance condition it will be identical to the original signal. If the magnetization relaxes fully before the return sweep, so that it is pointing in the original direction and antiparallel to $\overline{H}_e$, the return sweep will carry the magnetization around the back side of the coordinate frame (past -x') and the signal will be upside down if a detector is used which can distinguish between the magnetization along the x' and -x' axes. Such a detector is called a phase sensitive detector.

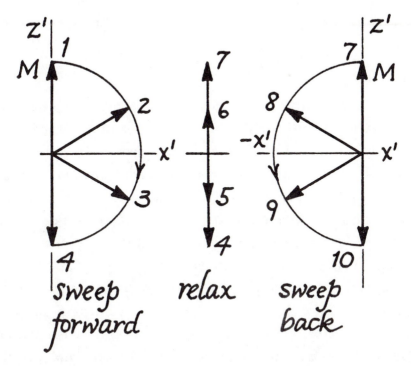

There are many reasons for detecting more than just the

amplitude of the magnetization as it precesses around the axes
not the least of which is the fact that knowing exactly where
in the rotating frame the magnetization is at a given time
(rather than just an amplitude away from the z' axis, for
example) represents more information.  A phase sensitive
detector, described in Chapter V, is used to gain information
about projections of the rotating magnetization along any axis
in the x'-y' rotating plane.  With phase sensitive detection, the
return sweep in the fully relaxed case will yield an inverted
signal so that there will be a special delay time $t_{\frac{1}{2}}$ between the
two sweeps for which the magnetization is a null.  This will
occur when $t_{\frac{1}{2}} = T_1 \ell n2$, where $T_1$ is the relaxation time of the
magnetization towards the positive z axis.

## I.C.      PULSE NMR

The first NMR experiments were the cw (continuous
wave) experiments by Bloch, Purcell, and their respective
colleagues.  The first pulse experiments followed shortly
thereafter and Hahn published his now famous spin echo
paper in 1950.  For the following decade and a half, pulse
NMR was used mostly for determinations of relaxation times $T_1$
and $T_2$.  In the meantime Lowe and Norberg proved that an
NMR frequency spectrum is related to an FID by Fourier
transformation and Cooley and Tukey rediscovered an efficient
Fourier transform algorithm.  Ernst and Anderson recognized
the potential for great efficiency in using Fourier transform
pulse NMR for multiline high resolution spectra and performed
the first successful demonstration.  More recently, line narrow-

ing experiments with solid samples have become possible due to the pioneering efforts of Hahn, Mansfield, Waugh, and their colleagues, and the overwhelming majority of these utilize pulse techniques. In the next two sections, we introduce you to the pulse NMR experiments without getting into the hardware or even the Fourier transformation process.

## I.C.1. INTRODUCTION TO PULSE NMR

In pulse nuclear magnetic resonance (NMR) experiments, an intense rf pulse in the laboratory frame, with an amplitude $H_1$ larger than that used in cw experiments, is applied to the sample. At resonance, $\overline{H}_e = \overline{H}_1$, say along the rotating x' axis, and the magnetization will rotate about $\overline{H}_1$ in the rotating frame y'-z' plane with an angular velocity $\gamma\overline{H}_1$. If such an $\overline{H}_1$ field is turned on at resonance for a time $\tau$, the magnetization vector will be rotated by an angle $\gamma H_1 \tau$. An $H_1$ field of the correct amplitude and duration to produce a rotation of 90° is called a 90° (or a $\pi/2$) pulse. If the $H_1$ field is

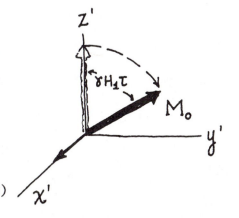

applied for twice as long, this results in a 180° (or a $\pi$) pulse. Similarly one could obtain a 270° pulse, a 360° pulse, etc.

Consider the 90° pulse. Immediately following it, the magnetization lies along the rotating y' axis. This magnetization will decay in time as the system comes back to equilibrium due to two mechanisms. The magnetization dephases in the x'-y' plane because the field inhomogeneity, for example, will cause some of the nuclei to precess faster than others but it also tends to regrow along the z'=z direction because thermal equilibrium is being re-established. After the 90° pulse, the magnetization will induce a decaying sinusoidal voltage in a pickup coil in the x-y plane. For a single Lorentzian line the decay of the magnetization in the x-y plane will be exponential with the time constant $T_2^*$ and the output of the pickup coil will be a sinusoidal wave whose amplitude is decaying at the rate $\exp(-t/T_2^*)$. If one waits long enough, the magnetization will re-establish itself along the z direction in equilibrium with the applied field with a time constant called the spin-lattice (or longitudinal) relaxation time $T_1$, i.e., it will recover as $M_z = M_0[1-\exp(-t/T_1)]$. The decay rate for the magnetization in the x-y plane is usually larger than or equal to the decay rate for the recovery of the magnetization along the z direction.

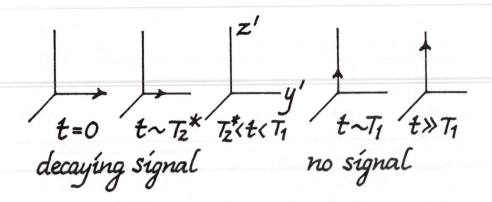

Because the pickup coil is sensitive only to the component of magnetization in the x-y plane, the magnetization being re-established along the z' axis due to the $T_1$ process is not detectable until it is rotated away from the z' axis. Therefore, a second pulse is always necessary to measure the spin-lattice relaxation time $T_1$. The basic idea is to perturb the magnetization from its equilibrium state with one pulse and examine its recovery along the z' axis after a variable delay with another pulse, for example, a 90° pulse. Several methods for measuring $T_1$ are described in III.D.

The effect of a 90° pulse can be duplicated in cw NMR although with more difficulty than in pulse NMR. When a resonance line is swept rapidly in the conventional cw high resolution experiment, a ringing pattern follows the line if phase detection is used. Such a ringing pattern occurs in the following way: As one sweeps into resonance, the magnetization is tipped away from the applied magnetic field as in adiabatic fast passage, as discussed in I.B. However, if one sweeps through resonance too rapidly, that is, non-adiabatically, the magnetization gets left behind by the effective field $H_e$. (The best chance of this happening is near resonance because the precession frequency $\gamma H_e$ about $H_e$ is the smallest and the rate of change of the $H_e$ direction is the greatest there.) After the passage, the field has been increased way beyond the resonance value for that line and the magnetization will freely precess. If a phase sensitive detector is used, the signal from the pickup coil will get in and out of phase with the reference signal in the detector giving rise to a decaying ringing pattern. The length of time during which the ringing pattern persists is the time during which there is a net x-y component of the magnetization. Therefore, under conditions in which the $H_1$ field does not exert a torque on the magnetization, the ringing pattern will be observed to decay in time as $\exp(-t/T_2^*)$. The $T_2^*$ time constant here is identical to the $T_2^*$ observed following a 90° pulse in a pulse experiment.

The magnetization decay rate constant in the x-y plane, $1/T_2^*$, consists of a term related to the recovery rate constant for the magnetization along z, $1/T_1$, as well as other mechanisms which give rise to a dispersal of the magnetization only in the x-y plane. Two examples of the latter are $1/T_2'$ which results from dipolar processes (to be discussed later) and $\gamma \Delta H_0$ which is due to the applied magnetic field inhomogeneity $\Delta H_0$. In this case the relaxation rate is approximately

$$1/T_2^* \cong 1/(2T_1) + 1/T_2' + \gamma \Delta H_0$$

although usually one of these mechanisms will dominate. In many high resolution spectra, $T_2^*$ is determined by the spin lattice relaxation time $T_1$ which, in turn, is affected by motional parameters of the molecule such as characteristic times for reorientation and diffusion. In a typical solid, on the other hand, the dipolar contribution $T_2'$ is much more likely to determine $T_2^*$ because the effects of these fields from the neighboring spins do not cancel as they do in liquids wherein the molecular tumbling averages out the dipolar interaction. Strictly speaking, the line will not even be Lorentzian in most solids but $T_2'$ and $T_2^*$ still can characterize some decay or dephasing time. For very narrow lines such as for natural abundance $^{13}$C nuclei without proton and fluorine neighbors, the magnetic field inhomogeneity can be the main contribution to the line width and the relaxation time $T_2^*$. This is usually the least interesting and informative contribution to the $T_2^*$ process and there are ways such as the CPMG method discussed in the following section to get around this problem.

## I.C.2.  PULSE NMR EXPERIMENTS

At this point, we study in more detail the behavior of the magnetization following an rf pulse. Consider a sample placed between the poles of an electromagnet. In any real magnet, different parts of the sample experience different fields owing to magnet imperfections. In NMR spectroscopy it is important to minimize this inhomogeneity in the applied static field for maximum resolution, although the required homogeneity varies widely depending on the experiment.

Suppose a 90° pulse is applied to a sample in an inhomogeneous magnetic field. If the pulse is very intense, the entire magnetization will be rotated through the appropriate angle in a very short time. Shortly thereafter, however, the magnetization amplitude will decrease due to the field inhomogeneity in the following way. The total magnetization vector is the sum of smaller magnetization vectors each arising from a small volume experiencing a homogeneous field. After a 90° pulse, each of these components of the magnetization will precess with its own characteristic Larmor frequency. Therefore, the magnetization from those portions of the sample with slightly larger magnetic fields will precess faster than those which are in a smaller field. As a result, the different contributions of the magnetization will get out of phase with each other. The contribution to the magnetization which arises from one such small segment of the sample, experiencing a homogeneous applied field, is called a spin isochromat. The term isochromat is used because it implies a constant frequency (i.e., color) and the implication is that all the nuclear moments in that segment of the sample will precess with the same Larmor frequency.

In the following figure, the results of applying a 90°

pulse in an inhomogeneous field is seen in the rotating reference frame.    In the rotating frame the magnetization immediately following the pulse starts out along the rotating y'
axis.    Shortly thereafter, some of the components of the
magnetization, i.e., some of the spin isochromats, start
getting ahead of the average and some start getting behind.

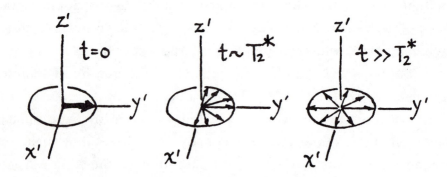

The result is that the net magnetization along the y' axis
decays to 0 as the spin isochromats fan out in x'-y' plane.
If there is a pickup coil in the x-y plane, the voltage induced
in the coil will decay as the spin isochromats fan out.    This
signal following a pulse is the free induction decay (FID).
[See the last figure in Section I.A.]    As a quantitative example, consider protons in a field inhomogeneity $\Delta H$ of 1 gauss
across the sample.    [That's quite a poor magnet -- or a huge
sample.]    Since the angular velocity is given by $\gamma H$, its spread
in this sample is $\gamma \Delta H = (4.3 \text{ kHz}) \times 2\pi$.    This means that in a time
of the order of $1/(4.3 \text{ kHz})$ the fastest isochromats would have
crossed the slowest isochromats.

As mentioned in the previous section, the time constant
which describes the decay of the magnetization in the x-y
plane is $T_2^*$.    We define an intrinsic relaxation time which is
characteristic of the magnetization decay in one of the spin

isochromats without any field inhomogeneity effects and this is the spin-spin or the transverse relaxation time $T_2$. Thus, $T_2$, $T_2^*$, and the spread in Larmor speed $\gamma\Delta H_0$ due to field inhomogeneity $\Delta H_0$ are related by $1/T_2^*=1/T_2+\gamma\Delta H_0$. If the magnetic field homogeneity is improved, $T_2^*$ will increase until it becomes equal to $T_2$. When $T_2^*$ is dominated by magnetic field inhomogeneity, it will give us little information about the sample. However, the $T_2$ processes which cause the magnetization decay within a spin isochromat, will be due to fundamental molecular processes and are likely objects of the experiment. The time constant $T_2^*$ is related to the line width in the frequency spectrum as discussed in III.B.

While all this is going on, spin-lattice relaxation is also taking place. The spin-spin relaxation does not involve any exchange of energy with the world outside the spin system, whereas the spin-lattice relaxation depends on an outside agent (which is called the lattice) accepting energy from the spin system so that the latter can relax towards the thermal equilibrium state given by the Boltzmann populations discussed near the beginning of this chapter. As we will discuss in III.A., the lattice can be of many possible forms such as molecular rotation, diffusion, or lattice vibration. We only note here that the spin-lattice relaxation process can be quite weak in the absence of molecular motion or paramagnetic ions as in many rigid solids but it can be very strong in many solutions and in some solids which exhibit molecular motion. The spin-spin relaxation process contains a contribution from the spin-lattice process so that $T_1 \gg T_2^*$ in most solids while $T_1 \sim T_2^*$ in most liquids.

Ingeneous experiments have been designed to remove the effect of the applied field inhomogeneity. They are called spin-echo techniques and have application in the realm of

Fourier transform NMR as well as pulse NMR (Hahn, 1950).
Suppose that a 90° pulse is applied along x' at time 0 to a
spin system for which magnetic field inhomogeneity is the major
contribution to $1/T_2^*$.  Shortly thereafter, the spin isochromats
will have dephased in the x'-y' plane.  As a result, there is
no net magnetization component in the x'-y' plane, although
the individual spin isochromats have not yet dephased.
Suppose at a time τ later, a 180° pulse is applied along the
x' axis.  Any magnetization along the z direction would simply

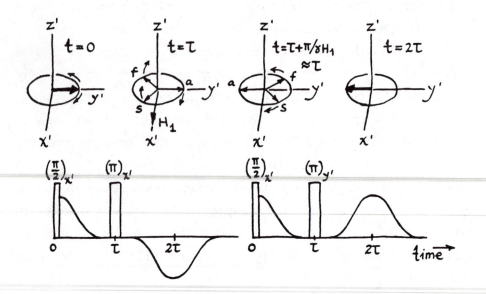

be inverted to the -z direction and be of no consequence.  Of
the magnetization remaining in the x'-y' plane each one of the
spin isochromats would be rotated 180° about the x' axis.  As
a consequence, those spin isochromats which had gotten ahead
of the average spin isochromats by a certain angle are now
behind the average of the pack of spin isochromats by the
same amount.  Those spin isochromats which were going slower

than average and had gotten behind the rotating y' axis are now ahead of the rotating y' axis by the equivalent amount. Therefore, following the 180° pulse, the spin isochromats begin to rephase to form a net magnetization as the rapid isochromats catch up with the slow ones. The result is that the magnetization becomes refocussed along the -y' axis at time 2τ and it will cause an inverted spin echo, as shown in the sketch at left. The spin echo consists of two FID's back-to-back. It is also possible to get spin echoes by applying the 180° pulse along the y' axis in the rotating frame. The refocussing will then take place along the y' axis so the echo will have the same sign as the FID.

At this point we pause in our discussion of pulse experiments in order to give an application of spin echoes. One of the problems in accumulating FID's is that enough time must elapse between the FID's in order for the magnetization to build up along the z'(=z) axis so that the signal is reasonably large. If $T_1 \cong T_2^*$ as in many liquids this is not a serious problem since the recovery along the z axis will already have taken place by the time the FID has been recorded. If, on the other hand, $T_1$ is long compared with $T_2^*$, time intervals of the order of $T_1$ are still needed between the FID's even though the FID has disappeared long before.

One way to increase the efficiency of data taking in such a case is a sequence called DEFT for driven equilibrium Fourier transform (Becker, et al. 1969). The idea of this experiment is to force the system back to equilibrium more rapidly after each FID so that it can be pulsed more frequently. In a DEFT sequence, a 90° pulse is applied to the system, and the FID is recorded. A 180° pulse is applied at a time τ when the FID has disappeared due to the dephasing of the spin isochromats to form an echo at time 2τ. At the peak of the spin echo, when all the magnetization in the x-y plane that can be refocussed will be along the -y' axis, a 90° pulse is applied to rotate that magnetization back to the z

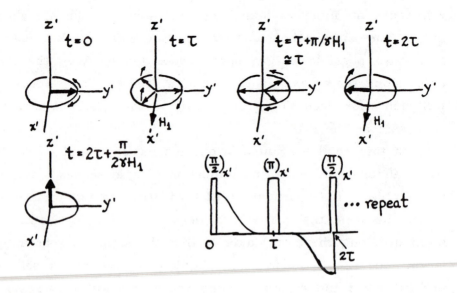

direction. In other words, the system has been forced back close to thermal equilibrium and we are prepared to repeat the experiment by applying a 90° pulse and recording the FID again.

We shall now return to spin echoes by expanding the discussion to include echo trains. For simplicity, let us assume that we have only one line in a high resolution spectrum. On exact resonance in an inhomogeneous field, the 90° pulse yields an FID with a time constant $T_2^*$. At a time $\tau$ later, a 180° pulse is applied and the echo maximum occurs at time $2\tau$, since the time required for rephasing the spin isochromats equals the time it took for dephasing. Because the spin echo arises from magnetization that has regrouped along the -y' axis, the echo will be inverted compared to the FID. Another 180° pulse at time 3τ will result in another echo at time 4τ and this will be right side up. One can continue to apply 180° pulses with a spacing of 2τ with echoes occurring between each of the pulses. This spin echo train is called a Carr-Purcell (CP) spin echo train (Carr and Purcell, 1954).

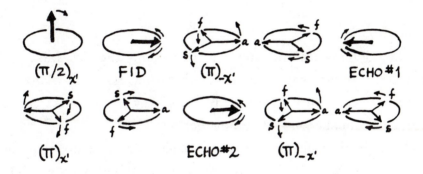

The echo amplitude maxima should decay with the time constant $T_2$, the intrinsic spin-spin relaxation time, which is the time it takes for the magnetization to decay in the x-y plane in the absence of any external field inhomogeneity.

In practice, Carr-Purcell echo trains usually result in measured $T_2$'s that are too short because of cumulative errors of each pulses not being exactly 180° pulses and of $H_1$ inhomogeneity which spreads out the magnetization in a plane containing $H_0$ and $H_1$. One way to compensate for these errors is to alternate the phase of each 180° pulse by 180° phase shifts as shown below. The first 90° pulse occurs with the rotating $H_1$ field along the rotating x' axis. The first 180° pulse, how-

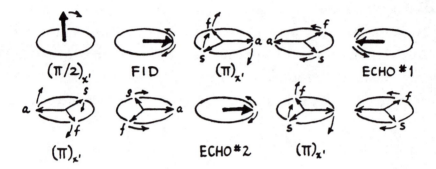

ever, would have its rotating $H_1$ field along the -x' axis. The second 180° pulse would have the rotating $H_1$ field along the rotating x' axis and so on. In this way, any pulse length errors are cancelled on alternate echoes.

A slightly simpler and more common spin echo sequence for measuring long $T_2$'s is the Meiboom-Gill modification of the

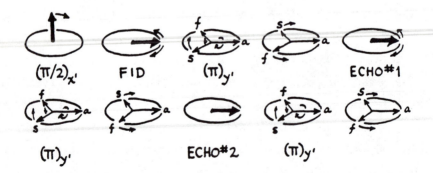

Carr-Purcell sequence (CPMG) in which all the 180° pulses in the train are phase shifted 90° with respect to the inital 90°

pulse, i.e., if the 90° pulse is along x', the 180° pulses are along ±y' (Meiboom and Gill, 1958). Now all the echoes form along $H_1$ for the 180° pulses in the rotating frame regardless of the exact tip angle, and this is shown at left (bottom).

CPMG echoes also differ from CP echoes in their having the same sign since the echoes are always formed along the same direction in the rotating frame. This is a slight inconvenience compared to the bipolar CP echoes which have no baseline ambiguity. An example of a CP signal with the corresponding CPMG envelope indicated is shown below.

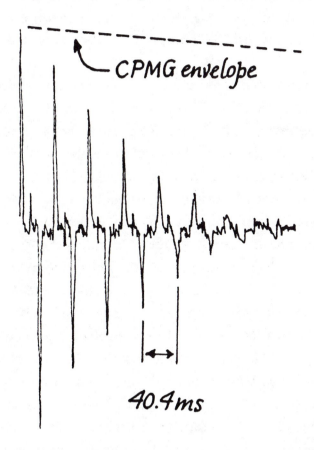

This example is taken from $^{13}C$ $T_2$ measurements in liquid CO at 77K at 8.9 MHz where $T_2$ was 5.8 seconds as determined by the CPMG data.

Suppose we had decided to measure this $T_2$ without using the CPMG method. From the relation

$$1/T_2^* = 1/T_2 + \gamma\Delta H_0$$

we know that $\gamma\Delta H_0 \ll 1/T_2$ in order for the measured $T_2^*$ to accurately reflect $T_2$. Because $T_2 \sim 6$ s, let us say that $\gamma\Delta H_0$ must be less than $10^{-2}$ radians/sec. For $^{13}C$, $\gamma \sim 2\pi$ kHz/G so that $\Delta H_0$ must be less than $(10^{-2}/2\pi \times 10^3)$ G or about 1.5 μG! In a field of about $10^4$ G, that is an inhomogeneity of $1.5 \times 10^{-10}$ which is well beyond the capability of the usual NMR magnets.

It is obvious that the CPMG sequence is particularly useful for determining the spin-spin interaction time $T_2$. Long $T_2$'s requiring CPMG measurements usually contain information about very slow motions of the molecules containing the nuclei. See Sections III.E.1. for details.

The spin echo train has an additional use in NMR. Remember that a spin echo consists of two FID's back to back. Suppose we co-add the echoes. Out of a spin echo train, one could get a fairly large number of FID's which could be added together to produce an enhanced S/N. This could be done simply by triggering the digitizer at each echo. With more sophistication, S/N can be further improved by taking the left-hand half of each echo, reversing it left to right, and adding it to the right-hand half. Using spin echo trains to improve the S/N in Fourier transform NMR is called SEFT (Spin Echo Fourier Transform) (Allerhand and Cochran, 1970). It is clear that SEFT will be advantageous under the same conditions that DEFT is advantageous, namely when $T_2^*$ is much smaller than $T_1$. Neither DEFT nor SEFT have found widespread application in NMR but are discussed here because they are techniques which add to the understanding of pulse NMR.

## REFERENCES

A. Allerhand and D. W. Cochran, "Carbon-13 Fourier-transform nuclear magnetic resonance. I. Comparison of a simple spin-echo procedure with other methods," J. Am. Chem. Soc. 92, 4482-4484 (1970).

Edwin D. Becker, James A. Ferretti and T. C. Farrar, "Driven equilibrium Fourier transform spectroscopy. A new method for nuclear magnetic resonance signal enhancement," J. Am. Chem. Soc. 91, 7784-7785 (1969).

H. Y. Carr and E. M. Purcell, "Effects of diffusion on free precession in nuclear magnetic resonance experiments," Phys. Rev. 94, 630-638 (1954).

E. L. Hahn, "Spin echoes," Phys. Rev. 80, 580-594 (1950).

S. Meiboom and D. Gill, "Modified spin-echo method for measuring nuclear relaxation times," Rev. Sci. Instrum. 29, 688-691 (1958).

## I.D.   INTRODUCTION TO FOURIER TRANSFORM NMR

In the following three sections, we will get our feet wet with simple concepts of Fourier transform NMR and the time and frequency domains which are related by the transform. Detailed treatments will be given in Chapter II.

## I.D.1.  THE DRIVEN VS. PULSED SIMPLE HARMONIC
### OSCILLATOR:  An Introduction to
### Fourier Transform Pairs

A nuclear spin system behaves much like a damped
harmonic oscillator.  Suppose a mass is attached to one end of
a spring and this mass-spring system is lowered into a liquid
which provides damping.  Attach the upper end of the spring
to a motor which will vibrate the mass-spring system.  As the
frequency of the motor is varied through the natural
resonance frequency of the system the response is a
resonance curve as shown.

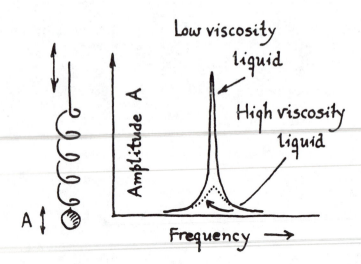

The amplitude of the oscillation is a maximum when the driving
frequency equals the natural resonance frequency and the
width of the resonance line depends on the degree of damping
of the system.  If the liquid in which the system is immersed
is not viscous, the line will be narrow.  On the other hand,
the line will be quite broad for a viscous liquid.  The lineshape produced by such a damped harmonic oscillator is a
Lorentzian lineshape.  The full width at half height of such a

line is $2/T_2^*$, where $T_2^*$ is defined in the following paragraph.

If we take the same damped mass and spring system and, instead of driving the system with a motor, simply displace the mass and let it go, we find that it will oscillate with its natural resonance frequency and the amplitude of the oscillation will decay in time. The decay will be exponential with the amplitude proportional to $\exp(-t/T_2^*)$ where the viscosity of the liquid will determine $T_2^*$. It is interesting to note that if you pluck the system, its oscillation and decay contains all the information of the system's behavior just as the frequency dependence of the amplitude in the driven simple harmonic oscillator experiment does. In the one case, the full width at half height of the resonance curve is $2/T_2^*$, while in the other, the decay time constant is $T_2^*$.

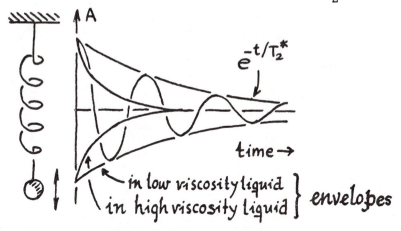

To summarize, all of the information of the system is available from either means of exciting the resonance, driving it and sweeping the frequency, or hitting it with an impulse for its time response. The first experiment is performed in the frequency domain and the second in the time domain. The mathematical transformation of one representation into the other is the Fourier transform. The time domain response and the frequency domain response are called Fourier transform

pairs.    Next we explore the nature of the Fourier transform
and how it can be applied to NMR.

## I.D.2.  THE NATURE OF FOURIER TRANSFORM NMR

Before discussing Fourier transform NMR, the nature of
Fourier transforms shall be explained.  We have already noted
that a representation in the time domain can be transformed
into an equivalent representation in the frequency domain
and vice versa.   Consider the two different representations

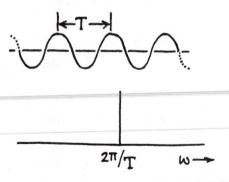

of a sine wave.  One is an amplitude vs. time graph as shown.
An alternative representation would be a spectrum of amplitude
vs. frequency which amounts to a histogram of the frequencies
in the sine wave.   Since a sine wave is monochromatic by
definition, the spectrum would be a single peak located at the
sine wave frequency.   The width of the peak is inversely
proportional to the duration during which the sine wave
exists since the longer it exists the more precisely the
frequency can be determined.

Now suppose that we have two such sine waves superimposed with different amplitudes and different frequencies.
In the next figure, the superposition of two sine waves of
different frequencies is represented.  The result is an interference pattern with beats in it.  In the representation of
amplitude vs. frequency there are two spikes with horizontal
coordinates corresponding to the appropriate frequencies.
One can continue to add sinusoidal waves to both diagrams
and see that the pattern in the amplitude vs. time diagram
becomes increasingly complex, while in the amplitude vs.
frequency diagram one simply adds more spikes to the curve.

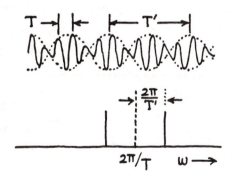

These two representations, amplitude vs. time and amplitude vs. frequency, both contain the same information.  One
representation can be transformed into the other by a mathematical technique called a Fourier transform represented by

$$A(t) = \frac{1}{2\pi} \int_{-\infty}^{\infty} A(\omega) e^{i\omega t} d\omega$$

and

$$A(\omega) = \int_{-\infty}^{\infty} A(t) e^{-i\omega t} dt.$$

In the transformation from the frequency to the time domain (the top equation), the amplitude of a wave $A(\omega)$ is multiplied by a sinusoidal wave of unit amplitude, $\exp(-i\omega t)$, and this product is added over all frequencies at time t. The translation from time space to frequency space is the inverse of this process. Consider now the application to nuclear magnetic resonance. A spectrum is a representation in the frequency domain. The individual lines represent the sinusoidal waves occurring at those frequencies. A given NMR line has a certain characteristic shape and a certain characteristic frequency. Consider a frequency domain spectrum consisting of two lines with finite widths. If the individual lines are Lorentzian in shape (which is the common lineshape observed in liquids) each one would correspond to an exponentially decaying sinusoidal wave. If all these individual decaying sinusoidal waves are algebraically summed the result is a very complex beating pattern called an interferogram or a free induction decay (FID) which is what is recorded in an actual experiment. The Fourier transform yields the spectrum corresponding to such an FID.

Next consider how the Fourier transform NMR experiment is performed. Consider a molecule with only chemically equivalent protons in it (for example benzene), resulting in a large magnetization vector in equilibrium with the applied field. A 90° pulse applied at time t=0 will rotate this magnetization into the x-y plane at which point it will precess in time at the Larmor frequency and will decay exponentially with the time constant $T_2^*$. A coil mounted in the laboratory x-y plane will have voltage induced in it which is a decaying sinusoid. If the output of the pickup coil were sent to an analog-to-digital converter, and the digitized decaying sinusoidal wave is Fourier transformed in a digital computer, the output would be the NMR spectrum of the molecule. In this case the spec-

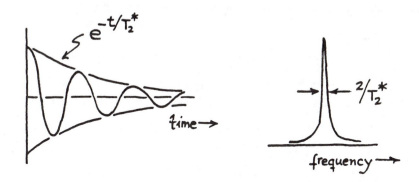

trum would be a single Lorentzian line with a full width at
half height of $2/T_2^*$.

If the sample contained molecules with several chemically
different groups of protons (e.g., ethyl alcohol) then the
90° pulse would rotate all contributions to the magnetization
by approximately 90° and all of these contributions would
precess in the x-y plane with their own characteristic Larmor
frequencies so that various contributions to the magnetization
would get in and out of phase with each other. The resulting
FID would be a complex beating pattern of different sinusoidal
waves all starting with the same phase. If this complex interferogram were digitized and then Fourier transformed, the
output would be the spectrum of this molecule. In this way
we see that in pulse NMR the different contributions to the
magnetization precess at different frequencies, producing a
characteristic interferogram which has all the spectral information in it.

One advantage of Fourier transform NMR becomes immediately obvious. Consider an analogy: Suppose there were a
series of bells, each with a different pitch or frequency, and

suppose that one were asked to characterize the loudness and frequency of each of these bells. One might ring each bell and plot the amplitude against the frequency. For N bells this would require performing N different experiments. This would be comparable to the cw experiment in which each of the contributions for the magnetization at different frequency is sampled, one by one. On the other hand, all the bells can be struck simultaneously and the resulting ringing, which is a complex interferogram, can be digitized and Fourier transformed to yield the spectrum of all the bells. It is clear that this experiment is far more efficient than the first.

In an NMR experiment a typical cw proton spectrum might take 500 seconds to record. If 1 Hz resolution is required, the interferogram could be gathered for one second after a $\pi/2$ pulse and the computer would then proceed to generate the high resolution spectrum. This means that in terms of data gathering time the Fourier transform experiment has a factor of 500 advantage in this example. Another way of looking at this experiment is to consider the high resolution spectrum to be broken up into resolution elements. With a cw experiment each resolution element is examined one at a time. With a Fourier transform experiment all the resolution elements are sampled simultaneously. The application of this "multiplexing" advantage to high resolution NMR was pioneered by Ernst and Anderson (1966).

If there are two different contributions to the magnetization which are very close to each other in frequency, they have to be observed for a long enough time for them to get out of phase so that they can be distinguished. As a result, the ability to distinguish two lines close together (called the resolution) is limited by how long the FID is sampled. If the data is sampled for one second total duration, the resolution of the resulting spectrum is 1 Hz. If it is sampled for 5 seconds, the resulting resolution of the spectrum becomes 0.2 Hz.

The signal-to-noise ratio (S/N) of an accumulation of spectra is the S/N from one spectrum times the square root of the number of scans added together as mentioned in the introduction. If, for example, one were to take 500 seconds to record a high resolution spectrum in the cw mode and were satisfied with 1 Hz resolution (so that the Fourier transform system could be pulsed once every second), one could take 500 scans by the pulse technique in the time taken to take a scan by the cw technique. The S/N improvement following those 500 scans would be $\sqrt{500}$ or about 22. The advantage of the pulse method in terms of saving time or in terms of improving S/N in a given period of time is even greater when the spectrum range is very wide as it often is for carbon-13 NMR. Consider an example of a carbon-13 spectrum which is 1 kHz wide with the FID accumulated once a second. This leads to a saving of a factor of 1,000 in time or a S/N improvement of about 30. The S/N improvement and the saving in time is quite dramatic and this fact enables many experiments which otherwise would not be possible. Even if the time saving is only 500 it is possible to do a pulse FT experiment in 24 hours which would have required 1-1/3 years by cw. Aside from the intolerably long wait, the cw experiment simply is not practical because the electronics most likely will not be stable for that period.

## REFERENCE

R. R. Ernst and W. A. Anderson, "Application of Fourier transform spectroscopy to magnetic resonance," Rev. Sci. Instrum. _37_, 93-102 (1966).

I.D.3.  COMPARISON OF FT AND CW NMR

Spectra acquired by FT techniques are nominally the same as those obtained by slow passage cw techniques but may differ in certain ways.  For example, the presence of ringing and other artifacts of scanning rapidly are absent in the FT spectrum.  On the other hand, the FT spectrum may have artifacts of its own, for example, the "feet" on spectral lines resulting from truncating the FID if one does not remove them by apodization routines (II.B.5.), baseline problems most probably due to improper values of FID near t=0, and "glitches" of varying origin.

Furthermore, there are important differences between FT and cw NMR in experimental methods and in the kind of information easily obtainable.  For example, FT NMR is much better suited for almost any relaxation time measurements as well as for experiments in which the sample is undergoing some transformation with time.  Furthermore, the breakthroughs in the field of high resolution NMR in solids (Section IV.E.) have involved pulse FT NMR.  Perhaps surprisingly, cw spectral acquisition has some advantages over FT as well.  A large but uninteresting resonance, such as a large solvent peak, can sometimes dominate a spectrum and therefore the FID.  Of course, FT techniques such as WEFT or homonuclear decoupling with FT do exist for the attenuation of such a resonance.  However, in the cw approach, one can simply stop the field scan short of such a large peak.  This is used most frequently in "rapid scan" NMR (Dadok and Sprecher, 1974; Gupta, et al., 1974).  In addition, cw may be the only feasible method for very broad lines for which the FID does not extend past the apparatus deadtime and for which it is not possible to get an echo.  If echoes can be formed, as in inhomogeneously broadened resonances, a line profile can be traced out by scanning the field or frequency and plotting the echo height in a method which is a combination of pulse and cw NMR. The echo shape itself is devoid of the desired spectral information because it mostly reflects the shape of the transmitter pulse if $T_2^* < \tau$ where $\tau$ is the pulse length.

Many sophisticated data manipulation techniques used in NMR spectroscopy were ushered in together with Fourier transform NMR. Specifically, FT introduced spectroscopists to the power of dealing with information in both the time and frequency domains. These data manipulation techniques are also available for use on cw spectra, which can be Fourier transformed into the time domain to carry out such operations, if necessary, and then transformed back to the spectral representation. Having the opportunity to manipulate the data both in the time and the frequency domain adds greatly to the ability to refine the data.

With many of these information manipulation techniques one is generally trading off one kind of information in order to enhance another. For example, as discussed in II.B.5., the FID can be multiplied with a decreasing exponential function to enhance S/N at the expense of resolution and knowledge of the lineshapes, or it can be multiplied with an increasing exponential function to increase resolution at the cost of poorer S/N. Similarly where the spectrum has narrow and broad lines, either component may be enhanced by emphasizing an appropriate part of the FID.

The advent of FT spectroscopy has made spectroscopists far more knowledgeable about dealing with spectral information in both FT and cw techniques and in both wide line and high resolution NMR.

REFERENCES

J. Dadok and R. F. Sprecher, "Correlation NMR spectroscopy,"
J. Magn. Resonance 13, 243-248 (1974).

R. K. Gupta, J. A. Ferretti, and E. D. Becker, "Rapid Scan
Fourier Transform Spectroscopy," J. Magn. Resonance 13,
275-290 (1974).

CHAPTER II

DETAILS OF PULSE AND FOURIER TRANSFORM NMR

## II.A.  FUNDAMENTALS

We now go into pulse NMR in more detail after learning a few Fourier transform theorems.  The second section considers the behavior of the magnetization in the rotating frame under pulsed irradiation and how the spectrum is affected, a very important topic.  The third section is a discussion of quadrature detection with an educational as well as a practical aim.

## II.A.1.  SOME USEFUL FOURIER TRANSFORM THEOREMS

There are several basic theorems of Fourier transformation which we have to know in order to understand much of

Eiichi Fukushima and Stephen B. W. Roeder, Experimental Pulse NMR: A Nuts and Bolts Approach

what happens in FT NMR.  In this section, we discuss only
four of them, namely, the addition theorem, the shift theorem,
the modulation theorem, and the convolution theorem.  The
reader is referred to treatises on Fourier transformation for
more details and more theorems.  One of the most useful and
readable is the book by Bracewell (1978) in which Chapter 6
is devoted to these and other theorems.

Before getting into the theorems, we have to remember
what the transformation equations look like.  They are reproduced here from Section I.D.2.

$$A(t) = \frac{1}{2\pi}\int_{-\infty}^{\infty} A(\omega)e^{i\omega t}d\omega$$

$$A(\omega) = \int_{-\infty}^{\infty} A(t)e^{-i\omega t}dt.$$

The addition theorem states that if f(t) and g(t) have
the Fourier transforms F(ω) and G(ω), then the function
f(t)+g(t) has the Fourier transform F(ω)+G(ω).  This follows
easily from the definition of the transform.

The shift theorem states that if the function f(t) has the
Fourier transform F(ω), then the function f(t-a) has the
transform F(ω)·exp(-iωa).  Its derivation is also quite simple.
We go through it here for illustration.

$$\int_{-\infty}^{\infty} f(t-a)e^{-i\omega t}dt = \int_{-\infty}^{\infty} f(t-a)e^{-i\omega(t-a)}e^{-i\omega a}d(t-a) = e^{-i\omega a}F(\omega)$$

This theorem means that if a time domain signal is shifted in
time, its Fourier transform remains identical except that it

gets multiplied by a phase factor which varies linearly with the frequency ω and is proportional to the time shift a. What the time origin shift is doing is to change the even/odd character of a sinusoidal component of the signal. (All signals can be decomposed in that way.) A Fourier transform of an even and real function is even and real while that of an odd and real function is odd and imaginary. Thus, shifting the time origin of a real function to change the even/odd character of the components can mix in varying amounts of imaginary components which is the same as introducing phase shifts.

For an example, consider the function $f(t)=\cos\omega t$. It is even and real. Its FT $F(\omega)$ is a symmetric impulse at $+\omega$ and another at $-\omega$, i.e., there is no phase shift between $+\omega$ and $-\omega$. For the function $g(t)=\sin\omega t$, the transform $G(\omega)$ is an antisymmetric imaginary impulse function at the same frequencies, so that there is a phase shift of 180° from $-\omega$ to $+\omega$.

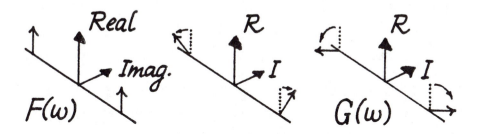

Since you can go continuously from $\sin(x)$ to $\cos(x)$ by shifting the time coordinate, you can introduce the phase difference continuously into the other domain.

The <u>modulation theorem</u> says that if the function $f(t)$ has the transform $F(\omega)$, then the function $f(t)\cos\omega_0 t$ has the transform $\frac{1}{2}F(\omega-\omega_0)+\frac{1}{2}F(\omega+\omega_0)$. For pulse NMR, $f(t)$ would be the FID envelope which contains the desired information while $\cos(\omega_0 t)$ is the carrier.

A convolution is the process whereby the effect of a broadening function can be included in an otherwise sharp NMR lineshape. A convolution of the functions f(t) and g(t) is $h(t_o) = \int_{-\infty}^{\infty} f(t)g(t_o-t)dt$ and is writtten as f(t)*g(t). The convolution theorem states that if f(t) and g(t) have transforms F(ω) and G(ω), then f(t)*g(t) has the transform F(ω)G(ω). The modulation theorem stated above is clearly a special case of the convolution theorem. The mechanics of the convolution process and its physical picture are treated in great detail in Chapter 3 of Bracewell's book. We also discuss a specific example of the theorem in II.B.6.


REFERENCE

Ron Bracewell, The Fourier Transform and Its Applications, 2nd edition (McGraw-Hill, New York, 1978).

## II.A.2  PULSE LENGTH, TIP ANGLE, AND ALL THAT
### (OR JUST WHAT IS THE RF PULSE DOING
### TO THE MAGNETIZATION?)


In Chapter I, we saw that a magnetic moment with gyromagnetic ratio γ in a magnetic field H precesses with angular frequency ω=γH. We also discussed the rotating frame concept with which it is possible to transform away all or a part of the static laboratory magnetic field. In section I.C.1., we introduced the 90° pulse as one with the property $\gamma H_1 \tau = \pi/2$ where τ is the 90° pulse length and $H_1$ is the magnitude of the magnetic field vector (originating in the applied rf magnetic field) which is stationary in the rotating frame. In pulse NMR, the length of the irradiation is an important parameter as we will show in this section.

We now consider how the magnetization is affected by the rf irradiation and how this interplay affects the resulting spectrum. The usual treatment of this problem is in terms of the power spectrum of the pulse by Fourier analysis (Shaw, 1976). Consider a very long pulse of rf with frequency $\nu$. As the pulse length $t_p$ approaches infinity, the signal becomes a pure sinusoid of frequency $\nu$ whose Fourier transform is an infinitely narrow contribution at the frequency $\nu$. The only way to do spectroscopy with an infinitely narrow frequency irradiation is to sweep the frequency $\nu$, in other words to do a cw experiment. If the pulse is made shorter, we will no longer have a truly monochromatic frequency spectrum even though the source (say, the signal generator) is still monochromatic. This is because many different frequencies have to be combined in order to form the rising and the falling edges of the rectangular pulse.

What, then, is the frequency distribution of a rectangular pulse? Suppose its duration is $t_p$, its area is normalized so that its height is $1/t_p$, and we take its start to be at $t=-t_p/2$ so that $t=0$ is at the middle of the pulse for convenience in performing the Fourier transform. From the definition of the transform in the previous section, the frequency spectrum is given by

$$F(x) = (2/t_p) \int_0^{t_p/2} \cos 2\pi xt \cdot dt = \frac{\sin \pi xt_p}{\pi xt_p} \, .$$

In this integration, we used the fact that the pulse was even in time so that only the cosine part of the exponential function

in the Fourier integral contributed.  The resulting function in
the frequency domain is, therefore, real and even.

  This function occurs often enough in different applications that a special symbol exists for it (Bracewell, 1978).
By definition, $\text{sinc}(x) = \sin\pi x/\pi x$ and it has the properties that

$$\text{sinc}(0) = 1,$$

$$\text{sinc}(n) = 0, \ n = \text{nonzero integer},$$

and

$$\int_{-\infty}^{\infty} \text{sinc}(x)dx = 1.$$

It looks like a sine wave that is damped as $1/x$ for large $x$
but, unlike a sine wave, approaches one for $x \to 0$ which is easy
to see by L'Hopital's Rule in elementary calculus.  Furthermore, it is an even function.  Therefore, it looks like what is
shown in the following figure with the nodes at x=1, 2, 3,...

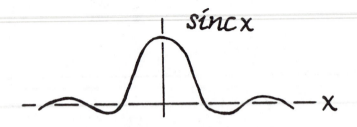

which means that for our rectangular pulse of duration $t_p$
nodes occur at values of x equal to multiples of $1/t_p$.

  Now what if we had taken the t=0 point to be at the

beginning of the pulse instead of in the center? Since this amounts to a time coordinate shift of $t_p/2$, the shift theorem (from II.A.1.) requires a multiplication of $F(x)$ by $\exp(-i\pi t_p x)$ so the resulting frequency spectrum is

$$\exp(-i\pi t_p x)\frac{\sin\pi xt_p}{\pi xt_p} = \frac{\sin 2\pi xt_p}{2\pi xt_p} - i\frac{\sin^2\pi xt_p}{\pi xt_p} .$$

Its magnitude is still $\operatorname{sinc}(xt_p)$, which can be verified by taking the square root of the sum of the squares of the real and the imaginary parts, but we can see that the phase has become frequency dependent just by a choice of a different origin. Furthermore, the real and the imaginary parts do not have the same periodicity and, in fact, the real part has twice as many nodes as the imaginary part. We will remember these observations when we consider an alternative formulation of how the nuclear spins respond to rf pulses close to the Larmor frequency.

For the time being, we note that this treatment of the Fourier components of the square pulse must only give an approximate result. For one thing, the pulses are never strictly rectangular so that its Fourier transform cannot be the same as that of the square pulse.

With this background, we now consider an alternative formulation of the problem of the response of nuclear spins to rf pulses. This formulation involves the motion of the macroscopic magnetization in the frame rotating with the applied rf field as discussed in I.B. In such a frame, the rotating component of the rf field $H_1$ is arbitrarily placed along the y' axis and any residual component of the static magnetic field $H_0 - \omega/\gamma$ will point along the z'=z' axis. The magnetization $\overline{M}_o$ in thermal equilibrium points along the z' axis and is rotated away from this axis by an effective field made up of $H_1$ and $H_0-\omega/\gamma$ which has a magnitude $[(H_0-\omega/\gamma)^2 + H_1^2]^{\frac{1}{2}}$ and an angle $\theta$ with the z' axis given by $\tan\theta = H_1/(H_0-\omega/\gamma)$. Whenever $\overline{M}_o$ is pointing away from the z' axis, it will induce an rf signal in a coil in the laboratory frame with its axis perpendicular to the z axis. With the use of a phase sensitive detector, we can observe any component of $\overline{M}_o$ in the x'-y' plane. It will turn out that in the coordinate frame we use here the component along the x' axis will be the absorption and that along the y' axis will be the dispersion component. The former is characterized by a maximum FID amplitude at t=0 while the latter will have zero amplitude at t=0.

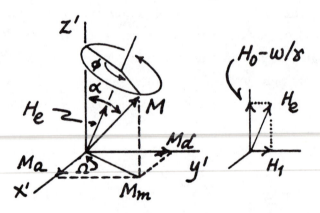

Let us define $x=(H_0-\omega/\gamma)/H_1$ as the offset parameter. Further, let $\theta$ be the polar angle of the effective field $H_e=H_1(1+x^2)^{\frac{1}{2}}$ so that $\tan\theta=1/x$ and $\sin\theta=(1+x^2)^{-\frac{1}{2}}$. Let $\phi$ be the nutation angle of the magnetization $\overline{M}_o$ about $\overline{H}_e$ in time $t_p$. If $\phi_o$ is the rotation angle of $\overline{M}_o$ in the same duration $t_p$ about $\overline{H}_1$, i.e., $\overline{H}_e$ with no offset, then $\phi=\phi_o(1+x^2)^{\frac{1}{2}}$. In terms of $t_p$, $\phi=\gamma H_e t_p$ and $\phi_o=\gamma H_1 t_p$. The various vectors and angles are shown in the figure.

In an NMR experiment, what is detected is the projection of $\overline{M}_o$ in the x'-y' plane which contains the axis of the receiver coil. Therefore, we now calculate the parameters defining this projection in terms of $\phi_o$ and x. The projection of $\overline{M_o}$ in the x'-y' plane has components along the x' and y' axis which correspond to the absorption and dispersion components $M_a$ and $M_d$, respectively. By straightforward geometry, we obtain

$$M_a = M_o \frac{\sin[\phi_o(1+x^2)^{\frac{1}{2}}]}{(1+x^2)^{\frac{1}{2}}}$$

and

$$M_d = M_o \frac{2x}{1+x^2} \sin^2\left[\frac{\phi_o}{2}(1+x^2)^{\frac{1}{2}}\right].$$

The projection can alternatively be specified by its magnitude $M_m$ and its phase angle $\Omega$ from the x' axis. The magnitude $M_m$ is the square root of the sum of the squares of $M_a$ and $M_d$ and is equal to

$$M_o\left\{\frac{1}{1+x^2}\left[\sin^2\phi + \frac{x^2}{1+x^2}(1-\cos\phi)^2\right]\right\}^{\frac{1}{2}}$$

whereas the phase angle $\Omega$ is equal to $\arctan(M_d/M_a)$.

The magnitude expression is especially important because it is a measure of the maximum signal possible for a given set of $\phi_o$ and x. It is customary to speak in terms of a tip angle $\alpha$ which is the angle $M_o$ makes with the z' axis after a pulse, and we see from the figure that it is defined by $\sin\alpha = M_m/M_o$. For a given value of $\phi_o$, $\alpha$ depends on x, i.e., for a given transmitter pulse the tip angle depends on how far off resonance the line is. The full expression for $\alpha$ is

$$\sin\alpha = \left\{ \frac{1}{1+x^2} \left[ \sin^2\phi + \frac{x^2}{1+x^2}(1-\cos\phi)^2 \right] \right\}^{\frac{1}{2}}.$$

We now present several figures which illuminate these relations. The first figure is a projection of the magnetization in the x'-y' plane as a function of the offset for a 90 degree transmitter pulse at resonance. In terms of the equations in this section, we have plotted $M_a$ and $M_d$ as a function of x with $\phi_o$ constrained to $\pi/2$ and $M_o$ chosen to be 1.0 although an entirely equivalent interpretation is to consider the plot as a polar coordinate plot of $M_m$ and $\Omega$ as a function of x. The dots on the curve represent x in increments of 0.25. The

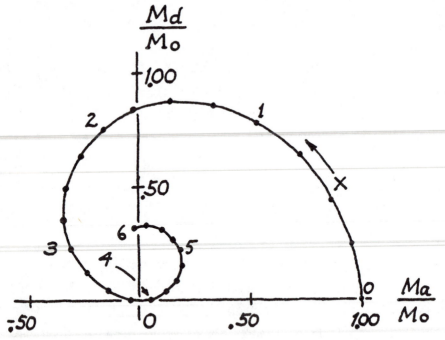

absorption component has a value of 1.0 at resonance where x=0 and oscillates about zero as it decreases with the first two nodes at about x=1.75 and x=3.9. The dispersion component starts at zero for x=0 and increases to a maximum of

about 0.9 for x of about 1.5 and decreases to zero at x close
to 3.9 where $M_a$ had its second null. The magnitude $M_m$
starts at 1.0 like the absorption component but decreases
more slowly and reaches zero for the first time for x around
3.9 like the dispersion component.

$M_a$, $M_d$, and $M_m$ are sketched in the next figure as a
function of x under these conditions. It is well to remember
that $M_a$ drops off twice as fast as $M_d$ and $M_m$ as a function of
x as this fact could bail you out when you are trying to
observe a wide frequency range in pulse NMR. The fact that

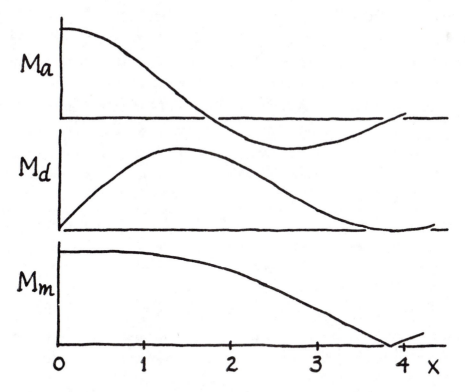

the magnitude of the spectrum is quite insensitive to the offset is illustrated by Meakin and Jesson (1973) in a nice parallel treatment to this section.

In the polar coordinate plot (p. 56), the angle between

the line joining the origin to a point on the spiral and the x'
axis is $\Omega$. Therefore, it can be seen that there is an approximately linear dependence of $\Omega$ on x for conditions not too far
from resonance and this fact is utilized in correction schemes
as described in II.B.4. There is a discontinuity at x near
3.9 because the trajectory of M is tangent to the x' axis at
that point.

It is important to realize that this approximately linear
dependence of the phase angle on the offset is an intrinsic
property of the experiment and has nothing to do with other
mechanisms which can give rise to phase shifts such as instrumental delays in filters or inappropriate choice of the time
origin in the Fourier transform process. In fact, since all
these effects are approximately linear in x, they are usually
lumped together and treated as one effect. Since the choice
of the time origin is arbitrary and is entirely under the
experimenter's control, all contributions to such frequency
dependent phase shifts can be cancelled by an appropriate
choice of the time origin and this is done quite often in FT of
very wide lines.

In 99% of FT NMR experiments the authors are aware of,
the quantity plotted as a function of x is $M_a$, $M_d$, or $M_m$.
The fourth quantity $\Omega$ can also be plotted as a function of x
and this is a very useful thing to do to check the phase
correction for broad lines. Although the S/N is very poor at
the discontinuities, it is still possible to adjust the reference
phase so that the phase plot is antisymmetric about x=0
(Gibson, 1978).

In the next figure which is a composite of the first two
figures of this section, we show a 3-D sketch of the magnetization as a function of x for $\phi_o = \pi/2$. Because of the difficulty in making a 3-D line drawing, we show $M_a$ and $M_d$ along

two perpendicular directions but do not show the resultant vector sum except as a projection in the x'-y' plane.

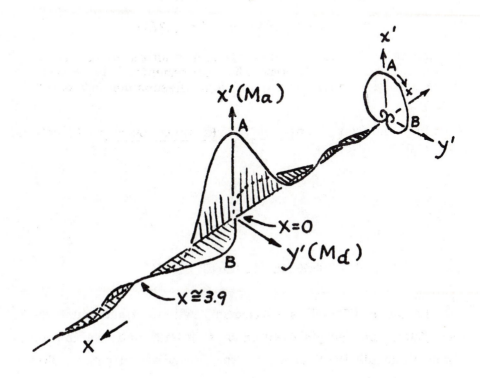

We note that in the limit of x>1, $M_a$, $M_d$, and $M_m$ become proportional to $\sin(\phi_o x)$, $\sin^2(\phi_o x/2)$, and $\sin(\phi_o x/2)$, respectively, in agreement with the earlier result for the frequency distribution of a rectangular rf pulse. Therefore, the FT of a rectangular pulse does approximate the response of the magnetization provided that the pulse parameters are chosen to yield the first maximum at the center of the frequency range of interest. It turns out that the first null as a function of x occurs at 87% of where the FT formulation predicts but the subsequent nulls agree almost exactly.

REFERENCES

R. Bracewell, The Fourier Transform and Its Applications,
2nd edition (McGraw-Hill, New York, 1978).

A. A. V. Gibson, unpublished results (1978).

P. Meakin and J. P. Jesson, "Computer simulation of multi-
pulse and Fourier transform NMR experiments.  I.  Simulations
using the Bloch equations", J. Magn. Resonance 10, 290-315
(1973).

D. Shaw, Fourier Transform N.M.R. Spectroscopy (Elsevier,
Amsterdam, 1976).

## II.A.3.   QUADRATURE DETECTION

In pulse FT NMR spectroscopy without quadrature detection (QD), the carrier frequency is intentionally set higher or
lower than all NMR lines in the immediate region.  This is
because the detector sees only one magnetization component in
the rotating frame and it is unable to tell if the magnetization
is precessing faster or slower than the carrier frequency.
Suppose we look at a nuclear resonance signal with a Larmor
frequency offset from the rotating frame frequency by 100 Hz.
The tip of the magnetization vector after a 90 degree pulse
describes a spiral in the rotating $x'$-$y'$ frame at 100 Hz with
the direction of rotation depending on whether the offset is
positive or negative.  Because there is only one detector,
only the $x'$ or the $y'$ component or a combination in between is
detected and clockwise and counterclockwise rotations cannot
be distinguished so that we are getting only a part of the
information that is available.  In Fourier transform language,

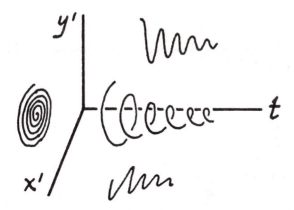

any signal or noise on one side of the carrier at $\nu_0$ is indistinguishable from the same signal or noise on the other side of
the carrier.  So, a single spectral feature can be detected at
two different frequencies symmetrically located about $\nu_0$ and
the process of reflecting the feature at, say, $\nu_0+\Delta\nu$ to $\nu_0-\Delta\nu$
is called folding.

Now we show you a sketch of what is happening in the
frequency domain.  The rf carrier frequency is at $\nu_0$ which is
chosen so that the entire spectrum of interest is bracketed
between $\nu_0$ and $\nu_0+\Delta\nu$ or $\nu_0-\Delta\nu$ (but not both) where $2\Delta\nu$ is
the digitizing rate.  In a single phase detection experiment
the folding problem is avoided by deliberately locating the
carrier frequency $\nu_0$ to one side of all spectral features and
ignoring the other side of $\nu_0$.  There are a couple of fairly
obvious but not fatal problems with this scheme, however.
One is that since everything gets folded over about $\nu_0$, the
noise to the left of $\nu_0$ in the figure will get folded into the
right half of the figure which doubles the noise power in the
spectrum.  A quick fix to this problem has been known and
that is to use a sharp crystal filter very close to $\nu_0$ to cut

out the noise from the left half (Allerhand, et al. 1973).

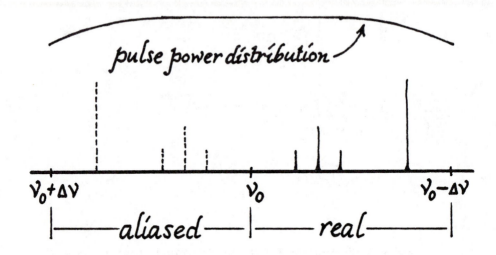

The frequency $\Delta\nu$, defined above as one-half the digitizing frequency, is called the Nyquist frequency.  Its significance is that any signal above $\nu_0+\Delta\nu$ cannot be represented
accurately.  Let us backtrack for a minute.  What does $\nu_0$
have to do with all this?  Not much, since we get rid of the
effect of this frequency $\nu_0$ by considering the FID in the
rotating frame with angular speed $2\pi\nu_0$.  So, $\nu_0$ is the origin
from which we measure $\Delta\nu$.  Any spectral information gets
folded (aliased) about the Nyquist frequency $\Delta\nu$.  Thus, $\Delta\nu$
must be set high enough, i.e., the digitizing rate must
exceed twice the highest frequency feature in the spectrum
or any spectral feature beyond $\Delta\nu$ will be aliased below $\Delta\nu$
and things will become very confusing.  There is noise being
aliased, of course, into the spectrum from beyond $\Delta\nu$ but this
can be filtered out with a low pass filter and is independent
of any reflections about $\nu_0$.

For the record, we point out that the Nyquist frequency
is occasionally taken to be the minimum allowable sampling
frequency rather than the highest frequency feature uniquely

reproducible with a given sampling rate. The latter is the generally acceped definition. For further details on sampling and the Nyquist theorem, refer, for example, to Cooper (1977).

Because the reference frequency for the detector $\nu_0$ cannot be set within the spectrum without aliasing real lines in a single phase detection experiment, it is difficult to perform selective saturation experiments wherein only a small part of the spectrum is irradiated. We will see that it will be trivial to irradiate anywhere in the spectrum with quadrature detection. For such selective irradiation experiments, usually to saturate an unwanted water line, the reader is referred to the papers by Redfield (1976) and Hoult (1976).

Another problem with not having QD is that the irradiation spectrum is centered at $\nu_0$ and the transmitter power to the entire left half of the spectrum in the last figure is wasted. Since transmitter power is a limited quantity, usually we cannot afford to waste it. Particularly for small gyromagnetic ratio nuclei, it often happens that the pulse length cannot be made short enough to irradiate the spectrum of interest uniformly (or in the rotating frame language of section I.B., the $H_1$ amplitude cannot be made large enough compared to the offset field $H_0 - \omega/\gamma$ with the result that different spectral lines are rotated by different amounts) and the resulting spectrum is distorted. If it is possible to center the irradiation on the spectrum rather than on $\nu_0$, a factor of two gain is possible in the spectral range irradiated. Since $H_1$ is proportional to the square root of the transmitter power, this is equivalent to quadrupling the transmitter power, a significant feat.

One way in which this might be implemented is to offset the transmitter frequency as well as the center frequency for the window provided by the receiver including the filter(s) from the reference frequency by, for example, a local oscillator output mixed into the reference signal so that it is at $\nu_0 \pm \Delta\nu/2$ while the reference stays at $\nu_0$ (Grutzner and Santini, 1976). This scheme is not implemented very often because the standard QD scheme is easier to implement. In fact, an equivalent scheme can be implemented through a variant of quadrature detection (Redfield and Kunz, 1975).

A simple but instructive implementation of QD (Pajer and Armitage, 1976) is shown in the next figure. With this setup, we can first accumulate the FID's in one memory block in the

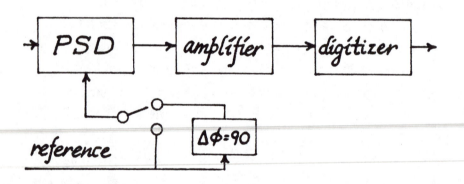

computer with the reference phase $\Delta\phi$ set, say, to 0°. Then the reference phase can be changed to 90° and an equal number of FID's accumulated in another memory block. Then the two blocks of data in quadrature are Fourier transformed from one block of complex time domain signal to an identical size block of complex frequency domain signal to create a spectrum. If the ADC conversion rate is $\nu$ Hz, the resulting spectrum width will also be $\nu$ Hz because digitizing two blocks at frequency $\nu$ is like digitizing one block at $2\nu$. A possible problem with this scheme is the difference in experimental parameters between the two accumulations caused by electronic instabilities which lead to distortions of the spectra as described later.

A more usual QD configuration is shown in the next figure. Reference signals in quadrature, i.e., 90 degrees out of phase with each other, are created with the use of a quad hybrid (discussed in V.C.10.) or a phase shifter. These reference signals go to two nominally identical phase sensitive detectors and their outputs are amplified, digitized, and stored in blocks of memory as real and imaginary components of the FID.

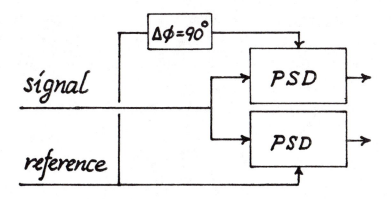

A complex (as opposed to real, not simple) fast Fourier transform (FFT) routine is needed to obtain the spectrum but this is not a hardship since the standard FFT algorithm is written for complex numbers. In fact, the data handling is easier with QD because the extra operations necessary to make the real single phase FID look complex to the FFT routine can be omitted. The main problems with the standard QD schemes are the difficulties in getting two reference signals exactly in quadrature and in the two amplifiers to have identical gain. Imperfect phase and gain adjustments lead to ghosts and glitches, i.e., unwanted phantom signals folded in from some other region of the spectrum.

Let us now consider how these unwanted features might

arise. We will do this by considering the QD experiment in more detail in terms of the symmetry properties of the various signal components. We begin by stating without proof that every well-behaved function can be separated into an odd and an even part. The oddness and the evenness is defined in terms of the symmetry property of the function about some chosen point, usually zero. An odd function in time $O(t)$ has the property that $O(t)=-O(-t)$ while an even function $E(t)$ has the property that $E(t)=E(-t)$.

Now what does all this have to do with an FID? It starts at $t=0$ and evolves until the NMR signal decays away. So it exists only for $t \geq 0$ and how can we tell what its evenness is? This question of what the FID is supposed to be for $t<0$ has to be addressed because the Fourier transform, as already stated in Chapter II, is an integral which goes from $-\infty$ to $+\infty$:

$$F(\omega) = \int_{-\infty}^{\infty} f(t)e^{-i\omega t}dt.$$

For a hint, we ask if the FID can be formed through some natural course of evolution rather than abruptly at $t=0$ by an rf pulse. Such an evolution does exist and it is the spin echo in which, under most circumstances, the positive time part after the center of the echo is the FID and the negative time part is a reflection of the positive time part in some way. In general, there will be both an even and an odd part to the echo unless the phase detector reference is adjusted to yield just the absorption or the dispersion components. The absorption component $u(t)$ of the echo is even and it has a maximum at $t=0$, namely at the center of the echo. The dispersion component $v(t)$, on the other hand, is odd and is equal to zero at $t=0$ since the signal should be continuous for all time and in particular at t=0. Another argument to support the notion of a signal null at t=0 for dispersion FID is that the disperson signal has a net area of zero in the frequency domain and this has to equal the initial FID value. So, the FID which gives rise to an absorption signal in the frequency domain is even in the time domain and that giving rise to dispersion is odd in the time domain. A general FID can be written as

$$f(t) = x_e(t) + x_o(t) + iy_e(t) + iy_o(t)$$

where x and y denote real and imaginary components and the subscripts e and o stand for even and odd. Its Fourier transform $F(\omega)$ is

$$F(\omega) = 2 \int_0^\infty x_e(t)\cos\omega t dt + 2i \int_0^\infty y_e(t)\cos\omega t dt$$

$$-2i \int_0^\infty x_o(t)\sin\omega t dt + 2 \int_0^\infty y_o(t)\sin\omega t dt \quad .$$

In QD, there are two signal channels, or their equivalent, and let us designate the two signals to be the real and the imaginary parts of the magnetization. Let us further define the absorption (even) FID u(t) as real and the dispersion (odd) FID v(t) as imaginary. Then since x(t)=u(t) will be even and y(t)=v(t) will be odd, the second and the third terms in the equation above are zero. Thus, the absorption FID contributes an even and real spectrum and the dispersion FID contributes an odd and real spectrum. The total spectrum is the sum of these contributions and we summarize this in a sketch.

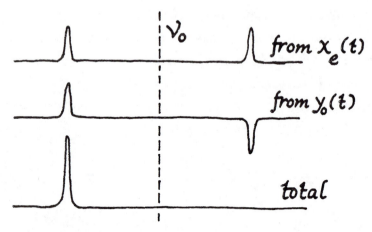

It is now clear how QD works and how ghosts may appear.  The reflected features, be they signal or noise, have opposite symmetries in the two spectra arising from the two FID's so they cancel while the true features add.  The cancellation is complete only if the amplitudes and the phases of the two FID's are correct.

This sketch also shows the origin of the $\sqrt{2}$ S/N improvement in a different way.  Because the noise associated with the two FID's in quadrature are also in quadrature, they are uncorrelated.  Therefore, when the two spectra are added, the signal is doubled while the noise increases by $\sqrt{2}$ resulting in a net $\sqrt{2}$ increase in S/N.

The assignment of which FID in QD is real and which is imaginary is arbitrary and, in fact, a continuous variation of the composition is possible by changing the phases of the detectors (while maintaining the 90° difference between the two detectors).  What happens if we shift the reference phases together by 90° so that the absorption FID is now considered to be imaginary and the dispersion FID real, that is, $x(t)=v(t)$ and $y(t)=-u(t)$?  The former will transform as an even-imaginary spectrum and the latter as an odd-imaginary one and the first and the fourth terms in the above equation will be zero. The total spectrum now displays a dispersion mode signal as shown. In general, x(t) and y(t) will be a linear combination of u(t) and v(t) so the total spectrum will have both real and imaginary parts and the usual software phase correction will be necessary to extract a pure absorption or dispersion mode spectrum.

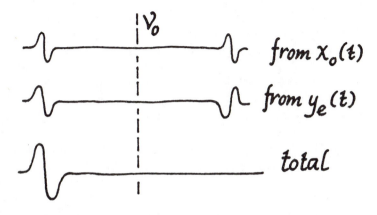

The common methods of phase and amplitude adjustments, for example, with an oscilloscope as monitor, can reduce the extraneous resonances to a few percent of the desired lines without undue difficulty. In order to cut down on the unwanted lines further, more sophisticated methods must be used. A popular solution is a scheme wherein the phase detected signals are alternately routed through the two channels and then combined in the memory in a certain way (Stejskal and Schaefer, 1974a, 1974c).

In this scheme, two sets of data are collected through the two channels. The reference phase is incremented 90° in both of them (which basically interchanges the absorption

and dispersion components coming through the two channels)
and the outputs are added to appropriate blocks of data from
previous FID's with suitable sign changes.  Suppose we start
the experiment with two channels as shown on p. 65, and that
the phases are adjusted initially to give a purely absorption
mode FID u(t) through channel A ($\phi$=0) and a dispersion
signal v(t) through channel B
($\phi$=90°).  Earlier in this section, we defined x(t) and y(t)
to be signals coming through
the    channels    A    and    B.
Following Stejskal and Schaefer
we can indicate the situation
with a figure where along the

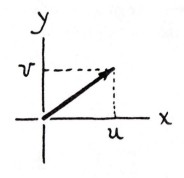

x and y axes lie the components of the magnetization coming
through the A and B channels which, in this case, are u(t)
and v(t), the absorption and dispersion components, respectively.

If  we  repeat  the  experiment  with  the  reference  phase
incremented by 90°, so that
$\phi$=90°  for  channel  A  and
$\phi$=180° for channel B, channel
A  now  sees  v(t)  and  channel
B sees -u(t) as shown.  These
two  sets  of  signals  can  be
combined  by  taking  the  first
pair  of  signals  through  the

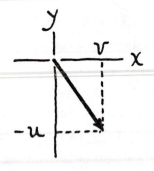

two channels (u,v), swapping  them  in  memory  to  (v,u),
changing  the  sign  of  the  second  block  to  (v,-u) and adding
the second pair of signals (v,-u) to form (2v,-2u).  If the A
channel gain exceeds that in B channel by 10%, say, then the
signals to be combined are (1.1u,v) and (1.1v,-u) resulting

in (2.1v,-2.1u) which has the correct phase. In a similar way, phase errors between the two channels can also be corrected.

Clearly, there are several ways of implementing the necessary data routing. The most primitive would be to accumulate one-half of the total FID's with one set of phase, perform the required swapping and sign change, and then accumulate the other half of the FID's on top. This method is susceptible to drifts in the electronics resulting in the two data sets being accumulated under different conditions possibly leading to imperfect cancellation of the phase and amplitude errors. A common way to avoid this problem is to alternate the two different sets of conditions every other pulse with the swapping and sign change operations performed either with the use of buffer memory or in the data blocks.

Another method to correct imperfect phases and amplitudes of the two supposedly quadrature signals is to do it mathematically. The basic idea is to construct a linear combination of the two signals to create a third which is truly in quadrature with one of the original signals (Stejskal and Schaefer, 1974b; Parks and Johanneson 1976).

We now consider quantitatively a few representative ways in which QD can be implemented. Suppose we want to record a 500 Hz wide spectrum. In a non-QD experiment, the ADC should operate at 1 kHz and the carrier is offset from the center of the spectrum by 250 Hz as shown in the figure below. This system can be converted to a QD system by running the ADC at the same 1 kHz rate, alternating the input between the two phase detectors whose references are in quadrature, and setting the carrier in the center of the spectrum. Then each FID is recorded at 500 Hz and the complex FT yields a 500 Hz wide spectrum with identical resolution as before. If the digitizer could store 2048 points so that each component of FID is defined by 1024 points, the

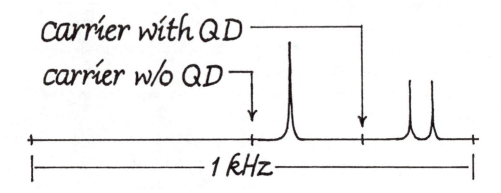

transformed spectrum will have 1024 points with or without QD (but without zero filling). As stated earlier, the advantages gained (over a system with same digitizing rate and number of data points but without QD) would be a $\sqrt{2}$ improvement in S/N due to elimination of noise from the left half of the spectrum of the figure and a factor of two reduction in the spread of the magnetization about $H_e$ or, alternatively, in the power fall-off across the spectrum.

Suppose we choose to implement QD by purchasing another digitizer, instead. It is clear that two digitizers running at 500 Hz is equivalent to the previous example of one digitizer running at 1 kHz and alternating between two FID's in quadrature. Having two 500 Hz digitizers would simplify the operation slightly because they can be digitizing the FID's simultaneously rather than alternately.

Finally, again consider the example by Pajer and Armitage on page 64. If the digitizing rate is set to 500 Hz, the result would be identical with the example of the 1 kHz digitizer working alternately on the two FID's in quadrature except that it will require two tries to digitize both FID's so the $\sqrt{2}$ advantage of the alternate sampling method is lost and the S/N is identical with the single phase experiment. (All other advantages of QD still remain, however.) But if we had a 1 kHz digitizer to start with, we could recover that factor of $\sqrt{2}$ by making an alternate sampling scheme by toggling the switch in this example with the phase alternating between every digitization. A typical electronic phase shifter responds in a few microseconds so that is perfectly capable of performing this function in high resolution NMR.

We now mention one last advantage of QD over single phase detection which is significant under some circumstances. Consider a spectrometer with a deadtime of 30 µs with which we want to record a 25 kHz wide spectrum. With single phase detection, the digitizer has to run at 20 µs per point so that the first two points will be within the deadtime. On the other hand, with a QD scheme in which the two signals are digitized simultaneously by two ADC's, each FID is recorded at a rate of 40 µs per point so that only the first point for each FID is lost in the deadtime. Now, because of the finite time resolution of the digitizing process, the apparent deadtime in the digitized FID may not be the actual deadtime. In

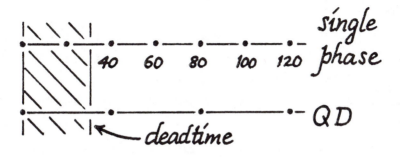

this example, the Fourier transform program perceives the deadtime to be 20 μs in the case of the single phase experiment and just a single point in the QD experiment. Since the FT of a single point is a constant, the QD version of this experiment should be free from baseline distortion due to deadtime effects whereas the single phase version should suffer some distortion.

Under what conditions does this advantage for QD over single phase detection hold? Suppose we take a spectrometer with a fixed deadtime and examine the relative merits of the two detection schemes only with regard to this deadtime effect as a function of the digitizing rate. Clearly, there is no advantage either way when the deadtime is shorter than the digitizing period for the faster digitizer. As the digitizing rate is increased, there will be a time when the second data point in the single phase FID will occur within the deadtime. This is the case discussed above and there will be a baseline distortion only in the single phase experiment. As the digitizing rate is increased further, the third and the second points, respectively, of the single phase and QD FID's will simultaneously become part of the deadtime. It can be shown that the two experiments are equivalent under these conditions. In the next increment of the digitizing rate, we add one more point only in the single phase experiment to the deadtime and there will be a slight advantage to QD but in the following increment both FID's get one more point into the deadtime regime and QD's advantage is lost again. So, in each successive increment of the digitizing rate, the advantage enjoyed by QD, if any, diminishes and in the limit of a large number of points within the deadtime, when the baseline problems are fierce, QD does not enjoy any advantage over single phase detection insofar as the deadtime effect is concerned. This limiting behavior is consistent with the notion that the frequency dependence of the distortion should depend only on the duration of the deadtime and not on how many data points were taken within that deadtime.

In summary, quadrature detection is an extremely advantageous detection scheme for FT NMR and should be incorporated as a standard item in any FT spectrometer. Without it, 1) the transmitter power is used inefficiently leading to larger frequency dependent tip angles, 2) less than optimum S/N, the amount of which depends on the method of implementation; and 3) inconvenience in doing experiments requiring irradiation of a part of the spectrum such as selective saturation experiments.

## REFERENCES

A. Allerhand, R. F. Childers, and E. Oldfield, "Carbon-13 Fourier transform NMR at 14.2 kG in a 20mm probe," J. Magn. Resonance 11, 272-278 (1973).

J. W. Cooper, The Minicomputer in the Laboratory (John Wiley & Sons, New York, 1977), p. 225.

J. B. Grutzner and R. E. Santini, "A broadband system for the observation of NMR spectra of any resonant nucleus," J. Magn. Resonance 22, 155-160 (1976).

D. I. Hoult, "Solvent peak saturation with single phase and quadrature Fourier transform," J. Magn. Resonance 21, 337-347 (1976).

R. T. Pajer and I. M. Armitage, "Single-channel quadrature FT NMR," J. Magn. Resonance 21, 365-367 (1976).

S. I. Parks and R. B. Johannesen, "An algorithm for automatic correction of quadrature phase detector outputs," J. Magn. Resonance 22, 265-267 (1976).

A. G. Redfield, "How to build a Fourier transform NMR

spectrometer for biochemical applications" in Introductory Essays, edited by M. M. Pintar (Springer-Verlag, Berlin, 1976).

A. G. Redfield and S. D. Kunz, "Quadrature Fourier NMR detection: Simple multiplex for dual detection and discussion," J. Magn. Resonance 19, 250-254 (1975).

E. O. Stejskal and J. Schaefer, "Data routing in quadrature FT NMR," J. Magn. Resonance 13, 249-251 (1974); "Comparisons of quadrature and single-phase Fourier transform NMR," J. Magn. Resonance 13, 249-251 (1974); "Comparison of quadrature and single-phase Fourier transform NMR," ibid 14, 160-169 (1974).

## II.B. THE MECHANICS OF FT NMR

The following short sections deal with how the FT NMR experiment is performed in terms of setting the correct parameters. They dovetail into many of the sections in Chapter VI. Aside from the specific references listed in each section and the general references in Appendix A, there are other good reviews of the how and why of FT NMR. For example, there is a compact elementary treatment by Van Hecke (1977), which can be read quickly to get an idea of the field. A different kind of a reference we cite here is an in-depth article specifically on digital data acquisition and processing in FT NMR by Lindon and Ferrige (1980).

## REFERENCES

J. C. Lindon and A. G. Ferrige, "Digitisation and data processing in Fourier transform NMR," Progress in NMR Spectroscopy <u>14</u>, 27-66 (1980).

P. Van Hecke, "The Fourier transform in NMR" in <u>Nuclear Magnetic Resonance in Solids</u>, edited by L. Van Gerven (Plenum Press, New York, 1977), pp. 201-228.

## II.B.1. DIGITIZING RATE AND THE SIZE OF THE TRANSFORM

The sampling theorem states that to record a spectrum unambiguously with frequencies as high as $\nu$, one must sample the FID at a rate of at least $2\nu$ (Nyquist frequency). For example, a 5 kHz wide $^{13}C$ spectrum must be digitized at

10 kHz, so that the duration between points (called dwell time) will be 100 µs. For protons in the same magnet, the spectral width might be 1 kHz so the digitizing rate must be 2 kHz and the dwell time, 500 µs. Failure to sample at a high enough rate results in the system being unable to distinguish frequencies X Hz higher than 2ν from those X Hz lower. For the same number of data points in a given time, the spectrum width is independent of whether or not QD is used. With QD, however, the window will be centered on the carrier whereas, without it, the window will have the carrier at one edge and the other edge will be one-half of the sampling frequency away.

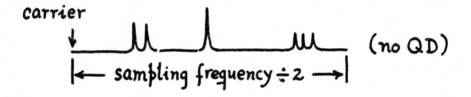

Any spectral components (including noise) lying outside either limits will be folded back into the spectrum. In order to prevent this, a low pass filter with corner frequency near the high frequency edge of the range of interest can be used.

Once the digitizing rate has been chosen, the size of the transform determines the resolution of the spectrum because the number of points times the dwell time is the duration of the FID observation. For example, if the FID is digitized at 10 kHz for 1 second, a 1 Hz resolution is the best that can be expected and the frequency window will be 5 kHz wide.

The number of points digitized in a run must be sufficient to define the narrowest feature in the transformed spectrum. In order to resolve a line which is 1 Hz wide, it must be defined by several points. If we say that five points are required to define a peak, the measurement must last for 5/(1 Hz) which is 5 seconds. If the desired spectrum width is 5 kHz, you need to digitize and record (5 kHz)×(5 s) points which is 25,000!

In order to create more points defining the lines without any more real resolution, we can use zero filling. For example, if 8K zeroes were appended to the extant 8K points of an FID, a subsequent 16K point transform will result in 8K points in the real spectrum and they will consist of the original 4K points with an additional 4K points interpolated between them. This doubles the number of points and makes the spectrum smoother. Note that you could have digitized the FID at 16K points in the first place, but without severe apodization the last 8K points would have consisted only of noise. Zero filling does the same job without the noise from the last 8K points.

Are the interpolated points telling you anything new? Not really except what you have assumed, namely that there are no signal components in real time after 0.8 sec in the above example. So, we have dictated the interpolation.

## II.B.2.  DYNAMIC RANGE AND DIGITIZER RESOLUTION

Dynamic range is the ratio of the biggest signal to the smallest significant feature within the data of interest be they FID's or spectra. The size of the digitizer word must be adequate to contain the most intense signal in the FID and also be able to resolve the smallest real signal feature. In

practice, the maximum voltage excursion is adjusted to fit
within the digitizer input range and the digitizer resolution,
e.g. 8, 10, or 12-bits, is chosen so that the smallest significant feature is larger than 1-bit.  The word "resolution" used
here is resolution in amplitude and not in frequency.  The
digitizer resolution should be dictated by the S/N because we
cannot obtain the information any more accurately than permitted by the noise.  However, the resolution must be fine
enough so that the least significant bit is smaller than the
noise amplitude.  Otherwise, the random character of the
noise will not show up and there will be discrete steps in the
data.

The choice of digitizer resolution depends in part on the
dynamic range of the signal and in part on the number of
transients to be sampled.  Choose a short enough digitizer
word so that all the transients desired can be added without
overflowing the word length allocated in memory.  We might
think that with a 12-bit digitizer and a 16-bit word length,
only $2^4 = 16$ transients can be accumulated without overflow.
Because the noise does not add up coherently, but rather as
the square root of the number of transients, far more transients can be added before the memory overflows than the
calculation above indicates if S/N is not large which is when
it is most necessary to signal average.  In particular the total
number of transients N without overflow is given by

$$N = \left\{ \frac{[1 + 4S(S + 1)2^{w-d}]^{\frac{1}{2}} - 1}{2S} \right\}^2$$

where S is the S/N, w is the memory word length, and d is
the number of digitizer bits (Cooper, 1976).  Thus, for a
signal with S/N=1, 32587 FID's can be accumulated in a 16-bit

computer with a 2-bit ADC for a signal enhancement of 180 whereas only 26 scans can be accumulated with a 12-bit ADC for a signal enhancement of 5!

These considerations are less critical in many commercial NMR spectrometers which are provided with computer word lengths of 20 or 24-bits and the problem is largely eliminated by using double precision signal averaging in which double length words are used for signal accumulation. Nevertheless, it is good to know how to manipulate the digitizer resolution according to the S/N in order to maximize the number of scans possible because the additional memory capacity neces- sary to do double precision averaging can be considerable for large data tables. Even more important is the extra computer time required to do double precision arithmetic which is a handicap under certain circumstances. On the other hand, if you are doing single precision signal averaging, the above relation is so handy that it will serve you well to make up a table of N as a function of S and w-d and keep it near the spectrometer.

REFERENCE

J. W. Cooper, "Computers in NMR--I: Signal averaging in Fourier transform NMR," Computers and Chem. 1, 55-60 (1976).

II.B.3.   POSITIONING THE CARRIER

The width of the frequency window in an FT NMR experi- ment is the digitizing rate if the folding problems are ignored.

If the carrier frequency were inside the spectrum, there could be magnetizations precessing faster and slower than the carrier. Since the detector, in the absence of quadrature detection (QD), cannot distinguish them, the resulting spectrum consists of one part of the spectrum folded over into the other part. This problem can be avoided by placing the carrier to one side or the other of the entire spectrum and using only half of the available frequency range although the best solution is to use QD as described in II.A.3.

Since the carrier is the reference frequency in the detection process, it becomes the origin for the frequency scale after the Fourier transform. It is strictly arbitrary whether the frequency scale runs left to right or conversely but the usual convention is to plot the spectra so that the frequency increases to the left. This is equivalent to the laboratory field increasing to the right in cw NMR, an even older convention.

We digress here to point out that describing spectral features in terms of upfield or downfield is a bit ambiguous. This nomenclature arose, of course, from the way the laboratory magnetic field was swept in cw NMR to detect spins resonating at a particular frequency. Even in cw NMR, the nomenclature is awkward because the "upfield" peak is actually experiencing a smaller total magnetic field than the "downfield" peak at a given laboratory field, thus requiring a larger applied field to resonate at the Larmor frequency.

The spins resonating upfield experience a smaller fraction of the applied field than those downfield because they are "shielded" better, i.e., their atomic electrons contribute induced fields opposed to the laboratory field. Thus, an alternative description for these cases is that the well-shielded spins are diamagnetically shifted and, conversely, the less-shielded spins are paramagnetically shifted with respect to each other. In addition, in most pulse NMR experiments, the laboratory field is fixed and there is a distribution of Larmor frequencies. Thus, it seems to us to be more logical to talk of up or down frequency, of paramagnetic or diamagnetic shifts, or of the amount of shielding, rather than down or upfield shifts. [The exception to this argument is the field scanning spin echo experiments commonly used to measure Knight shifts in metals.]

In any event, the spectrum moves back and forth in the window when the carrier frequency is changed. Assuming the conventional spectral representation discussed above, increasing the carrier frequency causes the window to move toward higher frequencies relative to the spectrum so that the spectrum will move to the right. Likewise, decreasing the carrier frequency will move the spectrum to the left.

If the window is not large enough to contain the entire spectrum, any lines which lie outside it will be folded back into the window an equal distance from the edge. Such folded lines can be so identified in two ways. First, they will not phase properly so that the correction giving the correct phase for the rest of the spectrum will not be correct for the folded lines. Secondly, the folded lines will move in the wrong direction when the carrier frequency is moved. Whenever there is a possibility of folded lines, increase the digitizing rate to increase the window width or change the carrier to move the window.

For fixed field FT NMR with QD, the above information is about all you need on this subject. However, without QD, there is one more decision to make and that is on which side of the spectrum to place the carrier. Here there is much less of a convention and you can put it on either side. With the above mentioned convention of having the frequency increase to the left, this means that when you offset the carrier so that it is less than the lowest frequency feature in the spectrum, the carrier will be at the right edge of the spectrum and vice versa. The opposite edge will be defined by the Nyquist frequency so that the spectral range will equal only one half of the digitizing rate. Note that in order to achieve the conventional spectral representation, the direction of increasing frequency in the memory after the Fourier transformation would depend on which end of the spectrum the carrier is located. Thus, depending on your software, you may have to reverse the abscissa before displaying the spectrum.

Although it makes no difference where the carrier is located, you should be aware of which edge of the spectrum is defined by the carrier so that you can take the appropriate action for lines which are folded over or aliased in the absence of QD. Be sure to label your spectra, especially for publication, as to where the carrier is and even as to which direction the shielding increases.

## II.B.4.   PHASE CORRECTIONS

In FT NMR, it is not trivial to set the detector phase so that it will be set to the absorption mode for all lines in the spectrum. In addition to a phase offset error due to improper setting of the detector for the absorption mode, there will be frequency dependent phase changes across the spectrum. Lines at the start of the spectrum might appear absorption-like while the percentage of dispersion-like character may increase with frequency. One possible reason for this is that the time taken as t=0 in the actual FID may not be the correct time origin because of either the finite width of the pulse itself (the correct time origin is in the middle of the pulse), the time required for receiver recovery, or any other delays in starting the digitizer. As can be seen from the following diagram, this produces a frequency dependent phase shift.

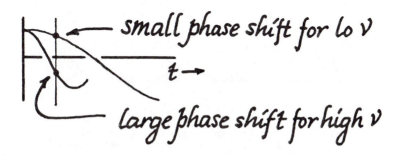

If the initiation of data collection is delayed, the effective phase shift at different frequencies becomes larger as the frequency becomes larger. For frequencies near the carrier the difference between the first point and the initial value of the FID can hardly be detected but for higher frequencies, starting late is equivalent to introducing a larger and larger phase shift. This result is a direct consequence of the shift theorem discussed in II.A.1. The exact amount of the phase shift due to missetting the time origin is 180 degrees per point, that is, the relative phase shift across the spectrum changes by 180 degrees for each shift in the time origin by the digitization interval regardless of the digitization rate.

The receiver system may also introduce phase shifts. The most basic cause for frequency dependent phase shifts, though, is that a finite $H_1$ rotates magnetizations at different offsets by different amounts (see II.A.2.).

All of these phase shifts are approximately linear in frequency for moderate frequency deviation. For example, it was pointed out in II.A.2. that the phase shift due to finite $H_1$ is quite linear for offsets up to $\gamma H_1$. Therefore, a linear phase correction of the form $\Delta\phi=(A\omega/2\pi)+B$, implemented by software, is quite effective.

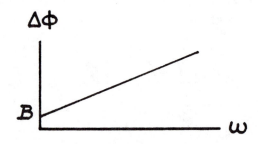

In this connection, we point out that determining the required phase correction parameters at two points may not be sufficient. For example, if there are only two peaks and they seem to be phased correctly, it could be a coincidence with a 360 degree phase shift between them. In fact, this situation has been deliberately created for a two-line spectrum by adjusting the time origin suitably to make the two lines have the same phase and thus allow an accurate determination of their separation (Canet, et al., 1976). Separations measured in this way will be correct as long as the line shapes are symmetric. The shapes and widths themselves will be distorted in such a scheme, however, since there is a dispersion of phase shifts within each peak.

The fact that a software linear phase correction is successful in correcting for the time origin in solution NMR is partly accidental. The exponential function which occurs in the majority of solution NMR FID's retains the same functional form when its front end is truncated. This means that when the time origin is shifted, the individual line shapes are not distorted. Only the relative weights of FID components having long and short $T_2$ are changed and this fact has been used in resolution enhancement (see II.B.5.).

On the other hand, most decaying functions change their character when their front ends are truncated. If the front end of a Gaussian FID is cut off, the remaining FID gains an exponential character so that no amount of phase correction will bring back the correct line shape. (See the figure in the next section.) Thus, for solids, it is better to first adjust the time origin to be approximately at the center of the transmitter pulse and then make software phase corrections. In fact, in the solids experiments we have performed, it seems to work well if nearly all of the phase correction is made through the adjustment of the time origin.

For manual phase correction, the value of the intercept and the slope of the phase correction or equivalent parameters such as the phases at the two ends are entered by the operator into the computer after the FT is performed and the resulting spectrum inspected for absorption mode lines everywhere. Alternatively, one can run a single line spectrum at various frequency offsets and empirically determine the phase correction for each location in the spectral window. Either of these alternatives must be performed with a model sample having narrow symmetric lines before a sample having broad and possibly asymmetric lines is looked at. As the phase shift is largely an instrument determined parameter, once the phase correction has been determined, it should remain the same for all spectra taken under the same conditions except for minor adjustments. The correction can be stored in the computer and applied automatically. Metallic samples are exceptions to this rule because the electrical conductivity and thus the skin depth affects the rf phase.

REFERENCE

D. Canet, C. Goulon-Ginet, and J. P. Marchal, "Accurate determination of parameters for $^{17}O$ in natural abundance by Fourier transform NMR," J. Magn. Resonance 22, 537-542 (1976).

## II.B.5.   DIGITALLY MASSAGING THE FID

Different parts of the FID contain different kinds of information.   By manipulating the data in the time domain, we may choose to emphasize different parts of the FID and thereby emphasize different kinds of information.   Naturally, when we do so, we sacrifice other kinds of information.   The most common of such manipulations is to trade off resolution for better S/N or vice versa.

The FID contains signals from different magnetization components precessing at different rates and beating together. Information about the existence of magnetization components with large frequency differences will be apparent at the start of the FID but information about smaller frequency differences will not surface until sufficient time has elapsed to permit the slow beats to be seen.   The extent to which one follows the FID into this long time region to observe small frequency differences determines the ultimate resolution in the spectrum. Similarly, for a system consisting of narrow and broad lines, the nuclei having the narrowest line contribute more to the tail end of the FID than those having broader lines, which contribute mostly to the beginning.

Thus, there is some flexibility in choosing what spectral information to emphasize. The FID tail containing information about small frequency differences has a small amplitude and therefore has poor S/N, while the start of the FID has better S/N but it characterizes only the large frequency differences in the spectrum. Therefore, the start of the FID can be emphasized for optimum S/N while sacrificing resolution or the resolution can be emphasized at the expense of S/N by weighting the points at the tail relative to those at the start.

The weighting to emphasize either the S/N or resolution can be gentle or drastic. In the former category are multiplication of the FID by relatively slowly varying functions such as a decreasing exponential to emphasize the beginning and an increasing exponential to emphasize the tail.

A special form of such a convolution commonly used in broadline solid state work is closely related to zero filling discussed elsewhere. For a study of a solid line, the digitiz-

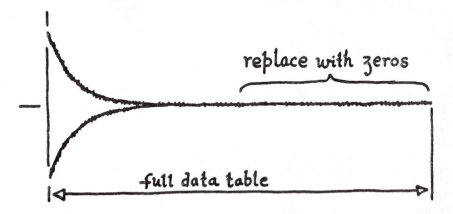

ing rate must be low enough so that there is a well-defined baseline in the frequency domain, i.e., the FID has to occupy a relatively small portion of the data table. In this situation, the majority of the data will be zero and you might as well

put in real zeros instead of using the noisy baseline you end
up with from the experiment. Therefore, it is common practice
to replace the last half, say, of the data table with zeros
before the Fourier transform is performed. Any zero filling
for interpolation would take place in addition to this operation.

In the category of more drastic weighting are multiplication of the FID by a faster varying function or, in an extreme
case, just dropping off data points. Suppose we just cut off
the tail of the FID. That is equivalent to multiplying the FID
by a square function before the transformation. From the
convolution theorem (Section II.A.1) we know that a Fourier
transform after such a multiplication leads to a convolution of
the Fourier transforms of the two functions. Since the transform of a square function is $sinc(x)=sin\pi x/\pi x$, the resulting
spectrum is the original one modified by each spectral feature

having the characteristics of the sinc function but at the
same time having better S/N. The better S/N has been
obtained not only at the expense of poorer resolution but, in
this case, also at the expense of putting the sinc function
wiggles on the spectral features. These wiggles can be
reduced or removed by cutting off the FID tail more gently
with a process called apodization (for cutting off the feet in

Greek).  Some of the obvious functions to use for apodization
are the linear ramp and the decreasing exponential (Cooper,
1977).  The latter is about the gentlest form of such an
operation.

Another technique for separating different contributions
to the FID which has been used sucessfully is to simply drop
off points at the beginning of the FID (Roth, 1980).  Suppose
the NMR line consists of a Lorentzian portion and a broader
non-Lorentzian portion which we can suppose to be Gaussian
without altering the argument.  As we discussed in the last
section, the Lorentzian line gives rise to an exponential FID
for which the shape is independent of the portion of the
signal considered.

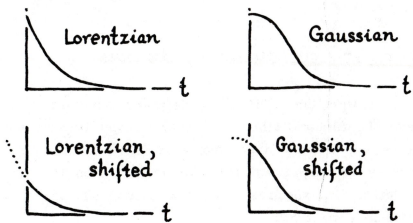

On the other hand, a Gaussian (and any other FID
shape usually encountered besides the exponential) FID starts
off flat, falls off, and levels off again so that different parts
of the decay can be identified by its behavior.  It turns out
that most, if not all, other common FID shapes besides the
exponential are quite flat at the beginning so that if the data
points are dropped off at the beginning, the non-exponential
character of the FID is reduced.  So, there are two effects of
this truncation.  One is to make the lines more Lorentzian

upon Fourier transformation and the other is to enhance the resolution at the expense of S/N. Needless to say, the frequency dependent phase correction must be changed because the time origin is, in effect, being changed.


REFERENCES

J. W. Cooper, The Minicomputer in the Laboratory: With Examples Using the PDP-11 (Wiley, New York, 1977), p. 300.

K. Roth, "A Simple resolution enhancement technique: Delayed Fourier transformation," J. Magn. Resonance 38, 65-70 (1980).

## II.B.6. FT SPECTROSCOPY OF WIDE LINES

For wide line NMR, it is especially important to understand the fundamentals of FT spectroscopy because some of the aspects which can be ignored in high resolution FT spectroscopy are significant for wide lines. Even for the high resolution NMR spectroscopist, this section will provide additional understanding.

Instrumentation is appropriately different for the wide line case. The demands on magnet resolution and stability are reduced but those on data acquisition and transmitter and receiver recovery are increased. While most high resolution FT systems use digitizers that can digitize at a maximum rate of 50 kHz or less, a MHz or greater rate may be required for the spectra of solids.

In high resolution FT, the long duration of the FID and the fact that the spectrum usually consists of Lorentzian absorption lines enable the operator to make simplifying assumptions. For example, it is common to slightly delay the data acquisition after the pulse to avoid pulse breakthrough in the FID. As this delay is normally insignificant in comparison to the total acquisition time, its effects on the spectrum are minimal except for the frequency dependent phase shift which can be adjusted computationally after the data gathering. The S/N is improved by applying a time window function to the FID in order to de-emphasize the tail of the FID in comparison with the early part of the FID where S/N is better. The most common time window function used is an exponential decay which is best suited for Lorentzian lines. Phase correction can be made empirically by examining lines in different parts of the spectrum and adjusting the phase until they are all in the absorption mode.

Contrast the above situation with that of wide line FT NMR. First, we need the instrumental capabilities mentioned earlier. Secondly, often we are trying to obtain the shape of a single line or those of a small number of lines which may be complex. Therefore, we cannot presume to know the line shape in order to phase it properly. Similarly, an exponential time window function may alter the lineshape information. Third, the delay time used to avoid the pulse breakthrough in the FID is almost certain to be a significant fraction of the total acquisition time and must be taken into account. Let us deal with each of these difficulties in order.

The phasing problem is usually overcome by a calibration of the required phase correction measured as a function of offset with a simple line from a liquid sample. This may not work for metallic samples which can have a signal with a phase different from non-metallic samples, not to mention the differences in the phase of the signal from different parts of the sample.  A possible correction procedure here, provided there is a good S/N, is to adjust the admixture of the components to minimize the ratio of signal below the baseline to that above.  A prerequisite to this is the elimination of frequency dependent phase shifts which can be accomplished with an adjustment of the delay in acquiring data and/or with a software correction.  The phase shift can be monitored by looking at the real and the imaginary components of the magnetization as suggested in II.A.2.

The deadtime problem is also not trivial.  Suppose the FID is set equal to zero during the deadtime.  (The argument will hold for any other constant.)  This is equivalent to taking an ideal FID $f(t)$ without deadtime problems and multiplying it by a function $\Pi'(t)$ which is zero from $t=0$ to $t_o$ and one for $t>t_o$.  Let us define a function $\Pi(t)$ to be equal to $1-\Pi'(t)$ which has the property that $\Pi(t)=1$, $t<t_o$, and $\Pi(t)=0$, $t>t_o$.  Then the actual FID is $f(t)\{1-\Pi(t)\}$.  Its Fourier transform can be calculated easily in terms of $F(x)$, the transform

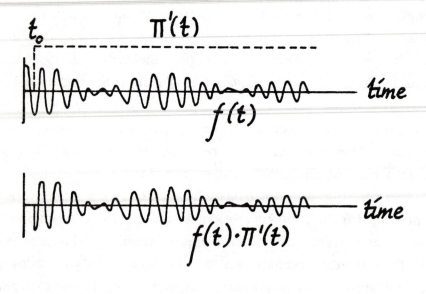

of $f(t)$, and $(\sin x)/x$, the transform of $\Pi(t)$, by the use of the convolution theorem described in section II.A.1. Since the Fourier transform of a produt is the convolution of the Fourier transforms and vice versa, the Fourier transform of $f(t)\{1-\Pi(t)\}$ is $F(x)*(-\sin x)/x$ where the symbol $*$ stands for a convolution. Thus, the spectrum obtained from the actual FID with the beginning missing is the ideal spectrum $F(x)$ with a $(\sin x)/x$ "folded" around each feature. That the baseline should be depressed around each line is clear from the fact that the initial value of the FID is proportional to the area under the spectrum. Since the initial value is forced to

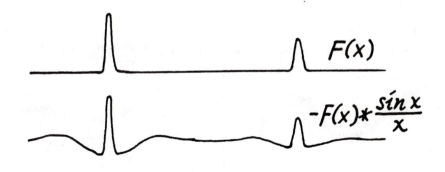

be zero, the baseline must be depressed to compensate for the peaks in order for the total area to be zero.

As it is clear from the above discussion, the deadtime problem is independent from the phase shift problem in their origins. They become related only when we attempt to eliminate the deadtime problem by adjusting the FID time origin so that $t_o=0$. This adjustment introduces a frequency dependent

phase shift by the shift theorem discussed in II.A.1. For
high resolution spectra, the phasing correction is easy so
that it is possible to correct for whatever small deadtimes
there may be in this way. For broad lines, the problem is
much more difficult and all possible combinations of adjustments
may be needed.

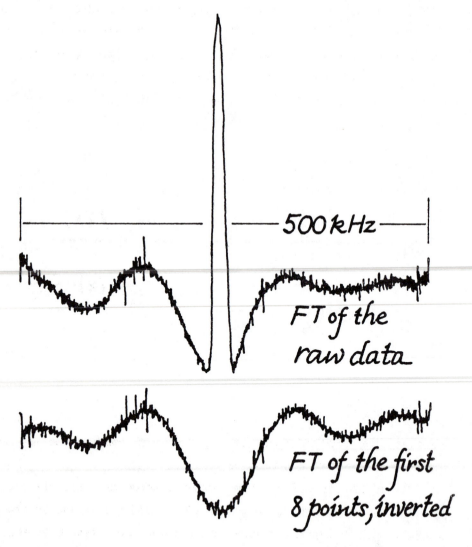

$500\,kHz$

FT of the
raw data

FT of the first
8 points, inverted

The figure above is an FT spectrum taken without QD of
protons in $KB_3H_8$ at room temperature. The FID was digitized

at 1 MHz and the deadtime corresponded to the first eight points which were set to zero. The lower trace shows an FT spectrum obtained from the same data by taking only the first 8 data points after the deadtime to approximate the function $\Pi(t)$ and inverting it. These traces do suggest a correction scheme for the deadtime problem but we have found it to be not straightforward because of phasing problems. In this example, the beginning of the digitizing process was adjusted to eliminate the frequency dependent phase shift in the upper trace but the lower trace required a frequency dependent phase correction due to the obviously different time origin. This example does serve to clarify the source of the oscillatory baseline seen quite often in FT spectra of broad peaks, however.

Although FT NMR can become a major tool for wide line NMR spectroscopy, cw spectrometers will still be necessary to record spectra which are so broad that the deadtime is comparable to or greater than $T_2$ from samples which do not give rise to echoes. In the frequency domain, this means that the $(\sin x)/x$ wiggles shown in the last figure will have frequencies and amplitudes comparable or greater than the peak itself, thus obscuring the desired spectrum. See IV.B.3. and VI.D.4. for other pulse techniques to overcome the deadtime problem.

## II.C.  DECOUPLING IN SOLUTION NMR

One of the great conceptual steps forward in NMR was the invention of multiple irradiation, and it came very early in the history of NMR. Even though the field of NMR is innundated with clever multiple irradiation schemes of major importance, for example, cross-polarization and multiple quantum coherence to name just two, the most common use for multiple irradiation is still the double irradiation we know of as decoupling and which produces nuclear Overhauser enhancement (NOE). We now discuss decoupling as it applies to high resolution solution NMR but not in great detail. The readers are referred to discussions and references such as those in Chapter 6 of Martin, et al. (Appendix A) for more details. Decoupling as it applies to solids is discussed in sections IV.E.2. and IV.E.3.

In high resolution NMR, there are spin multiplets most commonly due to electron coupled interactions between a pair of nearby spins. For example, a carbon-13 nucleus may have a neighboring proton whose possible spin-up or -down orientations are reflected in the slightly different resonance frequencies resulting in a doublet for the NMR signal. In a complex molecule with a multitude of such interacting sets of nuclei, it is often desirable to suppress the multiplets, at least temporarily, in order to simplify the spectra. One way to do this is to simply delete the offending nuclei which are giving rise to these multiplets by isotopic substitution. This is an expensive and difficult solution although it is a very powerful tool, especially if the substitution is performed selectively. The more practical method, and the one used most of the time, is to "decouple" the offending spins from the spins of interest with a second rf irradiation at or near the second Larmor frequency.

From the point of view of the actual methods used in the decoupling process, we may make a distinction between heteronuclear and homonuclear decoupling. The former is much easier because the decoupling irradiation frequency is significantly different from the NMR carrier frequency and there is little chance for interference. The example we cited of decoupling the effects of protons from carbon-13 spectra obviously is that of heteronuclear decoupling with the decoupling frequency approximately four times the Larmor frequency, as determined by the gyromagnetic ratios.

Now, how does an irradiation at the proton frequency decouple the protons from the carbon-13 nuclei, for example? In a qualitative way, we can see that the irradiation at the Larmor frequency of the proton can induce transitions between the two allowed proton levels corresponding to the parallel and antiparallel orientations with respect to the static field. If such a transition between the two orientations occurred fast enough, the carbon nucleus will simply see an average field from the protons. Thus, in the presence of a decoupling field the carbon resonance will be a singlet at the average position of the doublet.

## Broadband heteronuclear decoupling

Within the realm of heteronuclear decoupling, we have a choice of irradiating all of the lines in the proton spectrum (in the proton-carbon system we are still considering as an example) or just some of them. In order to irradiate all of the proton lines, we need a scheme for distributing the irradiation over the appropriate chemical shift range. Most commonly, noise decoupling is used wherein the decoupler output is modulated in a random way to make its spectrum uniform over a sufficient band of frequencies to reach all the proton lines.

The first generation of noise decouplers generated the decoupling irradiation by gating a coherent source on and off for randomly varying periods. The more recent decouplers have the rf on all the time but alternate the phase of the rf. This method has the advantage that, when necessary, higher average power is available than with the on/off modulation scheme. Typical proton decouplers need to put out a few to a few tens of watts at the present time.

Grutzner and Santini (1975) pointed out that coherent broadband decoupling by square wave phase modulation with about 50% duty cycle is much more efficient than noise decoupling in many cases most notably proton decoupling for $^{13}C$. The method has another advantage in that it is very simple to implement and is inexpensive. They find that there is a marked S/N improvement, typically a factor of two, especially for the carbons whose attached protons are at the ends of the chemical shift range. Noise decoupling still seems to be the method of choice for nuclei with large chemical shifts and small spin-spin coupling.

## Selective and off-resonance heteronuclear decoupling

Irradiating a selected line or a portion of a spectrum yields different information from noise decoupling. It is simpler in that the second irradiation need not be modulated and the total power required is less than noise decoupling. On the other hand, it is not as trivial to set up because a careful adjustment of where to put the coherent decoupling irradiation in the proton spectrum is necessary. In this regard, a most helpful capability for the spectrometer is that of being able to obtain the proton spectrum during the $^{13}C$ experiment without having to change the probe, sample, etc., but with some manipulation of cables, or in the case of one commercial spectrometer, simply with a throw of a console switch. The probe may be built with a special proton observe coil or the decoupling coil can be used as the observe coil for this purpose.

In coherent decoupling, if the second irradiation frequency is deliberately misadjusted, we will have a case which is called off-resonance decoupling. For example, in $^{13}C$ NMR, one might make the proton decoupler coherent and move it approximately 2 kHz off the center of the proton spectrum. The result is a carbon spectrum in which the coupling effects of the directly attached protons is manifest (but with reduced coupling constants J) and all the other couplings to protons are suppressed. One can immediately determine how many protons are directly bonded to each carbon. Its usefulness decreases with increasing frequency because the heating effects (to be described on pp. 104-105) are greater at higher fields and can induce shifts to confuse the assignments.

Aside from the effect on the line position mentioned in the last paragraph, decoupling irradiation, especially coherent irradiation, can also affect $T_1$, $T_2$, and $T_{1\rho}$. The artificial shortening of $T_2$ under decoupling actually can be beneficial for $T_1$ measurements by the steady state method described in III.D.2., however, because the undesirable spin echoes are attenuated for shorter $T_2$'s. Proton decoupling has been known to cause substantial errors in $T_{1\rho}$ measurements under certain circumstances (Freeman and Hill, 1971; Doddrel, et al. 1979).

## Gated decoupling

A subject closely related to decoupling is NOE, mentioned at the beginning of this section. This, simply put, is an effect on one nuclear spin population, say the carbon-13, induced by a change in the population caused by an irradiation of another spin system in contact with the first, say the protons. For our purpose, we will note only that NOE is brought on by the decoupling irradiation under favorable conditions and refer the readers to texts such as that by Noggle and Schirmer (1971). (For the record, we point out that some nuclei, notably nitrogen-15 and silicon-29, exhibit negative NOE's under proton decoupling.)

Gated decoupling is useful when we want to separate the decoupling effects from NOE effects. For example, we may want a decoupled spectrum without NOE for careful intensity studies. (NOE is usually not uniform across the spectrum.) In this case, we could turn on the decoupler only during the FID acquisition and off at other times as shown.

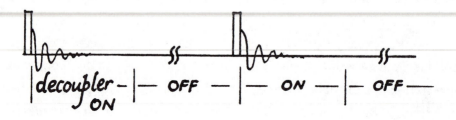

On the other hand, we may want a coupled spectrum with NOE, for example, for purely S/N reasons. In this case, the decoupler should be turned off only during the FID acquisition. It has been pointed out that accurate NOE determinations from gated decoupling measurements require waiting times between pulses much longer than for $T_1$ measurements. Canet (1976) and Harris and Newman (1976) recommend a wait of at least 9 or 10 $T_1$'s.

Both of these gated schemes are easy to implement because the timing requirements are not at all stringent in high resolution solution NMR. For a scheme with a more stringent timing requirement, we now turn to the subject of homonuclear decoupling.

### Homonuclear decoupling

In contrast to the case of heteronuclear decoupling which we have discussed for most of this section, homonuclear decoupling has associated with it the possibility of the decoupling irradiation interacting with the NMR receiver. This is especially acute in proton NMR where neither the couplings nor the chemical shifts are very large. (Incidentally, this problem is more acute in pulse FT NMR than it is in cw NMR for the very same reason that pulse FT NMR is much more efficient with its multiplexing advantage over the cw experiments.) In this case, we clearly should go to some scheme of time-sharing so that the receiver is off when the decoupler is on and vice versa. From the above discussion on gated decoupling, though, it is clear that those schemes will not work for homonuclear decoupling.

The solution is a time-shared decoupling scheme in which the "sharing" takes place within each digitizing period of the FID (Jesson, et al., 1973). The decoupler is turned on for as long as possible between digitizations by the ADC and is turned off during the digitizations. If the decoupler is not on long enough, possibly limited by the ADC requirements, the decoupling power may be inadequate. The timing requirements are still moderate compared with solid state experiments because the period between digitization in high resolution NMR is of the order of $10^{-3}$ seconds or longer. Time-share schemes are discussed further in V.A.5.

Decoupling power

From the discussion above, we can estimate the necessary decoupling field intensity. If the multiplet (doublet in this example) splitting were J-Hz, the proton transition must take place at a frequency J or greater. Since the precession frequency about a decoupling field $H_2$ perpendicular to the static field is $\gamma H_2$ where $\gamma$ is the proton gyromagnetic ratio, and this frequency must be comparable to or exceed $2\pi J$, the necessary field is approximately $H_2 > 2\pi J/\gamma$. Since $\gamma/2\pi$ is the usual 4kHz/gauss for protons, $H_2(mG) > J(Hz)/4$. This very crude estimate leads to a few tens of milligauss for the field intensity necessary to decouple protons from carbon under ordinary conditions.

It is an altogether different story for broadband decoupling, for which the decoupling power is spread out over a wide frequency range. In principle, the required total power goes up as the square of the frequency range irradiated. Furthermore, in contrast with the main NMR irradiation, the decoupler is required to run at a fairly high duty cycle. This creates problems not only because of the need for hefty hardware but because much of that power goes into heating up the probe and the sample. The probe heating with its attendant degradation of the probe performance is a serious problem but not nearly so compared to the heating of the sample. It can be alleviated by retuning the probe during the experiment after the probe has attained its steady state temperature. On the other hand, significant sample heating can make the temperature determination difficult, introduce temperature gradients into the sample, and cook biological samples. The need for more decoupling field $H_2$, and thus power, is much greater when the coupling J is large so that it is the least critical when decoupling protons.

Thus, efficient decoupler operation is crucial especially at the higher frequencies where certain samples may absorb more power than at the lower frequencies. Sample heating effects for ionic solutions have been discussed by Led and Petersen (1978) and by Bock, et al. (1980) and a more efficient coil design by Alderman and Grant (1979). A key fact to remember is that ionic heating is an electric field effect like the piezoelectric resonance discussed in VI.B.5. so that some sort of an electrostatic shielding as described there should work here, too.

## REFERENCES

D. W. Alderman and D. M. Grant, "An efficient decoupling coil design which reduces heating in conductive samples in superconducting spectrometers," J. Magn. Resonance 36, 447-451 (1979).

K. Bock, B. Meyer, and M. Vignon, "Some consequences of high frequency in NMR spectrometers," J. Magn. Resonance 38, 545-551 (1980).

D. Canet, "Systematic errors due to improper waiting times in heteronuclear Overhauser effect measurements by the gated decoupling technique," J. Magn. Resonance 23, 361-364 (1976).

D. M. Doddrel, M. R. Bendall, P. F. Barron, and D. T. Pegg, "Use of $^{13}C$ $T_{1\rho}$ measurements to study dynamic processes in solution," J. Chem. Soc., Chem. Comm., 77-79 (1979).

R. Freeman and H. D. W. Hill, "Fourier transform study of NMR spin-spin relaxation," J. Chem. Phys. 55, 1985-1986 (1971).

J. B. Grutzner and R. E. Santini, "Coherent broad-band decoupling -- an alternative to proton noise decoupling in carbon-13 nuclear magnetic resonance spectroscopy, J. Magn. Resonance 19, 173-187 (1971).

R. K. Harris and R. H. Newman, "Choice of pulse spacings for accurate $T_1$ and NOE measurements in NMR Spectroscopy," J. Magn. Resonance 24, 449-456 (1976).

J. P. Jesson, P. Meakin, and G. Kneissel, "Homonuclear decoupling and peak elimination in Fourier transform nuclear magnetic resonance," J. Am. Chem. Soc. 95, 618-620 (1973).

J. J. Led and S. B. Petersen, "Heating effects in carbon-13 NMR spectroscopy on aqueous solutions caused by proton noise decoupling at high frequencies," J. Magn. Resonance 32, 1-17 (1978).

J. H. Noggle and R. E. Schirmer, The Nuclear Overhauser Effect (Academic, New York, 1971).

## II.D.1.   PULSE RESPONSE IN THE PRESENCE OF QUADRUPOLE SPLITTING

A nucleus with nuclear spin I greater than ½ has an electric quadrupole moment in addition to a nuclear magnetic moment. An electric quadrupole moment arises from an asymmetry of the distribution of the electrical charges in the nucleus and does not depend on the net charge. It is often viewed as two electric dipoles displaced from one another and pointing in opposite directions. It has the unit of charge times area although it is usually given in units of $10^{-24} cm^2$ ($\equiv 1$ barn, for big as a....), the electronic charge having been divided out. In this discussion, we will only be concerned with the response of the quadrupole moment to pulsed irradiation.

The electric quadrupole moment becomes important only when the nucleus is in an electric field gradient (EFG), either static or dynamic. Any environment that does not have cubic symmetry will provide such an EFG. Suppose we consider a nucleus with a quadrupole moment near a positive charge. If the distribution of the net positive nuclear charge is prolate like a cigar, then the nucleus wants to orient perpendicularly to the direction towards the external charge because of simple electrostatic forces as shown in the figure below. Therefore, the interaction of the moment and the EFG

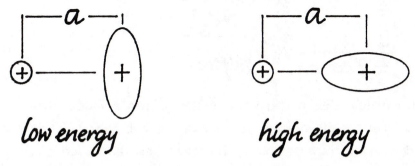

clearly augments that between the magnetic dipole moment and the magnetic field which gives rise to the Larmor precession. In this section we consider the case where the electric quadrupole effects are small compared to the magnetic effects. The opposite case falls into the domain of nuclear quadrupole resonance which we will not discuss.

In a stationary EFG, the quadrupole interaction shifts the Zeeman levels according to the square of the quantum number m to a first approximation. Thus, for $I=3/2$, for example, m can be $3/2$, $1/2$, $-1/2$, or $-3/2$, and the $m=\pm 3/2$ levels shift identically while the $m=\pm 1/2$ levels shift identically.

The splitting shown below can be quite large and the irradiation and the detection of such transitions can be difficult, as discussed below. If, due to the molecular motion, the EFG at the nucleus is averaging rapidly and isotropically on the time scale of the experiment, for example, for nitrogen-14 in an ammonia molecule dissolved in water at room temperature,

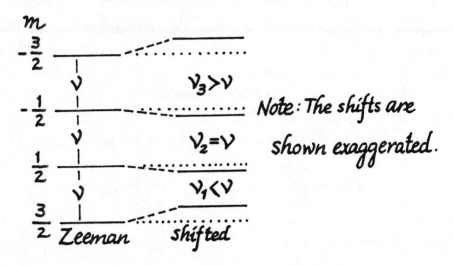

the multiple line pattern will collapse into a single line. There may still be effects on the relaxation behavior of that nucleus and also of its neighbors, however, as discussed in III.C.3. For now, we only consider the effects of pulse irradiation on the quadrupole split levels.

In Chapter I, we introduced the angular velocity of the magnetization around $H_1$ in the rotating frame as $\omega_1 = \gamma H_1$, in analogy with the Larmor frequency in the laboratory frame. This physical picture of the effect of $H_1$ breaks down for $I > \frac{1}{2}$ when the non-zero quadrupole interaction makes the energy levels unequal. The basic principles are treated in Chapter VII, Section II.B.(c) of Abragam, 1961 (Appendix A) but our discussion relies on some lecture notes by Schmidt (1972).

When there is quadrupole splitting, two parameters are affected in pulse NMR compared to the situation without any splitting or the cw case. The first parameter is the precession rate around $H_1$ in the rotating frame which depends on both I and m where the transition is, say, between the levels m and m+1 and where m can have any value between -I and I. The precession rate in the presence of the quadrupole interaction is <u>greater</u> and the required $\pi/2$ pulse shorter, than that for an equivalent (same $\gamma$) I=½ nucleus. The other effect is that the signal strength for the various split transitions are not those expected from the cw case, and become much weaker for larger I. The sum of all the quadrupolar satellite intensities add up to the unsplit line intensity in the absence of quadrupolar interaction in cw NMR but not in pulse NMR. Thus, the direct detection of quadrupolar satellites in solids is difficult by pulse NMR, although it is not exactly easy by cw NMR either.

The pertinent parameter is the magnetic dipole transition matrix element which is proportional to $A=\sqrt{I(I+1)-m(m+1)}$. In cw NMR the line intensity is proportional to the transition probability which, in turn, is proportional to $A^2$. For pulse NMR of quadrupole split lines, the peak magnetization is only proportional to A so that the satellites are much weaker, especially for large I and m. For spin 3/2, the sum of all the lines is 55% of the intensity of the unsplit line whereas for spin 7/2 it is only 28%.

The effective precession rate of the magnetization around $H_1$ in the rotating frame during the rf irradiation is proportional to A. Thus, for example, the $\pi/2$ pulse length for the $-\frac{1}{2}\leftrightarrow\frac{1}{2}$ transition for I=5/2 is 1/3 the length of a $\pi/2$ pulse for an equivalent I=½ nucleus with the same $\gamma$, or for that matter, for a I=5/2 nucleus with the same $\gamma$ in the absence of a quadrupole splitting as in liquids.

These results are given for $I\leq7/2$ in the table which is reproduced from the work of Schmidt (1972). The parameters 90° pulse length, peak pulse signal, and cw signal, are computed from 1/A, A/B, and $A^2/B$, where A and B are defined at the end of the table. The lines in the table labelled "all" refer to the case of simultaneously irradiating all the transi-

| I | Transition | $A^2$ | B | Values relative to those for unsplit spectrum | | |
|---|---|---|---|---|---|---|
| | | | | 90° Pulse length | Peak pulse signal | cw signal |
| 1/2 | 1/2↔1/2(all) | 1 | 1 | 1 | 1 | 1 |
| 1 | all | - | 1/4 | 1 | 1 | 1 |
| | ±1↔0 | 2 | $1/\sqrt{2} = 0.707$ | $\sqrt{2}/4 = 0.354$ | $2/4 = 0.500$ |
| 3/2 | all | - | 1/10 | 1 | 1 | 1 |
| | ±3/2↔±1/2 | 3 | $1/\sqrt{3} = 0.578$ | $\sqrt{3}/10 = 0.173$ | $3/10 = 0.300$ |
| | 1/2↔-1/2 | 4 | $1/\sqrt{4} = 0.500$ | $\sqrt{4}/10 = 0.200$ | $4/10 = 0.400$ |
| 2 | all | - | 1/20 | 1 | 1 | 1 |
| | ±2↔±1 | 4 | $1/\sqrt{4} = 0.500$ | $\sqrt{4}/20 = 0.100$ | $4/20 = 0.200$ |
| | ±1↔0 | 6 | $1/\sqrt{6} = 0.408$ | $\sqrt{6}/20 = 0.122$ | $6/20 = 0.300$ |
| 5/2 | all | - | 1/35 | 1 | 1 | 1 |
| | ±5/2↔±3/2 | 5 | $1/\sqrt{5} = 0.447$ | $\sqrt{5}/35 = 0.064$ | $5/35 = 0.143$ |
| | ±3/2↔±1/2 | 8 | $1/\sqrt{8} = 0.354$ | $\sqrt{8}/35 = 0.081$ | $8/35 = 0.228$ |
| | 1/2↔-1/2 | 9 | $1/\sqrt{9} = 0.333$ | $\sqrt{9}/35 = 0.086$ | $9/35 = 0.257$ |

| I | Transition | $A^2$ | B | Values relative to those for unsplit spectrum | | |
|---|---|---|---|---|---|---|
| | | | | 90° Pulse length | Peak pulse signal | cw signal |
| 3 | all | - | 1/56 | 1 | 1 | 1 |
| | ±3↔±2 | 6 | | $1/\sqrt{6}$ = 0.408 | $\sqrt{6}/56$ = 0.044 | 6/56 = 0.107 |
| | ±2↔±1 | 10 | | $1/\sqrt{10}$ = 0.316 | $\sqrt{10}/56$ = 0.057 | 10/56 = 0.179 |
| | ±1↔0 | 12 | | $1/\sqrt{12}$ = 0.289 | $\sqrt{12}/56$ = 0.062 | 12/56 = 0.214 |
| 7/2 | all | - | 1/84 | 1 | 1 | 1 |
| | ±7/2↔±5/2 | 7 | | $1/\sqrt{7}$ = 0.378 | $\sqrt{7}/84$ = 0.031 | 7/84 = 0.083 |
| | ±5/2↔±3/2 | 12 | | $1/\sqrt{12}$ = 0.289 | $\sqrt{12}/84$ = 0.041 | 12/84 = 0.143 |
| | ±3/2↔±1/2 | 15 | | $1/\sqrt{15}$ = 0.258 | $\sqrt{15}/84$ = 0.046 | 15/84 = 0.179 |
| | 1/2↔-1/2 | 16 | | $1/\sqrt{16}$ = 0.250 | $\sqrt{16}/84$ = 0.08 | 16/84 = 0.190 |

$$A^2 = I(I+1) - m(m+1)$$

$$B = \frac{3}{2I(I+1)(2I+1)}$$

tions with a $\pi/2$ pulse, as you would in the absence of quadrupole splitting.


REFERENCE

V. H. Schmidt, "Pulse response in the presence of quadrupolar splitting," in Pulsed Magnetic and Optical Resonance, Proceedings of the Ampere International Summer School II, Basko polje, 2-13 September 1971, edited by R. Blinc (University of Ljubljana, Ljubljana, Yugoslavia, 1972), pp. 75-83.


## II.D.2.   SELECTIVE EXCITATION

In section II.A.2. we discussed how the rf pulse acts on the nuclear spins. We now discuss this topic further and apply the results to several examples having to do with selective excitation. There are many reasons for wanting to perform selective excitation in NMR. Some of them are to simplify complex spectra, to do selective population transfer in order to get the sign of spin-spin coupling, and to study cross relaxation. An important application of what we might call selective de-excitation is a notch in the irradiation pattern to suppress an unwanted resonance such as the solvent peak in a complex proton spectrum. Morris and Freeman (1978) have reviewed many aspects of such experiments.

So far, we have only talked of applications where the word selective means with respect to locations of NMR lines in spectra. The selectivity can also apply to other properties such as $T_2$ and we now give some examples of this kind of experiment before continuing with the more common kinds of selective excitation.

Basically, it is possible to obtain FID's selectively according to $T_2$ because there will be a time during a pulse experiment when components of magnetization with short $T_2$ will have died out, whereas the long $T_2$ component will still be accessible. One of the earliest experiments of this kind that we know of was due to Goldman and Shen (1966) who studied $T_1$'s and $T_2$'s of inequivalent fluorine spins in $LaF_3$. They separated the two components with different $T_2$'s simply by waiting until the FID due to the shorter $T_2$ components died out and then applying a $-\pi/2$ pulse to the spins in order to return whatever magnetization there was in the rotating x'-y' plane back along the z' axis. Because of the $T_2$ difference, no short $T_2$ component was returned along the z' axis whereas a significant amount of the long $T_2$ component ended up along z'. Now other experiments can be performed with the long $T_2$ component, for example, to measure its $T_1$.

Another example of separating long $T_2$ components from short $T_2$ components is given by Rabenstein, et al. (1979). They named the sequence inversion-recovery spin-echo (IRSE) because it is literally just that. Instead of the $\pi/2$ pulse to monitor the magnetization recovering after the initial $\pi$ pulse, a $\pi-\pi/2$ echo is used. The separation of the two pulses in the spin echo sub-sequence is adjusted so that the short $T_2$ component does not contribute significantly to the echo. This sequence was successfully used to isolate the sharp NMR lines of small molecules from the broad lines of proteins in solutions.

As already mentioned, the overwhelmingly common excitation uses an approximately square pulse and such a pulse will excite the nuclei in only approximately a uniform fashion.

This is because $H_1$ is finite and there is a dispersion of

$$H_e = [(H_0 - \omega/\gamma)^2 + H_1]^{\frac{1}{2}},$$

which causes the spins at different offsets to nutate about $H_e$ by different amounts. In fact, for sufficiently large offset compared to $H_1$, the magnetization can make a full revolution (or more) about $H_e$ in which case it contributes no signal. This fact suggests a straightforward selective excitation (or rather de-excitation) experiment; namely, to adjust the $H_1$ (which is along the y' axis) and the offset in such a way that the unwanted magnetization executes a 360° rotation about $H_e$ while the desirable magnetization executes some other (and preferably uniform but otherwise known) amount of rotation. For a two line spectrum, this can be done trivially by nutating one component 360° and the other 180° about $H_e$ so that the latter ends up on its trajectory farthest from the z' axis (Virlet and Rigny, 1975, and references therein).

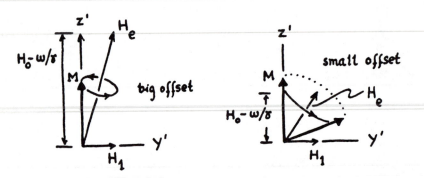

The most obvious use for such selective de-excitation is for solvent peak suppression wherein an overwhelming but undesirable solvent peak must be reduced or eliminated. Redfield (1976) has developed specialized techniques to selectively de-excite the solvent peak in a multiline spectra.

Since there will be a sizeable dispersion of $H_e$ across the spectrum due to the relatively long pulse necessary in his scheme, both phase and amplitude corrections must be performed from previous calibrations.

The next example we give is a selective excitation of a small part of a multiline spectrum by somehow exaggerating the spectral dispersion, i.e., enhancing the differences between the desirable and the undesirable spins. An elegant scheme to do this was proposed and first carried out by Bodenhausen, et al. (1976). They used a train of equally spaced "hard" pulses ($\gamma H_1 >>$ spectral width) with a flip angle much less than $\pi/2$. In the period $\tau$ between pulses, a line with frequency offset $\Delta\nu$ will precess through $2\pi\Delta\nu\tau$ rad. If $\tau$ is chosen so that the desired line precesses through an angle of $n2\pi$ radians in the rotating frame where n is an integer, all others precess through some other angle. The result is that the train of such pulses provides a cumulative tipping of the magnetization directly towards the x'-y' plane only for the magnetization component of interest because only that line has its magnetization precess back into the y'-z' plane before each subsequent pulse. The pulse train length is chosen to nutate this magnetization through 90° into the x'-y' plane and then the FID is collected. This scheme has been demonstrated to excite only a 10 Hz wide band at an offset of 500 Hz. This pulse sequence has the name DANTE because of the alleged similarity between purgatory in Dante's "Divine Comedy" and the paths of the magnetization components in the rotating frame under the influence of this sequence (Morris and Freeman, 1978).

In principle, any excitation pattern can be concocted by a Fourier transformation of the desired frequency distribution of the excitation and this is called tailored excitation.

The main problem is that such "odd" shape excitation is difficult to implement. For example, a flat distribution of excitation would result from a transmitter output having an envelope of the form $(\sin\omega t)/\omega t$ but you would first have to generate such a time dependence, either by analog or digital means and then have a transmitter which would respond linearly to such an input. A better alternative is to generate such an irradiation pattern by a series of equally spaced pulses whose widths or amplitudes are modulated appropriately (Tomlinson and Hill, 1973).

## REFERENCES

G. Bodenhausen, R. Freeman, and G. A. Morris, "A simple pulse sequence for selective excitation in Fourier transform NMR," J. Magn. Resonance 23, 171-175 (1976).

M. Goldman and L. Shen, "Spin-spin relaxation in $LaF_3$," Phys. Rev. 144, 321-331 (1966).

G. A. Morris and R. Freeman, "Selective excitation in Fourier transform nuclear magnetic resonance," J. Magn. Resonance 29, 433-462 (1978).

D. L. Rabenstein, T. Nakashima, and G. Bigam, "A pulse sequence for the measurement of spin-lattice relaxation times of small molecules in protein solutions," J. Magn. Resonance 34, 669-674 (1979).

A. G. Redfield, "How to build a Fourier transform NMR spectrometer for biochemical applications" in Introductory Essays, edited by M. M. Pintar (Springer-Verlag, Berlin, 1976), pp. 137-152.

B. L. Tomlinson and H. D. W. Hill, "Fourier synthesized excitation of nuclear magnetic resonance with application to homonuclear decoupling and solvent line suppression," J. Chem. Phys. 59, 1775-1784 (1973).

J. Virlet and P. Rigny, "$^{19}F$ nuclear magnetic resonance and relaxation study of hexafluoride solid solutions $(UF_6)_x(MoF_6)_{(1-x)}$ and molecular motion rates near the phase transitions in molecular crystals," J. Magn. Resonance 19, 188-207 (1975).

## II.D.3.  TWO DIMENSIONAL NMR

A branch of NMR offering unique capabilities and great promise is two dimensional (2-D) NMR.  We will just scratch the surface of this new field pioneered by Jeener and Ernst and leave the rest to current reviews such as the one by Freeman and Morris (1979).  Undoubtedly many new developments are forthcoming in this rapidly developing field.

Two dimensional FT NMR is possible when a time varying signal (like an FID) arises from a magnetization which has evolved in the presence of an interaction different from that which is present during the data acquisition.  Consider an actual example in a simplified form.  Waugh and co-workers (Hester, et al., 1976; Rybaczewski, et al., 1977) have performed the following experiment.  They take a single crystal containing natural abundance carbon-13 in the presence of abundant protons like in the CP/MAS experiments discussed in IV.E.3.  The carbon-13 FID is observed in the presence of proton decoupling.  Let us call the time scale during this FID $t_2$.  These FID's are accumulated as a function of time delay $t_1$ during which the carbons evolve without being decoupled from the initial condition of full magnetization.  Thus, the amplitude of the FID depends on $t_1$ because the magnetization has evolved in the presence of the proton-carbon dipolar interaction, even though the FID shape is independent of $t_1$.

We schematically show the carbon-13 signal during the two time periods in the following figure.

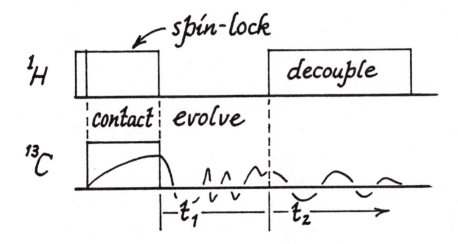

In practice, the first FID is accumulated in the $t_2$ time frame with a fixed value of $t_1$. Then it is Fourier transformed to yield a frequency domain spectrum, say, $F(w_2)$ for a particular $t_1$. Another FID is accumulated with a different frequency domain spectrum and as many such spectra are to be acquired as needed at different values of $t_1$ to form an array of $F(w_2, t_1)$. Now a Fourier transform is carried out with the independent variable $t_1$ for each value of $w_2$. For example, 128 different values of $t_1$ can be used to accumulate 1024 point FID's in which case the first set of transforms involve 128 1024-point transforms and the second set involve 1024 128-point transforms.

The resulting two dimensional plot will have the decoupled chemical shift spectrum along the $w_2$ axis (just like the ordinary proton decoupled carbon spectrum) and the dipolar split spectrum along the $w_1$ axis. Thus, the net effect of this experiment is to separate the chemical shift effects from the dipolar effects. Not only does this separation simplify the assignments of the spectral lines but it allows much better resolution of the dipolar lines (because the chemically inequivalent lines do not overlap) from which it is possible, for example, to make very accurate determination of the carbon-proton (in this example) bond distances and angles.

Any pairs of interactions may be separated in this way provided suitable signals are obtainable. To date, such pairs include chemical shift/spin-spin coupling, proton/carbon chemical shifts, as well as dipolar/chemical shifts discussed above. A most promising 2-D plot represents the correlation between proton and carbon shifts. It is also possible to look for polarization transfer only between interacting carbon-proton pairs. This last experiment is otherwise done with fairly complex methods such as spin tickling or off-resonance decoupling, in many cases 2-D NMR can do it in a straightforward manner. Finally, it is possible to detect forbidden transitions by 2-D NMR, but we will not go into it in this book.

As the example above implied, one of the difficulties with 2-D NMR is the sheer number of data points to be handled. This is troublesome in at least two ways. First, it is time consuming to do the experiment because the data acquisition, the transformation, and the output take much more time than they do in the usual FT experiment. In the above example of taking 128 different FID's each consisting of 1024 points, we might guess that the 2-D experiment will take approximately 256 times longer than the 1-D experiment with just a 1024 point FID. (We get that estimate by saying that there are 128 times the FID's and then there is another factor of roughly two because the FT process has to be repeated in the other parameter.) It turns out to be worse than that because of the required data manipulation. There is so much data that a mass storage device such as a hard disk must be used to store the data outside the computer and this causes additional delays because data transfer is slower between a computer and a disk (or any other peripheral) than within the computer memory. In addition, there is just a lot of data juggling needed because the $F(\omega_2, t_1)$ spectra at the end of the first FFT are arranged all wrong (on the disk) for the second FFT over the parameter $t_1$. The result of all this is that just the data processing for a 2-D NMR can take several hours or more. Even plotting the resulting 2-D spectrum could take many times longer than the usual spectrum.

The current practice is to rely on minicomputers with more efficient programs and faster mass storage devices to speed up the data handling and processing. (There is not much we can do about data acquisition.) Since, if the 2-D experiment is performed in any detail, the computation (and plotting) time could take many hours even with a modern minicomputer, we should rethink about how the experiments should be performed. Basically, we are back to the pre-FFT situation when it was unthinkable to do an on-line FT experiment. We believe that it makes much sense to process 2-D NMR data off-line on large computers designed for ultra fast computations and let the minicomputer concentrate on data acquisition.


## REFERENCES

R. Freeman and G. A. Morris, "Two-dimensional Fourier transformation in NMR," Bull. Magn. Resonance 1, 5-26 (1979).

R. K. Hester, J. L. Ackerman, B. L. Neff, and J. S. Waugh, "Separated local field spectra in NMR: Determination of structure in solids," Phys. Rev. Letters 36, 1081-1083 (1976).

E. F. Rybaczewski, B. L. Neff, J. S. Waugh, and J. S. Sherfinski, "High resolution $^{13}C$ NMR in solids: $^{13}C$ local fields of CH, $CH_2$, $CH_3$," J. Chem. Phys. <u>67</u>, 1231-1236 (1977).

## II.D.4.    CPMG J-SPECTRA

In I.C.2. we saw how useful the spin echo train, particularly the CPMG train, can be for obtaining the spin-spin relaxation time $T_2$ without the effects of magnetic field homogeneity. Echo trains like the CPMG not only cancel the effects of field inhomogeneity but also cancel other effects such as the chemical shifts and the quadrupolar splitting. On the other hand, homonuclear spin-spin coupling and homonuclear dipole-dipole coupling are not cancelled because the $\pi$ pulse inverts both of the coupled spins so the effects of either of these couplings continue unaltered by the pulse.

Consider an unresolved spin-spin multiplet in a high resolution spectrum. It turns out that the spin-spin coupling, resolved or not, will modulate the echo maxima. By this we mean that the maximum height of each echo will not decrease monotonically but rather will oscillate according to the spin-spin coupling. Of course, the shape of each echo will only tell you about the spectrum in which the spin-spin coupling under question is not resolved. Such echo envelope modulation has been known in certain areas of magnetic resonance. See, for example, the photograph in Brookeman, et al. (1972), and references therein.

The echo envelope modulation occurs because, since the spin-spin coupling is not reversed by the $\pi$ pulses, the situation is like superimposing an irreversible angular velocity proportional to the coupling constant to the spin components in the rotating frame. Thus, if J is in frequency units (like Hz), the modulation will have the frequency J/2. The coupling can be measured from the modulation pattern or the pattern can be Fourier transformed to yield the "J-spectrum" after it is digitized in such a way that only the echo maxima are sampled. Resolution in excess of that determined by the magnetic field inhomogeneity can be obtained in this way. In complete analogy with the usual FID, the duration of the echo train sampled determines the resolution. For example, two lines 0.05 Hz apart will be resolved by an echo train persisting for at least 20 seconds.

In order for the spin-spin coupling to be simply obtainable from the spin echo spectrum, the spectrum must be first order, the pulse repetition rate be slow compared to the smallest chemical shift difference, the $\pi$ pulses be set accurately, and no other source of modulation must exist. The method becomes considerably more versatile when the first two conditions, above, are relaxed at the cost of increased complexity in the analysis. See the articles by Freeman and Hill (1971), Vold and Shoup (1972), Vold and Vold (1974), and references therein, for details.

## REFERENCES

J. R. Brookeman, P. C. Canepa, and A. S. DeReggi, "Slow beats in the NQR echo envelope of solid $N_2$," Physics Letters 39A, 415-416 (1972).

R. Freeman and H. D. W. Hill, "High-resolution study of NMR spin echoes: 'J spectra'," J. Chem. Phys. 54, 301-313 (1971).

R. L. Vold and R. R. Shoup, "Proton spin echo NMR spectra

of 1,1-difluoro-2,2-dichloroethane: Subspectral analysis," J. Chem. Phys. 56, 4787-4799 (1972).

R. L. Vold and R. R. Vold, "Spin-echo measurements of long-range coupling constants in some simple esters," J. Magn. Resonance 13, 38-44 (1974).

## II.D.5.   ECHOES IN LIQUIDS

If you sit at a pulse NMR machine and try various pulse combinations with a liquid sample, you will find that it is easier to form echoes than not. Almost any two pulses will form an echo and, in particular, even two $\pi/2$ pulses (with the same phase) will form an echo which may seem a bit surprising at first. Historically, the echo from two $\pi/2$ pulses was the first published by Hahn (1950) even though it is easier, in retrospect, to understand the $\pi/2-\pi$ echo. In addition, the $\pi/2-\pi$ echo can be about a factor of two bigger than the $\pi/2-\pi/2$ echo [see eq. (6) in Hahn's article]. The formation of the echo with a pair of $\pi/2$ pulses is best understood with the help of good rotating frame diagrams and we can do no better than to refer you to figure 1 of the same article.

Even though it is amusing to form echoes practically at will, these echoes are not always a blessing. For many of the $T_1$ pulse sequences to be described in III.D.2., the steady state sequence, for example, the echoes must be suppressed before the application of the following pulse. This is usually accomplished by a weak pulsed field gradient called the homospoil pulse applied between the rf pulses (and after the FID has been recorded) or by reversing the phases of the pulses, usually in pairs. See III.D.3. for more details.

In the article already referred to, Hahn goes on to discuss the multiplicity of echoes which are induced upon application of three $\pi/2$ pulses. If we include the echo induced by the pair of pulses already discussed, there will be five(!) echoes. We will mention only one of them, the stimulated echo, because of its special characteristic that its lifetime is determined by $T_1$ rather than by the diffusion effects which rapidly dephase the usual echos according to eq. (2) of III.E.1. If the first pair of $\pi/2$ pulses are separated by $\tau$, and the third pulse occurs at time T after the first pulse, the stimulated echo occurs at T+$\tau$. We again refer you to Hahn's article for the explanation of the stimulated echo. See III.E.2. for an application of stimulated echoes.

These echoes formed by a set of $\pi/2$ pulses differ in a significant way from the other echoes we discussed in I.C.2. The echoes such as the $\pi/2$-$\pi$ echo are, under ideal conditions, homogeneous in the sense of IV.A.1. whereas the $\pi/2$ echoes are not. We mean by this that different parts of the $\pi/2$ echoes arise from nuclei in different fields and the nuclei in some fields contribute more to these echoes than those in other fields.


REFERENCE

E. L. Hahn, "Spin echoes," Phys. Rev. 80, 580-594 (1950).

CHAPTER III

RELAXATION

If you go to a social gathering and announce that you are an expert in relaxation, you could receive responses ranging from curious stares to hearty approvals neither of which are probably justified. In this chapter, we discuss nuclear spin relaxation, a subclass of the larger subject, in a qualitative way as well as pulse NMR methods for measuring them.

## III.A. GENERAL REMARKS

### What is spin-lattice relaxation?

At the beginning of the book (in I.A.) we said that a spin-$\frac{1}{2}$ particle with magnetic moment $\bar{\mu}$ in a static magnetic

Eiichi Fukushima and Stephen B. W. Roeder, Experimental Pulse NMR: A Nuts and Bolts Approach

field $\overline{H}$ has two energy states separated by $\Delta E = 2\mu H$. A group of spin-½ nuclei placed in a magnetic field would distribute themselves between the two states according to the Boltzmann relation, so that the ratio of the number of spins in the upper energy states to that in the lower energy state is exp(-$\Delta E/kT$) at equilibrium. As a result, there will be more spins in the lower energy state than in the upper state and the population difference gives rise to a magnetization $\overline{M}_o$, the magnitude of which depends on the field H, the temperature T, the number of spins N, the angular momentum $I\hbar$, and the gyromagnetic ratio $\gamma$, according to Curie's Law

$$M_o = \frac{N\gamma^2\hbar^2 I(I+1)}{3kT}H.$$

A chunk of $CaF_2$ (which can be thought of as a simple array of fluorine-19 nuclear spins) outside of the magnet, will have negligible magnetization because the earth's field is very small compared to the field of any NMR magnet. What happens when we insert this sample into a field of a magnet? We already know that a macroscopic magnetization will be induced, but how? We define the process of growth towards the equilibrium magnetization given by Curie's Law as spin-lattice relaxation. Assuming, further, that this process is exponential, the magnetization will have recovered to within a factor [1-(1/e)]=63% of the equilibrium value at a time $T_1$, the spin lattice relaxation time.

How does spin-lattice relaxation take place?

Why does the induced magnetization not form instantaneously? If we can insert the sample into the field instaneously, the nuclear moments will find themselves with their orientations distributed equally between the two energy states. In order for a spin with its moment $\bar{\mu}$ antiparallel to the field (with a potential energy +μH) to turn parallel to the field (in which case its potential energy will be -μH), it has to give up 2μH of energy. Unless there is an agent to accept this energy, and a mechanism to transfer it, the spin cannot make this jump to the lower energy state.

Thus, the rate at which magnetization builds up in a static field depends on the mechanisms available for the spins to transfer energy to something else -- namely the other repositories for thermal energy such as the translations, rotations, and vibrations, collectively called the lattice.

A spin in a high energy state can make a transition to the low energy state via spontaneous emission or stimulated emission. We mean by this that the spin can spontaneously jump to a lower energy state while emitting a photon or that it can be stimulated to do so. However, the probability of spontaneous emission depends on the third power of the frequency and at radio frequencies this term is too small to be significant. Thus, all NMR transitions are stimulated. Stimulated by what? In order to undergo a transition, the nucleus needs to see fields fluctuating at its Larmor frequency.

What kinds of fields? Any kind which will interact with the nucleus strongly enough and there are two such randomly fluctuating fields effective in NMR relaxation. One is a time varying magnetic field which can interact with the nuclear magnetic dipole moments and the other is an EFG which can interact with an electric quadrupole moment of the nucleus if it has one. [Only those with $I > \frac{1}{2}$ can have electric quadrupole moments.] The magnetic field interaction is the same as that which is utilized in the NMR experiment to cause transitions, for example, by a $\pi/2$ pulse.

What are the sources of these fluctuating fields? There

are a variety of possible sources, each one of which gives rise to a specific spin-lattice relaxation mechanism. Suppose on a molecule there is an unpaired electron which has a large magnetic moment, about 1000 times larger than the nuclear magnetic moment of the proton. If the molecule tumbles randomly, this creates randomly varying magnetic fields at the site of the nucleus, and the component of this fluctuation at the nuclear precession frequency will be effective in stimulating transitions.

If an unpaired electron is causing the nuclear relaxation, this is called paramagnetic relaxation, an ambiguous terminology since all nuclei with spins as well as electrons are paramagnetic. If the nucleus has an electric quadrupole moment, it can interact with a randomly varying EFG in the molecule and the process is quadrupolar relaxation. If the source of the randomly fluctuating magnetic field is another nuclear magnetic moment, then the process is called nuclear dipole-dipole relaxation. The known mechanisms for spin-lattice relaxation through magnetic dipole transitions are dipole-dipole; scalar relaxation of the first and second kind, spin rotation, and chemical shift anisotropy relaxation. These mechanisms will be outlined in III.C.1.

The important point to get out of all this is that since the nuclear spin-lattice relaxation processes usually depend on the the existence of molecular motion to generate a randomly varying magnetic field or EFG, we can get valuable information about these motions from the relaxation rates.

Can you have spin-lattice relaxation in the absence of the usual sort of molecular motion? By the usual sort, we mean nearly free molecular rotations and translations in liquids and gases as well as hindered rotations and translations in solids but not those involving only parts of the molecule like librations and vibrations.

The answer to the above question is yes, but whether or not a particular mechanism is effective depends greatly on the situation. For example, the lattice vibrations (which are intimately tied with the molecular vibrations) in solids are not very effective in causing relaxation through the nuclear dipole-dipole interaction mainly because the vibration frequency is so much higher than the usual Larmor frequencies. However, if there are some electronic spins present in the lattice, they become quite effective simply because of the size of the electronic moments. In this category are the large class of situations encountered in NMR of metals wherein the conduction electrons provide an extremely strong relaxation mechanism as well as the cases of insulators containing paramagnetic impurities.

In fact, for many metals $T_1 \sim T_2$ as it is the case with most liquids. Unlike liquids, however, the relaxation times are so short that it is often not possible to see an FID. Fortunately, it is usually possible to induce spin echoes in these inhomogeneously broadened metallic compounds. The one last obstacle, that of the lines being much wider than the spread of the irradiation pulse is solved by sweeping the $H_0$ field and plotting the area under the spin echo as a function of $H_0$, like in a cw experiment. You do need fast digitizers for this experiment (see sections V.A.3. and V.B.1.).

The case of insulators containing paramagnetic impurities is also important in solids because at low enough temperature the molecular motions which control the relaxation times at the higher temperatures slow down and the impurity relaxation usually takes over. That this happens depends on the fact that the relaxation mechanism via the paramagnetic impurities may have a weak dependence on the temperature.

For more details on relaxation in metals, we refer you to the books by Abragam and Slichter listed in the general section of Appendix A as well as that by Winter (1971). For relaxation by spin diffusion to paramagnetic impurities, we refer you to the article by Noack (1971) and references contained therein.

## Other kinds of nuclear relaxation

Besides <u>spin-lattice relaxation</u> with its characteristic time constant $T_1$, we will encounter several other relaxation processes, the most prominent being <u>spin-spin relaxation</u> with its characteristic relaxation time $T_2$. $T_2$ is the time constant for the decay of the precessing x-y component of the magnetization following a disturbance. We refer you to sections I.C.1. and III.B. for more discussion. When molecular motions are very fast, as in non-viscous liquids, $T_1=T_2$ for most interactions and $T_2$ offers no additional information. However, in solids typically $T_1 \gg T_2$ and then $T_2$ does offer additional information.

Other relaxation processes mentioned in this book are those characterized by $T_{1\rho}$ and $T_{1D}$. The former is called <u>spin-lattice relaxation in the rotating frame</u> and is discussed in IV.C. while the latter is called <u>dipolar spin-lattice relaxation</u> and is discussed in IV.B.

## The spectral density

If we take a pair of spins, originally infinitely far apart from each other, and bring them together to a normal internuclear distance in a molecule, we will have an energy of interaction which is a constant for a particular molecular arrangement. If we now make this energy of interaction time dependent by causing the spins to move randomly with respect to each other, the interaction energy will be distributed in frequency and time. The frequency dependence of the power is called the spectral density $J(\omega)$ and the spin-lattice relaxation rate $1/T_1$ is directly related to the value of this function at the Larmor precession frequency of the nucleus.

What might be the qualitative behavior of the spectral density and how might the spin-lattice relaxation behave because of it? The calculation of the exact spectral density for any real molecular motion has never been solved. The basic difficulty is that the molecular (or atomic) motions are extremely complicated. Although we said that the motion is random, it is not completely so and may even be anisotropic because of interactions between the atoms and molecules.

How would the spectral density look? We state without proof that a typical term in it looks like $\tau_c/(1+\omega^2\tau_c^2)$ where the quantity $\tau_c$ is called the correlation time and is a measure of the time between the field fluctuations. This particular form of the spectral density term is contingent on a truly random fluctuation which leads to a physical picture where the chances of not experiencing a fluctuation for a time t decreases monotonically according to $\exp(-t/\tau_c)$. The model used for the usual calculations assumes the actual change in the molecular state, be it rotation or translation, to take place infinitely quickly and then to stay in that state for $\tau_c$ until the next fluctuation occurs.

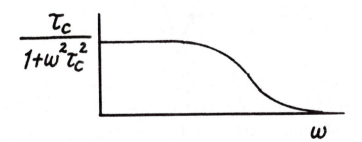

The typical spectral density term $\tau_c/(1+\omega^2\tau_c^2)$ has the following characteristics. In a particular system and for a given $\tau_c$, meaning given temperature and pressure, the spectral density is a constant at small $\omega$ and falls off as $1/\omega^2$ at large $\omega$, as shown above. There are no components at frequencies significantly exceeding $1/\tau_c$.

As implied earlier, the integral over the spectral density with respect to frequency $\int J(\omega)d\omega$ is a constant and represents the energy of interaction that is modulated by the motion. Thus, the temperature will change the shape of the curve $J(\omega)$ but not the area under it. At low temperatures, one expects that the spectral density will be bunched at low frequencies and as the temperature is raised, the spectral density will become spread out over a larger and larger frequency region. Note that if we sample this spectral density at a specific frequency, as the relaxation process does, we expect a smaller, then a larger and finally a smaller value of $J(\omega)$ as the temperature is raised from very low to very high temperatures. Therefore, a graph of $1/T_1$ will have a maximum at some temperature, i.e., there will be a minimum in the graph

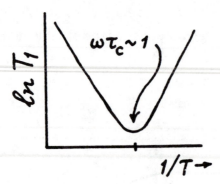

of $T_1$ vs. T and this will occur at a temperature at which the spectral density contribution at the Larmor frequency is a maximum. Given the usual form for the J's with terms like

$\tau_c/(1+\omega^2\tau_c^2)$, the $T_1$ minimum will occur at lower temperatures (where $\tau_c$ is longer) and will be deeper for smaller $\omega$ and $H_0$.

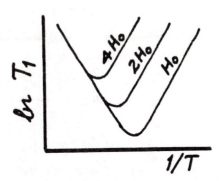

We close this subsection with an assurance that the qualitative arguments we have presented for a typical term in the spectral density holds for the total spectral density consisting of several such terms.

### Additional Observations

In all of our discussions of spin-lattice relaxation, we assume that the spin system has a very small heat capacity so that the spin energy cannot disturb the lattice temperature.

The relaxation is commonly described by an exponential function and so we characterize its rate by the reciprocal of the first order rate constant, a relaxation time. Even though the relaxation process may appear as an exponential process, we should not be mislead into believing that it is a real first order process similar to radioactive decay. The underlying process is generally not first order and in many cases, oftener than we like to think, this appears as a non-exponential relaxation, especially if the data were taken carefully enough.

The spin-lattice relaxation time can be viewed as a lifetime of an average chemically equivalent spin in the system. (In fact, the spin state half life is $T_1 \ln 2$.) One should not

confuse the mean lifetime of a spin with the correlation time
of the motions responsible for relaxation and a large number
of fluctuations may occur before a spin relaxes -- a situation
called the weak collision limit.  For example, the relaxation
time of water at room temperature is about 3 seconds but the
molecular correlation time is about $10^{-12}$ seconds.  On the
other hand, there are cases in which virtually all motions of
the spins result in relaxation -- the so called strong collision
limit.  See section IV.C.2 for further discussion.

It is important to note that the term describing the relaxation should only contain that part of the interaction which
is time varying.  For example, in the case of dipole-dipole
relaxation, if the total interaction contains terms that are
static as well as terms that are time varying, only the latter
contribution should be included.  This fact, while rarely
alluded to, becomes important in certain cases, for example,
rotating frame relaxation, which explicity depend on how much
of the local field is time varying.

The spin-lattice relaxation rates discussed so far are for
a single pair of protons.  For multiple interactions, contributions must be summed.  For example, if spin A interacts with
both spins B and C, then the total interaction for spin A
must be the sum of the two contributions.

In the above discussion, we assumed that the interaction
between a pair of dipoles is independent of interactions between all other pairs.  This means, for example, that the
relative motions of A and B are not dependent on the relative
motions of spins C and A.  This assumption would not be
true in a case such as a methyl group where the motions are
correlated.  This lack of independence of the relative motions
of pairs of spins is called cross-correlation and generally
results in a retarded and non-exponential relaxation.  Such

cross-correlation effects are considered in a series of papers by Hubbard and his colleagues, referenced in his 1969 paper, and the papers by Runnels (1964), Mehring and Raber (1973), Cutnell and Venable (1974), and Werbelow and Grant (1977).

We are only scratching the surface of the topic of relaxation in this book because most of it is inappropriate for a book on how to do NMR. For comprehensive treatments, we refer you to the general references (Appendix A) and the modern, comprehensive treatments and summaries of relaxation theories provided by Noack (1971), Werbelow and Grant (1977), and Spiess (1978).

# REFERENCES

J. D. Cutnell and W. Venable, "Nonexponential nuclear spin-lattice relaxation in polycrystalline dimethyl sulfone," J. Chem. Phys. 60, 3795-3801 (1974).

P. S. Hubbard, "Nonexponential relaxation of three-spin systems in nonspherical molecules," J. Chem. Phys. 51, 1647-1651 (1969).

M. Mehring and H. Raber, "Nonexponential spin lattice relaxation and its orientation dependence in a three-spin system," J. Chem. Phys. 59, 1116-1120 (1973).

F. Noack, "Nuclear magnetic relaxation spectroscopy" in NMR, Basic Principles and Progress, vol. 3, edited by P. Diehl, E. Fluck, and R. Kosfeld (Springer Verlag, Berlin, 1971), pp. 83-144.

L. K. Runnels, "Nuclear spin-lattice relaxation in three-spin molecules," Phys. Rev. 134, A28-36 (1964).

H. W. Spiess, "Rotation of molecules and nuclear spin relaxation," in NMR, edited by P. Diehl, E. Fluck, and R. Kosfeld, (Springer Verlag, Berlin, 1978) pp. 55-214.

L. G. Werbelow and D. M. Grant, "Intramolecular dipolar relaxation in multispin systems" in Advances in Magnetic Resonance, vol. 9, edited by J. S. Waugh (Academic Press, New York, 1977), pp. 83-144.

J. Winter, Magnetic Resonance in Metals (Oxford, London, 1971).

## III.B.   SPIN-SPIN RELAXATION

Spin-spin relaxation, also called transverse relaxation, is a process in which the magnetization in the x-y plane, perpendicular to the static laboratory field, decays. In section III.C.1., we will write down expressions for $1/T_2$ as they relate to the corresponding $T_1$'s for dipolar and chemical shift anisotropy relaxations. It will be pointed out that, in the presence of extreme narrowing due to rapid molecular motion, $T_1 \sim T_2$, with the consequence that these two relaxation rates do not contain different information in this limit, which almost always occurs in liquids.

When $T_2 \neq T_1$ so that $T_2$ and $T_1$ do contain different information, it becomes useful to measure $T_2$ for its own sake. There are many examples of $T_1$ and $T_2$ yielding completely different information and we cite only a case most familiar to us, namely solid CO in the β-phase where $T_1$ is determined by the spin-rotation interaction and increases for decreasing T while $T_2$ is determined by the molecular diffusion and decreases with decreasing T (Fukushima, et al., 1977; 1979). So here is a case where $T_1 = T_2$ in the liquid but $T_2$ suddenly diverges from $T_1$ when the material freezes, as shown below.

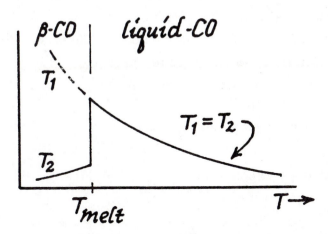

$T_1$ and $T_2$ are usually different in solids, although this can happen quite often in liquids as well. (This should be considered as a blessing rather than a nuisance since nature is trying to give us additional information.) In IV.A.2. we say that solids usually have a lineshape close to a Gaussian while most liquids have Lorentzian lineshapes. Since we have so far treated $T_2$ as a time constant for an exponential decay, there is an inconsistency here. How should we define a $T_2$ for a Gaussian or some other non-Lorentzian line?

A Gaussian function has the property that its Fourier transform is also a Gaussian. Thus, a function

$$f(\omega) \propto \exp(-\omega^2/2\sigma^2)$$

has a Fourier transform

$$F(t) \propto \exp(-\sigma^2 t^2/2)$$

where the full width at half height in the $\omega$ domain (which is the frequency domain) is equal to $2.36\sigma$. We can define a characteristic decay time for $F(t)$ as the time $T_2$ required for

F(t) to decay to $1/e$ of F(0).  Then $T_2 = \sqrt{2}/\sigma$ so that the full width at half height in the $\omega$ domain is $3.34/T_2$.  Since we will note in IV.A.2. that $S_2 = \sigma^2$ for a Gaussian function, $T_2 = \sqrt{(2/S_2)}$, where $S_2$ is the second moment.

In contrast, a Lorentzian function

$$g(\omega) \propto (1/T_2)/[(1/T_2)^2 + \omega^2]$$

transforms into an exponential

$$G(t) \propto \exp(-t/T_2)$$

and it is easy to prove that the full width at half maximum is $2/T_2$.

In any discussion involving relaxation times, remember that $1/T_1$, $1/T_2$, etc., have the units of angular velocity (radians/second) and that the usual NMR spectral display is in units of frequency (hertz).  Thus, in frequency units, the full widths at half maximum for the Gaussian and the Lorentzian lines are $3.34/(2\pi T_2)$ and $1/\pi T_2$, respectively.

Measurement of $T_2$ presents different problems depending on the magnitude of $T_2$.  If it is very long, say, greater than ten seconds, then the magnet inhomogeneity will usually make the spins dephase faster than the real $T_2$ and the CPMG sequence already discussed in I.C.2. should be used.  As with all multiple pulse sequences, there are many adjustments (such as tip angle, rf phase, and the resonance condition) which are all quite critical to the correct operation of the sequence.  We will not go into these details here but instead refer you to the articles by Vold, et al. (1973), Hughes and Lindblom (1977) and references therein, and sections 7.5.2. and 7.5.3. of the book by Martin, et al. (Appendix A).  We stress that when you are trying any complicated sequence for the first time, play around with it until you are convinced that you can get reproducible and correct results. Although that should go without saying for any NMR sequence, it is especially true for the multiple pulse sequences which are, after all, very complex.

If $T_2$ is very short, it is difficult to see the signal because the FID will decay before the instrumental dead time is over. This situation will usually occur only for solids, and one way to deal with it is to try to form an echo which occurs later than the end of the dead time. Such echoes are fairly easy to form in the presence of a large inhomogeneous broadening such as in metals and with many quadrupolar nuclei in NQR. You will need a high speed digitizer because the echoes will be very sharp. For some other solid state echoes, see IV.B.3. and IV.B.4. Another possible solution is to use the zero time resolution method described in VI.D.4.

## REFERENCES

E. Fukushima, A. A. V. Gibson, and T. A. Scott, "Carbon-13 NMR of carbon monoxide. I. Pressure dependence of translational motion in β-CO," J. Chem. Phys. 66, 4811-4817 (1977).

E. Fukushima, A. A. V. Gibson, and T. A. Scott, "Carbon-13 NMR of carbon monoxide. II. Molecular diffusion and spin-rotation interaction in liquid CO," J. Chem. Phys. 71, 1531-1536 (1979).

D. G. Hughes and G. Lindblom, "Baseline drift in the Carr-Purcell-Meiboom-Gill pulsed NMR experiment," J. Magn. Resonance 26, 469-479 (1977).

R. L. Vold, R. R. Vold, and H. E. Simon, "Errors in measurements of transverse relaxation rates," J. Magn. Resonance 11, 283-298 (1973).

## III.C.1.   SPIN-LATTICE RELAXATION MECHANISMS

In this section, we survey some of the spin-lattice relaxation mechanisms but not for completeness because there are many good references on the subject. We particularly recommend the books listed under general references in Appendix A. In addition, Noack (1971) has all the relaxation mechanisms (except in metals) tabulated and Spiess has a complete discussion of relaxation mechanisms arising from molecular motion including the interesting effects of $T_1$ on the solid state powder pattern. Noggle and Shirmer (1971) as well as Lyerla and Levy (1974) have written readable reviews of relaxation mechanisms in solution, and Winter (1971) augments Slichter's (Appendix A) discussion on NMR of metals. Needless to say, there are other good articles and books on spin-lattice relaxation but the above list is certainly a good start.

Believing that this book should concentrate on what is not generally known, we begin with a brief discussion of spin-lattice relaxation to paramagnetic impurities in insulating diamagnetic solids. Such paramagnetic impurities were recognized as an important relaxation mechanism as early as 1949 by Bloembergen, and it has been found that a rather small concentration of impurities can determine the $T_1$ of a solid if there are no other competing mechanisms. An important question to be asked at this stage is how we can talk about $T_1$ when there might be one impurity per hundreds or thousands of nuclei which need to be relaxed so that the nuclei are all at different distances from the electron whose field drops off as one over the cube of the distance. The answer is that, in many but not all systems, the strong dipole-dipole coupling maintains communication between all the spins so that all the spins can relax together. This is the reason why all the protons in, say, an organic solid tend to have the same $T_1$ as measured by high resolution NMR of solids. The magnetization diffuses among the spins to the impurity and this diffusion (not of the nuclei but just the polarization) is called spin diffusion.

Let us consider a thermal analogy. For this example, we define a local spin temperature which has a one to one correspondence with the spin polarization. Thus, if there is a small group of spins with no net polarization, we will say that its local spin temperature is infinite whereas if the small group of spins are all lined up (polarized) along the field we will say that its local spin temperature is zero. Then the spin diffusion problem can be recast as the following heat transfer problem. (Spin temperature is treated in IV.D.)

Suppose that we are going to measure the $T_1$ of the system by measuring the magnetization recovery after saturation (say by a comb as described in III.D.2.). Then immediately after the comb (t=0), the spin temperature $T_S$ is infinite throughout the sample. The lattice temperature $T_L$, however, is much colder, say, equal to room temperature, and for all practical purposes we can take it to be zero for this discussion. The paramagnetic impurity provides the thermal link between the hot spins and the cold lattice.

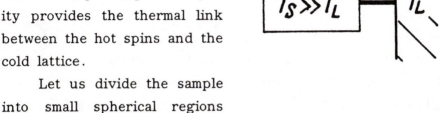

Let us divide the sample into small spherical regions each with a paramagnetic impurity at its center. All nuclei within each sphere will be relaxed by the impurity at the center which is, in effect, a heat sink through which the excess spin temperature drains to the cold lattice. The impurity will interact directly with and cool those nuclei which are close to it while those farther away will cool by diffusion to those closer which are already cold. This is analogous to Newton's law of cooling and the net result is an exponential decay of the average temperature.

From this qualitative picture, we would expect the $T_1$ to be proportional to the number of spins per each impurity so that $1/T_1$ must be proportional to the impurity concentration. $1/T_1$ must also be proportional to some parameter having to do with the strength of the paramagnetic impurity as a heat sink. The paramagnetic impurity interacts with the nuclei through the dipole-dipole interaction with its own correlation time determined by its interaction with the lattice. For the usual form of spin diffusion, $1/T_1 \propto [\tau/(1+w^2\tau^2)]^{\frac{1}{4}}$ so that the field dependence is very small. The temperature dependence, however, could be fairly substantial depending on the temperature dependence of $\tau$ which is simply the spin-lattice relaxation time of the paramagnetic ion. Finally, $1/T_1$ should be proportional to some parameter having to do with the diffusion coefficient which, in turn, is related to internuclear spacing and, thus, to $T_2$. The diffusion coefficient is greater for spins closer together, i.e., for small $T_2$. These behaviors usually are true as long as the impurity concentration is not too high.

There is a conceptual difficulty with all this. In order for diffusion to take place you need a temperature gradient. Initially, the entire sphere is at infinite temperature except the heat sink in the center, so there cannot be any gradient. So, there must be a period during which a gradient is created throughout the sphere before the normal diffusion process can take over. If the sphere is large so the impurity concentration is small, a relatively few of the spins out of the total will have relaxed in this period and the relaxation is negligibly different from exponential. Only in samples with large impurity concentration, say in the percent range or more, and only near the $T_1$ minimum does the non-exponential nature of the decay become measurable.

Spin diffusion to paramagnetic impurity is an important mechanism, not only in solids but in some large molecules which exist in solution. In fact, there can be spin diffusion not only to impurities but to any center which has good coupling to the lattice and this could be a sub-molecular unit with its own motion conducive to rapid relaxation or a quadrupolar nucleus. Because of the sheer size of the electron moment, spin diffusion to paramagnetic impurities is more common than to these other heat sinks. See the review by Noack (1971) for references to this mechanism.

A relaxation mechanism which we will not really touch on is that due to conduction electrons in metals. We will merely say that the primary interaction with the conduction electrons is through their spins rather than their orbital motions the latter being the case for electronic effects in diamagnetic materials to be discussed below.

There is a large class of relaxation mechanisms which operate on molecules in motion in non-metallic samples. All but one, the spin-rotation interaction, depend on the fact that the change in the molecular orientation or translation modulates the field due to that particular interaction and creates a randomly varying field at the site of the nucleus in question. Any such random motion can have associated with it a special form of an autocorrelation function $G(t)$, expressed in terms of a scalar product of the local field $h(t)$ and the same local field at an earlier time $h(0)$, which is a measure of how rapidly the local field changes in magnitude and direction. If the motion and, therefore, the local field fluctuation is truly random, the autocorrelation function

$$G(t) \propto h(t) \cdot h(0) \propto \exp(-t/\tau_c)$$

is exponential and is independent of the time origin.   This statement says that the probability of finding a correlation between the local field at time t and at t=0 decreases as $\exp(-t/\tau_c)$ so that $\tau_c$ which is called the correlation time for the motion is the mean time past which you could say that there is not much (or any) correlation.

The spectral density $J(\omega)$ which we have already mentioned is the frequency spectrum corresponding to the auto-correlation function $G(t)$, i.e., it is the Fourier transform of it, so that

$$J(\omega) = \int_{-\infty}^{\infty} G(t)\exp(-i\omega t)dt.$$

Just as in the relation of the FID to the spectrum as discussed in I.D.2., $G(t)$ and $J(\omega)$ have an inverse relationship such that for slow molecular motion in which $\tau_c$ is long its $J(\omega)$ is all bunched up for small $\omega$ and vice versa.

Suppose we now consider the nuclear dipole-dipole interaction.   Assume that there are two identical spins a distance R apart on the same molecule which is rotating randomly. The spectral densities which arise from the motion of the molecule in this case are

$$J^{(0)}(\omega) = (24/15R^6)[\tau_c/(1+\omega^2\tau_c^2)],$$

$$J^{(1)}(\omega) = (1/6)J^{(0)}(\omega),$$

and

$$J^{(2)}(\omega) = (2/3)J^{(0)}(\omega).$$

The three different J's correspond to cases in which there is no net spin flip, one spin flip, and two flips, respectively. Soda and Chihara (1974) have pointed out that in the weak collision limit ($\tau_c \ll T_2$) the spectral densities have fixed ratios independent of the type of isotropic molecular motion for each relaxation mechanism. This is a nice rule to know because only one J needs to be calculated from scratch for any given mode of motion and the other J's can be calculated from the first simply by choosing the proper argument, i.e., 0, $\omega$, or $2\omega$. These authors also give a useful relation between the reduction factor of the second moment arising from a particular mode of molecular motion to the $T_1$ minimum due to that same motion.

The relationship between $T_1$ and $T_2$ and the J's can be derived through time dependent perturbation theory as was done originally by Bloembergen, Purcell, and Pound or by the use of the density matrix which turns out to be equivalent. See the general references in Appendix A, especially Chapter 5 of Slichter. For our case of two spins on a rotating molecule, the relaxation rates are given by

$$1/T_1 = (3/2)\gamma^4 \hbar^2 I(I+1) \; [J^{(1)}(\omega)+J^{(2)}(2\omega)]$$

and

$$1/T_2 = \gamma^4 \hbar^2 I(I+1) \; [(3/8)J^{(0)}(0)+(15/4)J^{(1)}(\omega)$$

$$+(3/8)J^{(2)}(2\omega)].$$

The expression for $1/T_2$ consists of terms similar to those in the $1/T_1$ expression as well as a unique term with $J^{(0)}(0)$. This is because components of the field contributing to the $T_1$ process can help reduce components of the magnetization in the x-y plane but, in addition, the z component of the local field can influence magnetization components only in the rotating x-y frame but not those along the z-axis with which it is collinear.   Therefore, the term $J^{(0)}(0)$ contributes to $1/T_2$ but not to $1/T_1$ and this is the dominant contribution to $1/T_2$ in solids where $1/T_1 << 1/T_2$.

The substituion of the J's given earlier (with the correct arguments in the appropriate J's) into the above expressions will yield the usual $1/T_1$ and $1/T_2$ expressions to be found in most books and which we will rewrite in terms of the second moment in IV.A.2.   For the dipole-dipole case

$$J^{(0)}(\omega):J^{(1)}(\omega):J^{(2)}(\omega) = 6:1:4$$

so that in the motionally narrowed limit, ($\omega\tau << 1$), it is easy to see that $T_1 = T_2$.

We digress here to mention that the frequency independent term $J^{(0)}(0)$ will make $1/T_2$ behave differently from $1/T_1$ at large $\omega$ and/or large $\tau_c$ in this case.   As the temperature becomes lower so that $\tau_c$ becomes larger, the significance of the $J^{(0)}(0)$ term becomes greater by default, i.e., by having the other J's contribute less, and at temperatures below that for which the $T_1$ minimum occurs, $T_2$ will continue to fall until it reaches the value for no rotation as sketched here.

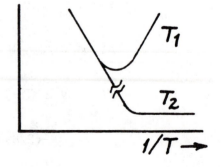

The usual way to plot the temperature dependence of the relaxation times when $\tau_c$ is determined by a thermally activated process obeying an Arrhenius relation, $\tau_c \propto \exp(-E/kT)$, is to plot $\ell n T_1$ [or $\ell n(1/T_1)$] against $1/T$ so that the sides of the $T_1$ dip yield the activation enthalpy $E$.

We now return to discussing some other relaxation mechanisms which involve random motions of molecules, namely the scalar relaxation of the first and second kinds, quadrupolar relaxation, chemical shift anisotropy relaxation, and spin rotation relaxation.

Scalar relaxation occurs when a spin $\bar{I}$ is coupled to a spin S through an electron moderated interaction $\bar{I} \cdot \underline{A} \cdot \bar{S}$. This so called spin-spin coupling is seen often in high resolution solution NMR wherein the coupling constant J is the average of the tensor components of $\underline{A}$ in frequency units.

Scalar relaxation of the first kind results when a nucleus $\bar{I}$ is jumping in and out of a site in which it is scalar coupled to a magnetically inequivalent neighbor $\bar{S}$. The modulation of the coupling may result in relaxation, in addition to the familiar collapse of the multiplet at high enough exchange rate. If the exchange correlation time $\tau_e$ is much shorter than the $T_1$'s for either spin,

$$1/T_1 = (2A^2/3)S(S+1)\tau_e/[1+(\omega_I - \omega_S)\tau_e^2]$$

where A is one-third of the trace of the tensor $\underline{A}$. This expression simplifies greatly when the $\bar{I}$ and the $\bar{S}$ spins are of the same species so that $\omega_I = \omega_S$.

Scalar relaxation of the second kind differs from the first kind only in that the modulation of the field at spin $\bar{I}$ arises from the relaxation effects of $\bar{S}$. A common example is a coupling to a quadrupolar nucleus undergoing rapid relaxation as decribed below.  The $T_1$ expression remains the same with a changed meaning for $\tau_e$ which now stands for the correlation time for $\overline{S}$ spin reorientation.

A nuclear electric quadrupole moment arises from a non-spherical distribution of electrical charges in the nucleus. Because the energy of an electric quadrupole moment depends on its orientation with respect to an EFG, a randomly varying electric field gradient associated, for example, with molecular rotation having a finite amplitude of its spectral density at the Larmor speed can induce effective relaxation of the quadrupolar nucleus.  In liquids, it is possible to write down expressions for $1/T_1$, only in special cases of $I=1$ or of motionally narrowed limit when $\omega\tau \ll 1$, while in solids it is usually not possible to define $T_1$.  For a further discussion of this topic see III.C.3. and the general references in Appendix A.

The chemical shift is due to the orbital effects of the nearby electrons.  When an atom or a molecule is placed in a static magnetic field, the electrons produce a small auxiliary field the direction and the magnitude of which depend in a complex way on the electronic structure of the atom or molecule.  Because the ability of the electrons to shield the nucleus depends in part on the direction of the magnetic field with respect to the electronic orbits, the chemical shift is a tensor quantity.  The shift observed in liquids as well as that in MAS experiments are motional averages of the tensor components.

Both the $T_1$ and $T_2$ relaxation rates due to chemical shift anisotropy coupling are proportional to the square of the anisotropy and the square of the static field.  For a nucleus at a site with axial symmetry

$$1/T_1 = (1/15)\gamma^2 H_0^2 \Delta\sigma^2 \tau_c/(1+\omega^2\tau_c^2)$$

and

$$1/T_2 = (1/90)\gamma^2 H_0^2 \Delta\sigma^2 [6\tau_c/(1+\omega^2\tau_c^2)+8\tau_c]$$

where $\Delta\sigma$ is the anisotropy of the shift. For nuclei at sites having less than axial symmetry, the expressions are more complex.

Notice that what counts is not necessarily the symmetry of the molecule but the symmetry of the site at which the nucleus is located. Thus, if we consider an octahedral metal hexafluoride $MF_6$ undergoing a random rotation, the metal will not experience any anisotropy relaxation effects but the fluorines will because they reside at sites having axial symmetry.

Note also that in the motionally narrowed limit where $\omega\tau<<1$, $T_1/T_2=7/6$ and this is the only case for which $T_1{\ne}T_2$ in this limit with random motion. Keep in mind, too, that the motional correlation time $\tau_c$ is the same for the chemical shift anisotropy, dipolar, and the quadrupolar interactions.

Finally, we consider spin rotation interaction which is important only for small spherical molecules under conditions in which they are relatively free to rotate in the absence of other relaxation mechanisms. It is most common in gaseous samples (and it is indeed a major source of relaxation in molecular and atomic beam experiments) but exists also in liquids and even some solids.

This interaction is between the nuclear spin and the molecular magnetic moment which arises from the angular momentum of the rotating molecule. When the molecule remains in a given angular momentum state, its electronic structure produces a constant electric current and this would not lead to

relaxation but collisions with other molecules can cause fluctuations of the angular momentum which, in turn, can cause
relaxation. This coupling is described in terms of a spin
rotation tensor $\underline{C}$. The spin-lattice relaxation rate is

$$1/T_1 = (2IkT/3\hbar^2)(2\pi C_e)^2 \tau_J$$

where $C_e/3$ is the average of the principle values of the tensor
$\underline{C}$ in units of frequency, I is the moment of inertia of the
molecule about the appropriate axis, and $\tau_J$ is the angular
momentum correlation time.

In general, for a liquid the density decreases as the
temperature increases and $\tau_J$ becomes longer due to less frequent collisions. $\tau_c$ usually has the opposite dependence as
the molecule changes orientation by large angles at each
collision. At the other extreme, collisions are frequent so
that $\tau_J$ is small and the molecular orientation can change by a
large angle only after many collisions. In this case, which is
the usual one, $\tau_J \ll \tau_c$ and the two correlation times are inversely related by the relation $\tau_J \tau_c = I/6kT$. Thus, as far as
our current discussion is concerned, the dependence of $T_1$ on
density or temperature for spin rotation interaction is opposite
to that due to other mechanisms already discussed which
depend on $\tau_c$.

## REFERENCES

J. R. Lyerla, Jr., and G. C. Levy, "Carbon-13 nuclear spin
relaxation" in Topics in Carbon-13 NMR Spectroscopy, vol. 1,
edited by G. C. Levy (John Wiley & Sons, New York, 1974),
pp. 79-148.

F. Noack, "Nuclear magnetic relaxation spectroscopy" in NMR,

Basic Principles and Progress, vol. 3, edited by P. Diehl, E. Fluck, and R. Kosfeld (Springer Verlag, Berlin, 1971), pp. 83-144.

J. H. Noggle and R. E. Shirmer, The Nuclear Overhauser Effect (Academic Press, New York, 1971).

G. Soda and H. Chihara, "Note on the theory of nuclear spin relaxation; exact formulae in the weak collision limit," J. Phys. Soc. Japan 36, 954-958 (1974).

J. Winter, Magnetic Resonance in Metals (Oxford, London, 1971).

## III.C.2.  HOW TO DIFFERENTIATE BETWEEN DIFFERENT RELAXATION MECHANISMS

In order to determine which relaxation mechanism is operating in a given case, you should rely on some estimates based on previous knowledge as well as doing experiments to sort out the different mechanisms.  Using common sense and chemical knowledge, you can significantly narrow the number of possible mechanisms so that the correct experiments can be chosen.

To estimate the strength of each interaction, you can rely on typical data for similar compounds from which extrapolation can be made.  Lacking data on similar compounds, you can try to estimate the potential strengths of different mechanisms indirectly.

Guessing at the relaxation mechanism will be easy when a general property of the sample makes it very likely for a single mechanism to dominate.  Some examples would be metals in which the conduction electrons will be the source of relaxation, paramagnetic samples in which the unpaired spin of the electron will be the relaxation mechanism, and nuclei with large quadrupole moments which will relax by the quadrupolar interaction in any surroundings but cubic sites.

In the case of paramagnetic impurities, remember that the relaxation rate is usually proportional to the impurity concentration. Just to give you an idea, the $T_1$ minimum for $CaF_2$ doped with 100ppm $Gd^{3+}$ at a flourine frequency of 20 MHz is well under 100 ms (Fukushima and Uehling, 1968). Of course, in most situations, you will not be close to the $T_1$ minimum near room temperature, especially for ions which are not in the S-state like $Gd^{3+}$ because the electron relaxation time will be much shorter than the Larmor period.

The quadrupole interaction depends not only on the size of the quadrupole moment, which is known for most of the nuclei of interest, but on the actual electric field gradient (EFG) at the nucleus. The latter is difficult to estimate because an atom in an EFG modifies the gradient seen by the nucleus in a complicated way. The usual effect is to make the field at the nucleus bigger than that felt by the atom by a factor called the Sternheimer antishielding factor. The Sternheimer factor generally increases with increasing atomic number so, in general, the quadrupole interaction becomes more important for the heavier atoms. In frequency units this interaction ranges between zero (for spin ½ nuclei) to hundreds of MHz. For further discussion, see II.D.1., III.C.3. as well as the general references in Appendix A and the articles by Cohen and Reif (1957) and Das and Hahn (1958). Quadrupole moments are listed, for example, in the NMR tables by Lee and Anderson referenced in Appendix A. There is also a fairly old tabulation of quadrupole couplings by Biryukov, et al., (1969).

The chemical shift anisotropy interaction is proportional to the squares of the anisotropy and of the Larmor frequency. Thus, it is much more important for nuclei having larger shifts like fluorine or carbon as opposed to protons and, at the same time, for nuclei at asymmetric sites like those on the ligands L as opposed to those on the metals M near the centers of the molecules like $ML_6$. Because of the field dependence, this interaction is relatively unimportant at low fields but must be seriously considered at high fields such as those of modern superconducting magnets. Typical range for this interaction might be from a few Hz to tens of kHz at currently typical fields.

The nuclear dipole-dipole interaction depends on the respective magnetic moments and the internuclear separation in addition to the existence of appropriate molecular motion. Thus, it is a very important mechanism for protons in, say, an organic molecule but not for naturally abundant carbon-13 $CS_2$. The maximum dipolar interaction is of the order of several kHz.

The scalar interaction, like the quadrupole coupling and the shift tensor, depends on the atomic number. There is a complete set of isotropic scalar coupling data from high resolution solution NMR. Typical proton coupling may be 10's of Hz while fluorine couplings can be as big as a few kHz, for example, in $TeF_6$.

Spin rotation interaction only occasionally affects the NMR relaxation times. It requires a rapid rotation of the molecule which limits its applicability to small or spherical molecules which are nearly free to rotate. Because it is relatively weak, it should be considered only after the other interactions have been considered. It is especially rare in solids where it is known to occur only in molecular crystals in their orientationally disordered phases (also known as plastic phases). Some of the better known examples include the $SF_6$ family (Blinc and Lahajnar, 1967), $Fe(CO)_5$, and $N_2$, the last two of which we will discuss further near the end of this section. The spin rotation relaxation has a unique signature, namely that its $T_1$ usually decreases with increasing temperature even at a fairly high temperature where the other mechanisms lead to increasing $T_1$'s.

There are a couple of special methods of separating the contribution of dipolar relaxation in solution. One is by the NOE factor which is the fractional difference in the signal intensity of one spin with and without irradiation applied to another spin system. For a sample containing protons and carbon-13 in the motionally narrowed limit, this factor should be 2 if the relaxation takes place through the dipolar and the scalar interactions. Thus, the departure from 2 of the NOE factor is an indication of other relaxation mechanisms. Clearly, any other pairs of spin systems with NOE's can be treated this way, with appropriate limiting NOE factors. See, for example, Noggle and Shirmer listed in Appendix A for more details.

Another way to separate the dipole-dipole interaction is to change the interaction while keeping everything else the same by making isotopic substitutions with nuclei having different magnetic moments. Deuterium for protons is a common substitution although $^{10}B$ for $^{11}B$, $^{15}N$ for $^{14}N$, etc., can be useful, too, and the relative mass change will not be as catastrophic thereby ensuring that the molecular motions, if any, will remain similar upon substitution.

The frequency variation of $T_1$ is a very good way to separate $T_1$ mechanisms. (See VI.D.1. for how to do such experiments.) For example, we have already said that $1/T_1 \propto H_0^2$

for chemical shift anisotropy relaxation. On the other hand, dipole-dipole relaxation in the regime $\omega\tau_c \gg 1$ has $1/T_1 \propto H_0^{-2}$. Other mechanisms are usually field independent.

The temperature dependence of $T_1$ is useful, too, in identifying relaxation mechanisms. We have already mentioned that spin rotation relaxation usually has an unusual dependence, namely that the $T_1$ decreases for increasing T in the region where the dipolar relaxation time usually increases with T. In general, the temperature dependence of $T_1$ (or other parameters, too, for that matter) is hard to use quantitatively because some other property of the sample, for example, the volume or the viscosity, is changing at the same time. Thus, in order to get a good thermodynamic parameter, say at constant volume as a function of temperature, you must make measurements as a function of pressure as well.

We finish this section with some examples of $T_1$ experiments showing different relaxation behavior depending on the experimental parameters. Maryott, et al. (1971), measured the $T_1$ for both $^{35}Cl$ and $^{19}F$ in $ClO_3F$ as a function of temperature. Because $^{35}Cl$

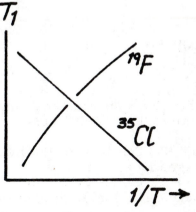

is a quadrupolar nucleus, it relaxes according to the correlation time $\tau_c$ whereas the $^{19}F$ relaxation is governed by the angular momentum correlation time $\tau_J$ and they have opposite temperature dependence in this temperature region which is above the $T_1$ minimum for $^{35}Cl$, that is $\omega\tau_c \ll 1$. An entirely similar example is the relaxation of $^{14}N_2$ and $^{15}N_2$ studied by Scott and co-workers, wherein the $^{14}N$ relaxes by quadrupolar and the $^{15}N$ by spin rotational interactions (Ishol, et al. 1976).

The final example is the temperature and field dependences of the $^{13}C$ resonance in Fe(CO)$_5$ by Spiess and Mahnke (1972). The temperature dependence of $T_1$ can be made to reverse upon application of a large field because the chemical shift anisotropy mechanism takes over at high field while the spin rotation is the main mechanism at low field.

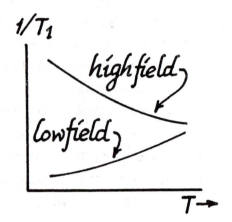

## REFERENCES

I. P. Biryukov, M. G. Voronkov, and I. A. Safin, Tables of NQR Frequencies (Daniel Davey and Co., Hartford, Connecticut, 1969).

R. Blinc and G. Lahajnar, "Nuclear-magnetic-resonance relaxation by spin-rotational interactions in the solid," Phys, Rev. Letters 19, 685-687 (1967).

M. H. Cohen and F. Reif, "Quadrupole effects in nuclear magnetic resonance studies of solids" in Solid State Physics, vol. 5, edited by F. Seitz and D. Turnbull (Academic Press, New York, 1957), pp. 321-438.

T. P. Das and E. L. Hahn, Nuclear Quadrupole Resonance Spectroscopy (Academic Press, New York, 1958).

E. Fukushima and E. A. Uehling, "Nuclear-spin-diffusion relaxation to a finite density of paramagnetic impurity ions," Phys. Rev. 173, 366-375 (1968); erratum, ibid 180, 632 (1969).

L. M. Ishol, T. A. Scott, and M. Goldblatt, "Nuclear-spin-lattice relaxation in solid and liquid $^{15}N_2$ and $^{14}N_2$," J. Magn. Resonance 23, 313-320 (1976).

A. A. Maryott, T. C. Farrar, and M. S. Malmberg, "$^{35}$Cl and $^{19}$F NMR spin-lattice relaxation time measurement and rotational diffusion in liquid ClOF$_3$," J. Chem. Phys. <u>54</u>, 64-71 (1971).

H. W. Spiess and H. Mahnke, "$^{13}$C anisotropic chemical shifts and spin-rotation constants in nickel tetracarbonyl and iron pentacarbonyl," Ber. Bunsen-Ges. Phys. Chem. <u>76</u>, 990-995 (1972).

## III.C.3.  RELAXATION OF QUADRUPOLAR NUCLEI

Spin-½ nuclei do not have quadrupole moments but nuclei with I>½ can have electric quadrupole moments and, in practice, always do. One consequence of I>½ is that there will be more than two allowed nuclear Zeeman energy levels because the number of levels goes as 2I+1. A static electric field gradient will perturb these levels according to the "m" labels (discussed in II.D.1.) so that the Zeeman levels (created by the magnetic moments interacting with the static magnetic field) will no longer be equally spaced. In section IV.D., we define a spin temperature and it will be quite clear from the discussion there, as well as that in II.D.1., that it is nearly impossible to maintain a uniform spin temperature in such a multilevel system upon the application of a pulse which causes transitions between some or all of the levels. Thus, in most solids with quadrupolar nuclei, it is impossible to define a unique $T_1$ because the nuclear relaxation will be a complex sum of relaxations between different levels.

If there is no electric field gradient (EFG) at the site of the quadrupolar nucleus, due to high site symmetry, there will not be any splitting and the relaxation will be exponential if there are no fluctuating EFG's as well as static EFG's.  If you want to know whether or not the relaxation is due to the quadrupole interaction, you might try to calculate what the relaxation time might be from the expression of the sort given below but it may be easier and more accurate to compare it with known systems relaxing via the quadrupole interaction. With certain nuclei, it is possible to isotopically substitute another nucleus with a different ratio of the magnetic dipole moment to the electric quadrupole moment and see whether $T_1$ scales with the square of the dipole or the quadrupole moments.

A modulation of the EFG at the site of a quadrupolar nucleus due to isotropic and sufficiently rapid molecular rotation in liquids (in the motionally narrowed case, where $\omega\tau \ll 1$) leads to the expression

$$1/T_1 = (3/40)[(2I+3)/I^2(2I-1)](1+\eta^2/3)(e^2qQ/\hbar)^2\tau_c$$

where $\eta$ is the asymmetry parameter and $e^2qQ/\hbar$ is the strength of the interaction between the quadrupole moment $eQ$ and the field gradient $eq$.  We parenthetically remark that $e^2qQ/\hbar$ is sometimes called the quadrupole coupling although it is more common to call its frequency equivalent $e^2qQ/h$ as the quadrupole coupling.  The quadrupole coupling ranges from very small to hundreds of MHz depending on both $Q$ and $q$.  If we take $I=3/2$ and $\eta=0$, the above equation reduces to $(e^2qQ/\hbar)^2\tau_c/10$, so that if $e^2qQ/h$ were 5 MHz, $(e^2qQ/\hbar)^2$ would be approximately $10^{15}$ and $T_1 \sim 10^{-14}/\tau_c$.  For a typical tumbling molecule in liquids, $\tau_c \sim 10^{-12}$ sec which leads to a $T_1$ of 10 ms!  Thus, generally speaking, quadrupole relaxation can devastate high resolution solution lines if the coupling is anything like hundreds of kHz or greater.  Even small couplings lead quadrupole nuclei to broader resonances in solutions than spin-½ nuclei. For further discussion of the above expression and its derivation, see Chapter VIII, section F(c), of Abragam (Appendix A).

There is one other special case of quadrupolar relaxation in liquids (or in solids with no net EFG which would lead to splittings of the lines) and that is when I=1. The relaxation will be exponential in this case with the rate given by an expression reminiscent of the dipolar relaxation (for a good reason) in which we have written $\tau$ for $\tau_c$.

$$1/T_1 = (3/80)(1+\eta^2/3)(e^2qQ/\hbar)^2[2\tau/(1+\omega^2\tau^2)+8\tau/(1+4\omega^2\tau^2)].$$

For further discussions, see the Abragam reference above or Carrington and McLachlan (Appendix A), Section 11.5.5.

We now finish this section with a discussion of what to do with quadrupolar relaxation in a solid. In a typical case, there will be a net EFG at the site of the nucleus in question and each nucleus will have 2I+1 allowable levels. Now a magnetic dipole transition caused by the interaction of a time varying magnetic field with a magnetic moment is limited to $\Delta m=\pm 1$. In addition to the magnetic dipole interaction, however, a quadrupolar nucleus can relax by its interaction with a time varying EFG accompanied by an emission of a single or a double quanta of energy, i.e., $\Delta m=\pm 1$ or $\Delta m=\pm 2$. So, the possible relaxation paths for a relatively simple quadrupolar system of spin 3/2 looks like the sketch at right. Not only is the system not likely to start with a unique spin temperature, the relaxation is bound to be very complex

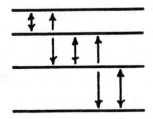

because of all these different possible relaxation paths.  Is there much hope in unravelling these different relaxation rates between different levels?  It certainly is a worthwhile effort since there must be quite a bit of information in these relaxation paths.

There have been relatively few experiments in this area and we make brief references to two of them.  Andrew and Tunstall (1961) considered quadrupole split systems in which the relaxation of the spins between the different levels takes place from two different initial conditions.  One is to start with equal populations in all levels which is the case if the magnetic field is turned on suddenly, a condition somewhat difficult to realize in an actual system because the relaxation rates between these quadrupole split levels tend to be very rapid except at very low temperatures.  The other initial condition is to equalize the $\frac{1}{2}$ and $-\frac{1}{2}$ levels without altering the populations of the other levels which is the case after a $\pi/2$ pulse on the central $\frac{1}{2} \leftrightarrow -\frac{1}{2}$ line.  You can imagine that the relaxation of the net polarization will be very non-exponential because the relaxation of the spin population for, say, the $\frac{1}{2}$ level will involve not only the $-\frac{1}{2}$ level but the other levels separated by $\Delta m = 1$ and 2.

Avogadro and Rigamonti (1973) saturated the central transition with a long train ($\sim 1800$) of pulses to adjust the populations of the other levels as if the $m = \pm\frac{1}{2}$ levels were collapsed.  Under this initial condition, the magnetization recovery is exponential for any $\frac{1}{2}$-integer spins with the relaxation time given by $[2(W_2 + W_m)]^{-1}$ where $W_2$ and $W_m$ are the double quantum electric quadrupole and single quantum magnetic dipole transition probabilities, respectively.  In order to separate the different W's (there is another one, $W_1$, for the single quantum quadrupole transition), other experiments such as Andrew and Tunstall's must be performed as well. Avogadro and Rigamonti's experiment is a good one to know because even if it proves impossible to separate these different W's, it can be used to quantify the relaxation rate because the recovery is exponential.

## REFERENCES

E. R. Andrew and D. P. Tunstall, "Spin-lattice relaxation in imperfect cubic crystals and in non-cubic crystals," Proc. Phys. Soc. 78, 1-11 (1961).

A. Avogadro and A. Rigamonti, "Nuclear spin-lattice recovery laws and an experimental condition for an exponential decay" in Magnetic Resonance and Related Phenomena, edited by V. Hovi (North-Holland, Amsterdam, 1973), pp. 255-259.

### III.C.4.          OXYGEN REMOVAL FROM SAMPLES

In the absence of a better competing mechanism, small amounts of paramagnetic impurities can determine the spin-lattice relaxation time in liquids and solids. The importance of the paramagnetic ion is greatest at temperatures below ambient where the electron correlation time (which is its spin-lattice relaxation time) $\tau_e$ can become of the order of the nuclear Larmor period, so that $\omega\tau_e \sim 1$, or any temperature regime where nuclear relaxa-

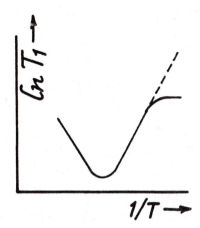

tion mechanisms which depend on molecular motions (section III.C.1.) are usually weak. The paramagnetic relaxation will often show up as a limit above which the temperature dependence of $T_1$ decreases noticeably as sketched.

Thus, in cases where long $T_1$'s are expected from intrinsic relaxation mechanisms, care must be taken during sample preparation to remove as much of these impurities as possible. An especially common paramagnetic impurity in liquid as well as condensed gas samples is molecular oxygen.

If the sample can exist in a gaseous state, it may be possible to pass it through a molecular sieve to remove molecular oxygen. (Such sieves are available commercially. See Appendix D.) It is also possible that the molecular oxygen may be removed from a gaseous sample by making it combine with something else to form a diamagnetic molecule. An example would be to pass CO over a hot filament to combine each $O_2$ molecule with two CO's to form two $CO_2$'s.

If you can work with the sample only in the liquid state, you must use the freeze-pump-thaw technique or some variant of it. In this technique, the sample chamber is pumped down and then the sample frozen usually with liquid $N_2$ to displace oxygen. The sample is then thawed, frozen, and pumped on again. Five or six such freeze-pump-thaw cycles is usually sufficient. Then the sample is sealed, while frozen, with a torch either under a vacuum or a partial atmosphere of an inert gas like nitrogen. A variation, useful for pure liquids, is vacuum distillation. The sample is distilled back and forth on a vacuum line under low pressure.

The freeze-pump-thaw technique is difficult to use with aqueous samples which commonly break the tube when they freeze. An alternative procedure is to bubble nitrogen through the sample to displace the oxygen and then to quickly seal the tube with an appropriate valve, stopper, cap, etc.

One may also combine the two methods and do a freeze-thaw technique on aqueous samples in a glove box. All the necessary materials are introduced into the box in which the air is replaced by an atmosphere of nitrogen. The freeze-thaw cycles can be performed with the sample in a plastic vial and then the deoxygenated material is sealed in an NMR sample tube. We have found it useful to reduce the nitrogen pressure before closing the tube to create a slight negative pressure. The task is considerably easier if the NMR tubes are fitted with ground glass stoppers or valves (available commercially) to ensure a good seal.

It may be difficult to determine whether sufficient oxygen has been removed. In general, there are no independent tests, but conscientious application of the above procedures should reduce the dissolved oxygen concentration sufficiently to lower its contribution to the relaxation rate below that of any competing mechanisms. Common sense clearly dictates that you should not stop whatever procedure you are using if the last application of it lead to an improvement, that is longer $T_1$. Keep doing it until the $T_1$ becomes independent of the treatment.

A check on the effectiveness of degassing for some nuclei may be possible by examining the nuclear Overhauser enhancement (NOE) factor. If you have independent knowledge of exactly what the NOE factor should be, then its measurement will reveal whether oxygen is playing a significant role in the relaxation by decreasing the NOE (Levy and Peat, 1975).

Note that because carbons generally lie on the inside of a molecule with protons and other functional groups protruding from them, they are predominantly relaxed by intramolecular interactions and the problem of dissolved oxygen may be less severe than it is for proton relaxation studies.   On the other hand, carbons can have very long T$_1$'s, say 50 seconds or more, so that they are particularly vulnerable to any additional contribution to its relaxation mechanism such as dissolved oxygen.   Equal care must be taken to prevent other paramagnetic impurities from entering the sample, for example, chromic ion from cleaning solution.


REFERENCE

G. C. Levy and I. R. Peat, "The experimental approach to accurate carbon-13 spin-lattice relaxation measurements," J. Magn. Resonance 18, 500-521 (1975).


III.D.   MEASUREMENT OF SPIN-LATTICE RELAXATION

The most often described spin lattice relaxation behavior is exponential, so that the magnetization M(t) recovers from zero (say, after a $\pi/2$ pulse) as

$$M(t)=M_o[1-\exp(-t/T_1)].$$

This does not at all mean that the majority of the spin-lattice relaxation processes in nature are exponential.   It is just that much NMR is done on samples in which most of the interactions which give rise to non-exponential effects are absent or else the non-exponentiality is ignored for one reason or another.   The spin-lattice relaxation time T$_1$ is defined only for exponential processes.   Therefore, any scheme for determining $T_1$ has to cope with the fact that the actual process may not be exponential and the value obtained may be meaningless, or worse, misleading.

## III.D.1.  STRATEGIES

$T_1$ determinations are generally time consuming because they are usually not "one shot" methods like the CPMG sequence for determining $T_2$. [There are certain cases in which one shot $T_1$ methods work. See pp. 183-184.] If many values of $T_1$ have to be monitored, for example, to monitor $T_1$'s which are changing in time, to map out the temperature dependence around a phase transition, or simply to measure long $T_1$'s, a possible shortcut is to measure relative $T_1$'s or to measure $T_1$'s from a limited portion of the data. A common technique for the former is to measure the time required for the magnetization to recover half way which is the desired $T_1$ times ℓn2.

Suppose we wish to measure $T_1$ by a $\pi$-$\pi/2$ sequence (to be described more fully later) for which the magnetization obeys the relation $M(t) = M_o[1-2\exp(-t/T_1)]$. We still have several choices of how the measurements should be made. One obvious way is to measure $M(t)/M_o$ and plot $1-M(t)/M_o$ against t on a semilog plot. Another way is to plot $M_o-M(t)$ against t on a semilog plot. Let us look at the two ways in which we can write the relaxation equation:

$$1-M(t)/M_o = 2\exp(-t/T_1)$$

and

$$M_o - M(t) = 2M_o \exp(-t/T_1) \ .$$

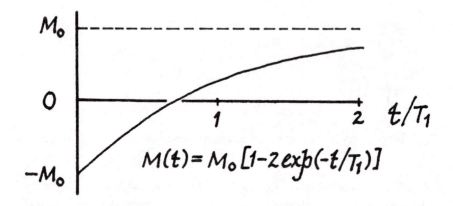

The practical difference is that any mis-estimation in the baseline for the signal cancels out for the second equation so that such errors do not lead to apparent non-exponentiality whereas they do not for the first. The penalty you pay for that is that the value of the function being plotted at t=0 is unknown, whereas for the first equation it is fixed at 2.00000 (assuming that the experimental parameters are set perfectly) which is very helpful to know especially if S/N is poor because it is a good data point, possibly the only one.

A related question is that of how to space out the sampling points in time for the optimum accuracy in a given time. Since the quantity plotted (for either equation) falls off with time while the noise stays the same size, clearly the S/N decreases as the delay increases. In fact, you can reach a point where the data contributes more noise than signal. There is also a strong desire to not measure these long delay values since they take more time. On the other hand, these are the points that tell you whether the relaxation is exponential or not and this is a very important fact to establish.

So, when it has been decided that the decay is exponential, the best procedure is to cut off the data at some maximum delay, beyond which the rms deviation for the estimated $T_1$ increases. A way to do this is to calculate the rms deviation as a function of the cutoff time for a trial run and decide when to stop taking data (like at 10% of full scale, for example). The only exception to this occurs when $M_o$ has to be determined directly.

The usual method of $T_1$ determination is to measure $M(t)$ at various values of t and fit them to one of the equations above either by a semilog plot or by a computer. The question, then, is how well should $M_o$ be determined? It must be determined very well for $T_1$ determinations using these methods, since the errors in $M_o$ count as much as those in each $M(t)$. This says that $M_o$ should be sampled as many times as $M(t)$. In fact, you can get away a bit easier since $M_o$ has S/N exceeding any value of $M(t)$ but the point is if $M_o$ is not determined very well, it doesn't do much good to determine $M(t)$ to great accuracy.

This consideration leads to a common scheme in which delays of the order of $7 \times T_1$ are used between the $\pi/2 - \pi/2$ clusters with the first and the second FID's (which are proportional to $M_o$ and $M(t)$, respectively) recorded and the quotient or the difference calculated, according to whether the first or the second of the above equations are being used. Thus, $M_o$ is determined each time $M(t)$ is determined and the above requirement is satisfied. This scheme has the added advantage that slowly drifting electronics do not affect the accuracy of $T_1$ because the drift should contribute nearly equally to $M_o$ and to $M(t)$. An obvious extension is to use clusters of more than two pulses and use the 1st and the last FID's. Such a sequence (discussed later) can be quite insensitive to the pulse lengths as long as they are close to $\pi/2$.

It is possible to determine $T_1$ from an exponential curve even if the asymptotic value of magnetization is not known at all. The brute force way is to simply fit it with an exponential function on a computer. There is also a very nice way to do this as you are taking data on a graph paper, too, and this seems not to be well known (or, at least, well used). These and other methods of $T_1$ determination are discussed in the following section. [We point out that it is most desirable to process as much data as possible as they are gathered so that the state of the experiment can be monitored. It is much easier to retake a questionable point or correct for some electronic drift at the time the experiment is being run than to do it later.]

## III.D.2.   SOME TYPICAL PULSE SEQUENCES FOR $T_1$ MEASUREMENTS

All pulse NMR sequences for measuring the spin-lattice relaxation time $T_1$ either initialize the spin system and monitor its evolution or achieve a steady state condition in which the signal size depends on $T_1$. The former scheme involves a pulse (or a group of pulses) to prepare the nuclear spins in some non-equilibrium configuration and then, after some waiting period during which the spins are allowed to relax, a pulse monitors the state of the spins. The recovery of the nuclear spin population is monitored as a function of the waiting time. The monitoring pulse need not be a special pulse; it may do double duty as part (or all) of the preparation sequence.

### Double pulse sequences (inversion and saturation recovery)

There are two popular sequences of this type: the $\pi/2$-$\pi/2$ (also called the 90-90) and the $\pi$-$\pi/2$ (also called the 180-90). In each of these sequences, the first pulse prepares the spins and the second pulse measures the magnetization after the waiting period. The $\pi$-$\pi/2$ sequence is called the inversion recovery sequence because the preparation pulse (being a $\pi$ pulse) inverts the spin population (and thus the magnetization) and the recovery therefore goes from $-M_0$ to $M_0$ where $M_0$ is the thermal equilibrium magnetization attainable only after waiting for a time much longer than $T_1$. Specifically, the magnetization after waiting for a time $\tau$ is given by

$$M(\tau)/M_0 = 1-2\exp(-\tau/T1).$$

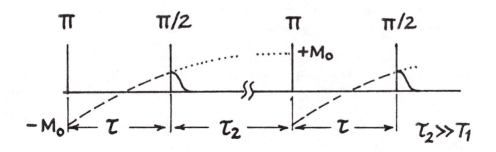

The popularity of the inversion recovery method is due to the fact that the range of magnetization is $2M_0$ as opposed to $M_0$ for the $\pi/2$-$\pi/2$ sequence to be discussed below. It will be shown, however, that the $\pi/2$-$\pi/2$ sequence has certain advantages over the inversion recovery method.

The $\pi/2$-$\pi/2$ sequence called the saturation recovery sequence with the magnetization given by

$$M(\tau)/M_0 = 1-\exp(-\tau/T_1),$$

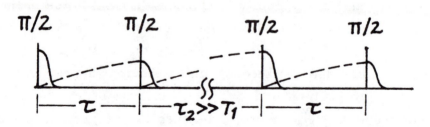

does not seem very different from the inversion recovery method.   One difference is that the data has only one-half of the
dynamic range of the inversion recovery data, as mentioned
above.   However, there are other significant differences with
practical consequences.   For a given transmitter power and
coil geometry the $\pi$ pulse is twice as long as the $\pi/2$ pulse so
that the frequency width of the preparation pulse power spectrum is only half as wide.   For a broad line or a broad spectrum, then, the $\pi/2$-$\pi/2$ sequence may yield much more accurate line shapes and intensities than the $\pi$-$\pi/2$ sequence especially if the magnitude of $H_1$ is limited due to, for example,
large sample size or small transmitter output power.

Equally as important, it is possible to perform the $\pi/2$-$\pi/2$
sequence with an arbitrarily short delay between the two
pulses whereas the $\pi$-$\pi/2$ pulse pairs must be separated by
several $T_1$'s so that the former is much more efficient than the
latter when $T_2 < T_1$.   This is because the z-component of the
magnetization is identically zero after any $\pi/2$ pulse regardless
of the state of the magnetization immediately before the pulse
so that nothing can be gained by waiting between the pulse
pairs.   On the other hand, the $\pi$-$\pi/2$ method relies on the fact
that the system starts with a full magnetization (the thermal
equilibrium value) to be inverted so that a long waiting period
is required to let the magnetization fully recover.   This need
for waiting several $T_1$'s more than offsets the wider dynamic

range of the inversion recovery signal so the apparent advantage of the inversion recovery method is not real, certainly in cases where $T_2 < T_1$, unless the $\pi/2$-$\pi/2$ sequence is used with a long waiting time. This discussion leads to two modifications of these sequences which will be considered a little later.

As discussed in III.D.1., if $M_0$ needs to be measured accurately, there is no way to get around long waiting times between sequences even with a saturation recovery sequence. In such cases, this particular advantage of saturation recovery does not exist. However, with long ($>> T_1$) delays between the pulse clusters, the saturation recovery is still more convenient than inversion recovery because a repetition of the simple $\pi/2$-$\tau$-$\pi/2$ sequence can yield both $M_0$ (after the first pulse) and $M(\tau)$ (after the second) with $\tau >> T_1$. With the inversion recovery method, a separate $\pi/2$ pulse must be used to get $M_0$ and this pulse destroys the magnetization waiting to be inverted.

It is worth pointing out that these sequences are often implemented with pulses which are not exactly $\pi/2$ pulses. Reasons for this include choosing shorter pulses to enhance the frequency coverage as well as the obvious cases in which it is difficult to set the proper pulse length due, for example, to very poor S/N even for a standard sample as with a low $\gamma$ nucleus. Provided that the magnetization recovery is exponential, no relaxation information is lost (except S/N) by slight missettings of the tip angle. The usual semi-log plot will still be straight but with an intercept that reflects the different tip angle. For a set of experimental data at constant tip angle setting,

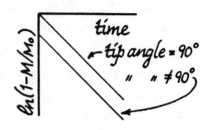

this intercept can be averaged for all runs to reduce random errors.

## Steady state sequence

Recalling the preceding discussion of the $\pi/2$-$\pi/2$ sequence, we now consider the limiting case where there is no delay at all between the pulse pairs. It is easy to see that we may, then, omit one of the two pulses since their functions may be performed by a single pulse, and each pulse can be a monitor pulse as well as a preparation pulse. The resulting sequence is an infinite train of $\pi/2$ pulses with a separation $\tau$ and the signal after each pulse (after a steady state condition is reached) is given by

$$M(\tau)/M_o = 1 - \exp(-\tau/T_1)$$

just like with the saturation recovery sequence.

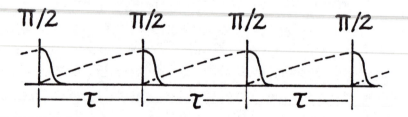

This is a very efficient sequence and its popularity dates back to the days when the signal amplitudes were read visually on oscilloscopes for $T_1$ measurements and the highest possible signal repetition rates were desirable. It has regained its popularity due to the need for the largest S/N in complex high resolution spectra obtained by FT NMR. Its use in high resolution NMR is limited, however, by the desired resolution which puts a lower limit on $\tau$ so that it is most suited for samples having long $T_1$'s relative to $T_2$'s. An additional practical limi-

tation of this sequence is on the repetition rate due to the
relatively high duty cycle requiring high transmitter output
power for short $T_1$ solid samples, although this should not be
a restriction for high resolution NMR.   Freeman and Hill (1971)
have a nice discussion of this method.

The method just discussed is a special case of an infinite
sequence of pulses not necessarily equal to $90°$.   If the tip
angle is $\theta$, the expression for the steady state magnetization
is

$$M(\tau)/M_0 = \{[1-\exp(-\tau/T1)]\sin\theta\}/[1-\cos\theta\exp(-\tau/T1)].$$

This means that the magnetization recovery shape is a unique
function of the tip angle so that both $\theta$ and $T_1$ can be obtained from the same run.   A quick way to implement this method
is to plot $M(\tau)/M_0$ vs. $\tau/T_1$ for various $\theta$'s, calculated from
the above equation, on a semilog paper as sketched.   Then a
plot of the experimental $M(\tau)/M_0$ vs. $\tau$ on a graph paper with

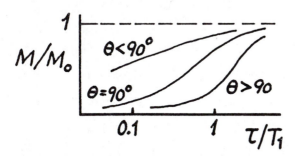

the same scale can be laid over the calculated plots.   The tip
angle $\theta$ can be determined from the best match for the shape
of the curve while the $\tau$ coordinate overlaying the $\tau/T_1=1$ abscissa yields $T_1$.   Random $T_1$ errors may be minimized by fix-
ing $\theta$ to some average value provided that the pulse power
and length are kept constant.

Progressive saturation with inversion recovery sequence

Another twist one can think of is the inversion recovery
($\pi-\pi/2$) without the adequate waiting period.  This is analogous to the generalized steady state method discussed above
(for arbitrary tip angle $\theta$) in that the amount of initial magnetization depends on the repetition rate.  This method is an
improvement over the $\pi-\pi/2$ method in terms of efficiency and
this has been demonstrated by Canet, et al. (1975).


Saturating comb

This method is a generalization of and functionally identical to the $\pi/2-\pi/2$ sequence in that a number of nearly $\pi/2$
pulses forming a "comb" sets the magnetization to zero and
the recovery is sampled at later times.  (In the $\pi/2-\pi/2$ sequence, only the z-component is initialized to zero but that's
equivalent to the total magnetization being zero for most experiments.)  The pulse spacing $t_1$ within the comb should be
set to $T_2^* < t_1 \ll T_1$ which restricts this sequence to solids and
liquids for which $T_2^* < T_1$.  Its virtue is that missetting the tip
angle so that it is not exactly $\pi/2$ does not affect the initial
condition of $M_z = 0$ provided that the comb length is sufficient.
In experiments where the magnetization recovery is not necessarily exponential, the knowledge of the initial magnetization is
most helpful.  The number of pulses required in a comb depends on how close each pulse is to being a $\pi/2$ pulse.  A

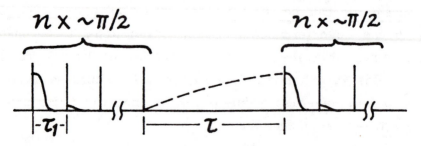

common comb length is between 4 and 16 pulses.

For the saturating comb, the deviation of the tip angle from $\pi/2$ does affect S/N by not rotating the magnetization exactly into the x-y plane which contains the receiver coil axis (but not very much for small deviation since the projection of M on the x-y plane does not depend very much on $\theta$ for $\theta \sim \pi/2$). The first pulse of the comb may be used as the detecting pulse for the magnetization initialized by the previous comb in which case the data gathering is very efficient just like the infinite train of $\pi/2$ pulses described above, under steady state sequence. The duty cycle, on the other hand, is likely to become quite large for short $T_1$ samples, especially if a large number of pulses are used in a comb. One can either live with this problem (by using a transmitter which is able to withstand the higher average power output), use another sequence if available, or use a dedicated monitoring pulse separated from the following comb and make this separation large enough to reduce the overall duty factor.

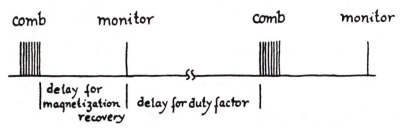

A special use for the saturating comb is in cases where a pulse might present different tip angles to different nuclei. This happens especially in metals in which the sample thickness exceeds the skin depth, although, this happens to some extent in most NMR experiments due to inhomogeneity of $H_1$ or to a distribution of $H_e$. Single coil probes inherently have inhomogeneous $H_1$ for those parts of the sample outside the coil but even if this were not so, $H_1$ is effectively inhomogeneous for a broad spectrum because what really counts in tipping the magnetization is not $H_1$ but $|H_e|=[(H_o-\omega/\gamma)^2+H_1^2]^{1/2}$ which depends on the offset.  (See the tip angle discussion in Section II.A.2.).  A saturating comb can be set up so that all parts of the spin system are saturated by it.  Bear in mind that the monitoring pulse still will not tip the nuclei at the different offsets by the same amount so the spectrum amplitude will still be distorted.  The relaxation behavior of each part of the spectrum will be accurate, however, and each part will have started (at t=0) without any magnetization.

Another use for a saturating comb is to initialize the populations of quadrupole perturbed Zeeman levels, when only one pair of levels can be irradiated.  As already discussed in III.C.3., Avogadro and Rigamonti used a very long (~1800 pulses!) saturating comb on the central $\frac{1}{2}\leftrightarrow-\frac{1}{2}$ transition of spin 9/2 niobium-93 in polycrystalline $KTa_{.63}Nb_{.37}O_3$ to obtain an exponential recovery of an otherwise non-exponential magnetization behavior for this transition.

## REFERENCES

D. Canet, G. C. Levy, and I. R. Peat, "Time saving in $^{13}C$ spin-lattice relaxation measurements by inversion recovery," J. Magn. Resonance 18, 199-204 (1975).

R. Freeman and H. D. W. Hill, "Fourier transform study of NMR spin-lattice relaxation by 'progressive saturation'," J. Chem. Phys. 54, 3367-3377 (1971).

## III.D.3.  $T_1$ MEASUREMENTS IN HIGH RESOLUTION NMR

We now consider the problem of determining the $T_1$'s of individual lines in a multi-line high resolution spectrum. The partially relaxed Fourier transform (PRFT) method, first performed by Vold, et al. (1968), works very well for this purpose. It is basically the inversion recovery sequence except that after the second pulse the partially relaxed FID is Fourier transformed to yield a partially relaxed spectrum.

In this scheme, a $\pi$ pulse at time zero inverts the magnetization. If a $\pi/2$ sampling pulse is applied immediately thereafter, the magnetization is rotated from the -z' direction to the -y' direction to induce an FID. When this FID is digitized and Fourier transformed, the result will be a spectrum upside down. The inversion occurs because the magnetization, before being sampled by the $\pi/2$ pulse, was along the -z' axis rather than along the +z' axis as it was initially. On the other hand, if we wait a while between the $\pi$ and the $\pi/2$ pulses, the magnetization will have partially relaxed back towards the +z' direction and a spectrum from such a partially relaxed FID will show peaks in between the fully inverted ones obtained with no delay between the pulses and the fully recovered ones with an infinite (compared to all the $T_1$'s) delay.

At some time following the initial $\pi$ pulse the magnetization for one of the peaks will have decayed to zero and this is called the null time $\tau_{null}$ for that peak. If the recovery of the magnetization is governed by $M(t)=M_o[1-2\exp(-\tau/T_1)]$, $\tau_{null}=T_1\times\ln 2$ provided that the first pulse was a true $\pi$ pulse for that peak. The general method of calculating $T_1$ from the entire recovery, unlike the $\tau_{null}$ method, works even for broad spectra in which there is a distribution of tip angles among the different peaks. This is because each line is assumed to recover exponentially to $M_o$ from whatever the initial magnetization might have been so that the recovery rates can still be determined.

Thus, from the spectrum as a function of the time between the two pulses, the spin-lattice relaxation times for each resolved line in the high resolution spectrum can be obtained. The relaxation behavior of each group of nuclei in the molecule characterized by different motions will be different and, the details of the relaxation behavior of the system can be very valuable, both in the study of molecular motions and in assigning lines.

There are two practical problems in the measurement of spin-lattice relaxation times in mutiline high resolution spectra that are not problems in measurements of broadline spectra. The first has to do with the effects of incorrectly set pulse lengths in the inversion recovery sequence. Suppose, for example, that the $\pi$ pulse is too short to invert the magnetization, resulting in a small component of the magnetization precessing in the x'-y' plane but that the nominal $\pi/2$ pulse is a true $\pi/2$ pulse. This can easily happen if the gating pulse length for the $\pi/2$ pulse was doubled for the $\pi$ pulse when both of the pulses are quite short. This is because the actual transmitter pulse is longer than the gating pulse by an amount independent of the pulse length. In the time between the $\pi$ pulse and the $\pi/2$ pulse this component of the magnetization will precess in the x'-y' plane by an amount determined by the time between the pulses and the precession rate in the rotating frame which, in turn, depends on how far its frequency is from the carrier.

The $\pi/2$ pulse will give rise to a magnetization lying somewhere on a cone about the negative y' axis, which should not give us any trouble only if we can look at the pure absorption mode signal. The dispersion signal, on the other hand, will be a decaying oscillation about zero because it is proportional to the deviation of the magnetization from the y'-z' plane. The magnitude which is the square root of the sum of the squares of the absorption and the dispersion signals will consist of a sinusoidal variation superimposed on the overall relaxation plot for those intervals between the $\pi$ and the $\pi/2$ pulses that are less than $T_2^*$ for the dispersal of the magnetization in the x'-y' plane.

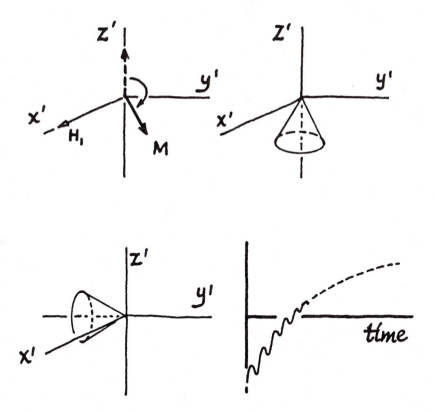

If you do the normal thing and arrive at the spectrum by a linear combination of the sine and the cosine transforms, you may very well get this sinusoidal oscillation on top of the relaxation curve, even for something that ends up as an absorption spectrum. Under these conditions, you may not safely assume that the initial slope of a semilog plot of the relaxation data correctly predicts the relaxation time.

A symptom of the above problem appears when we try to phase correct the spectrum after the Fourier transformation. In the absence of this problem, the phase correction required is independent of the interval between pulses. However, with this problem the projection of the magnetizaton on the x'-y' plane might lead or lag the rotating y' axis depending on its frequency in the spectrum and the time between pulses. So we experience an additional frequency dependent phase shift which depends on the interval between pulses in the range of intervals comparable to or less than T$_2^*$.

This problem was noticed in the earliest experiments in partially relaxed multiline spectra (Vold, et al. 1968). A simple solution is to alternate the phase of the $\pi/2$ pulse between sequences and appropriately add or subtract the FID's from memory (Demco, et al., 1974; Levy and Peat, 1975; Cutnell, et al., 1976). Circuits for alternating the phase are described in the above references. An alternative approach is to destroy these components precessing in the x'-y' plane following the $\pi$ pulse with a "homospoil" pulse. For example, a 50 ma dc pulse applied for 5 to 10 ms to the z gradient coils with an additional wait of 50 to 100 ms before the $\pi$ pulse for recovery of the homogeneity suffices (Demco, et al. 1974).

A second problem in getting accurate spin-lattice relaxation times in high resolution spectra has to do with the number of points Fourier transformed. It sounds trivial but there must be several points on each line to reasonably define the lineshape. If, for example, one has a 500 Hz carbon-13 spectrum stored as a 4K data table, each line with a typical linewidth of 1/4 Hz will be defined by about two points. This causes a lot of scatter in the relaxation plot. If digitizing more points is impractical, zero filling will help. In this procedure, already described in II.B.5., additional zeros are appended to the end of the FID before Fourier transformation, in order to interpolate between the original points thereby resulting in better definition of the lines and in the reduction of scatter in the relaxation plots.

Keep in mind, though, that interpolation does not mean the same thing as resolution enhancement. By zero filling, we constrain the transformed spectra to not have any signifiant features between the original points. Thus, for example, a line cannot be narrowed by zero filling in the time domain (Becker, et al. 1979).

# REFERENCES

E. D. Becker, J. A. Ferretti, and P. N. Gambhir, "Selection of optimum parameters for pulse Fourier transform nuclear magnetic resonance," Anal. Chem. 51, 1413-1420 (1979).

J. D. Cutnell, H. E. Bleich, and J. A. Glasel, "Systematic elimination of rf pulse defect errors in Fourier transform spin-lattice relaxation measurements," J. Magn. Resonance 21, 43-46 (1976).

D. E. Demco, P. Van Hecke, and J. S. Waugh, "Phase-shifted pulse sequence for measurement of spin-lattice relaxation in complex systems," J. Magn. Resonance 16, 467-470 (1974).

G. C. Levy and I. R. Peat, "The experimental approach to accurate carbon-13 $T_1$ measurements," J. Magn. Resonance 18, 500-521 (1975).

R. L. Vold, J. S. Waugh, M. P. Klein, and D. E. Phelps, "Measurement of spin relaxation in complex systems," J. Chem. Phys. 48, 3831-3832 (1968).

## III.D.4.   LONG $T_1$ PROBLEMS

The measurement of long $T_1$'s entails special problems. (We can arbitrarily call any $T_1$ exceeding one minute to be long.) Some of the more obvious problems such as the need for instrumental stability and minimal rf leakage from the transmitter during the "pulse-off" periods are mentioned elsewhere in the book.

Levy and Peat (1975) pointed out practical problems of sample geometry and sample state in measuring long $T_1$'s. The first problem is molecular diffusion in and out of the receiver coil during the course of the experiment. For a liquid sample in a single coil probe, a substantial fraction of the molecules in the sample coil at the start of the experiment can diffuse out (and back in) during the measurement if the $T_1$ is long enough and the diffusion rate great enough. The consequence is an artificially shortened (and non-exponential) relaxation process. This problem is not very severe for relaxation times less than, say, one minute and with sample lengths smaller than, perhaps, 30 mm in a 16-mm-long receiver coil. If it is a problem, use samples that are not much longer than the receiver coil with some sacrifice in spectral resolution and S/N. The problem is much less with a cross coil probe because the transmitter coil irradiates a region of the sample much larger than the receiver coil.

A second problem arises due to molecules leaving and entering the vapor phase to and from the liquid phase. Substantial exchange with the vapor phase can cause errors because the relaxation mechanism may be different in the two phases and because the molecular environment in the vapor phase has different physical characteristics. The severity of the problem is determined by the diffusion rate of the mole-

cules, the height of the liquid-vapor interface above the
receiver coil, the volatility of the material, the area of the
liquid-vapor interface, and the volume of the vapor region.
In $^{13}C$ NMR, this effect often leads to less than expected
nuclear Overhauser enhancement (NOE), although other
factors can also reduce NOE. One simple way to reduce the
problem is to have a constriction in the sample tube or a plug
at the interface.

Aside from these special problems, measurements of long
$T_1$'s are bothersome simply because they take so long. If
you choose a $T_1$ method which involves measurements of the
fully relaxed magnetization $M_o$, the total time required for the
experiment is on the order of the number of $T_1$ determinations times ten $T_1$'s. If ten delays are used and 10 pulses
are accumulated at each delay, the total time comes out close
to $10^3 T_1$. For a $T_1$ of one minute, this is about 17 hours.

For all methods mentioned in III.D.2., the above considerations hold because we start over every time after the
magnetization is sampled, that is, the measurement of the
magnetization destroys the magnetization. We now consider
some methods whereby the magnetization can be sampled
without being destroyed so that a series of magnetization
recovery points can be measured in a single recovery.

Basically, the "single-shot" $T_1$ measurement is similar to
the DEFT sequence described in I.C.2. We may start, for
example, with another $\pi/2$ pulse and monitor the recovery of
the magnetization. After waiting a time t, another $\pi/2$ pulse
will induce a partially recovered FID. As soon as possible
after this pulse, a $\pi/2$ pulse with the phase reversed 180
degrees (which is the same as a $-\pi/2$ pulse) returns the
magnetization to point along the static field. At time 2t, the
further recovered magnetization is again sampled by a $\pi/2$

pulse and then returned to point along the static field again with another $-\pi/2$ pulse. This process is repeated until the recovery curve is traced out. Such methods have been reviewed, for example, by Edzes (1975), referenced in the following (III.D.5.) section. Clearly, it cannot be used if T$_1$ is short enough so that significant magnetization recovery takes place during the sampling process.

We close this section with a consideration of when a single-shot T$_1$ method as we have described is useful and when it is not. We have seen that such a method can lead to a magnetization recovery curve in the time chosen to follow the recovery (apart from the need for signal averaging). As it has been pointed out in III.D.1., however, a T$_1$ method requiring the M$_o$ value is only as accurate as the determination of M$_o$. If, as advocated in that section, we measure an M$_o$ value for each partially recovered magnetization M(t), the single-shot methods offer no advantages whatsoever. In fact, a saturation recovery method with a determination of M$_o$ associated with each M(t) is better because it is less susceptible to the electronic (or other) drifts.

So, when is a single-shot method useful? It is useful, in fact essential, for long T$_1$'s but only for getting the recovery curve itself. If the recovery happens to be exponential, methods exist for determining T$_1$ as discussed in III.D.5. Even the recovery curve by itself will be quite meaningful in studying the relaxation mechanisms, especially if it is not exponential. We will continue the discussion of single-shot T$_1$ methods near the beginning of the following section.

REFERENCE

G. C. Levy and I. R. Peat, "The experimental approach to accurate carbon-13 spin-lattice relaxation measurements," J. Magn. Resonance 18, 500-521 (1975).

## III.D.5.  ANALYSIS OF EXPONENTIAL TIME CONSTANTS

Recently, strategies for accurate determination of exponential time constants in NMR have received increased attention. This is due to the increasing number of experiments in which $T_1$'s yield valuable information. Traditionally, the time constant has been evaluated by semilogarithmic plot of the decaying function versus time. This has the advantage that it is convenient and it gives the experimentor something to do between readings if the relaxation time is long or the signal is accumulated for a considerable period.

One commonly voiced objection to the semilog plot method, is that it is not quantitative because the fit to the time variation is performed by "eye." In our view, this objection is not as serious as it might seem. The fit by "eye" can be very good, depending on the eye, unless the scatter is so great that even a computer fit may produce a time constant which is not very meaningful. [It is true that because the scale is nonlinear, the weighting is also nonlinear, so you have to take this fact into account when you are eyeballing it. It may help to actually draw in the estimated error bars.] A more substantial objection is that the usual semilog plot method relies on an accurate determination of the value of the thermal equilibrium magnetization, to which the time dependent magnetization recovers.  Since the quantity plotted in this method depends on the ratio of the time-dependent magnetization to the thermal equilibrium magnetization, the latter should be determined as accurately as the former for optimum efficiency, a fact recognized a long time ago (Guggenheim, 1926).  Roughly speaking, this means that for each determination of the time dependent magnetization, there should be a determination of the equilibrium magnetization.  (Otherwise, the magnetization recovery can look very good with little scatter, but the resulting $T_1$ will be wrong.)  The result of all this is that the measurement takes a long time since it takes about five times the time constant for an exponential function to decay to about 1% of the initial value.  In this section, we will discuss some methods which yield the exponential time constant (such as $T_1$) without the need to determine the asymptotic value of the function.

Before we get started, we should point out another reason why the usual $T_1$ measurements cannot be made quickly compared, for example, to $T_2$ determination from the Carr-Purcell Meiboom-Gill data (see I.C.2.).  It is simply because in the former each data point is usually obtained after a fresh initialization of the magnetization while the latter is a "single shot" experiment wherein the entire decay curve is obtained in sequence with the measurement of each point not disturbing the evolution of the magnetization.  Under certain circumstances, single-shot $T_1$ measurements are possible.  They include cw methods like those due to Linder (1957) and Stepisnik., et al., (1970), fast passage experiments of Look and Locher (1968, 1969, 1970) and Heatley (1973), and pulse methods like those reviewed by Edzes (1975).  These methods were designed primarily for single line spectra and the measurements took place in the time domain.  The usual pulse

method involves reestablishing the value of the decaying magnetization after each point and this principle has been used in
high resolution multiple line spectra in methods like DEFT
(see I.C.2.).

Under ideal conditions, the magnetization recovery after
an exactly 90 degree pulse is $M(t)=M_o[1-exp(-t/T_1)]$ assuming
exponetial recovery. Unfortunately, in a real experiment
several things are not ideal. There are varying amounts of
noise which causes the points to scatter (usually in direct
proportion to the importance of the experiment). The noise
makes it difficult to estimate $M_o$, the thermal equilibrium magnetization, and also the initial value of magnetization which
should be zero after an exactly 90 degree pulse. This recovery can be recast in the form of a general exponential decay

$$Y(t) = Aexp(-t/T_1) + B \tag{1}$$

where Y represents the fraction of magnetization left to re-
cover, A contains information about initial magnetization, and
B about the thermal equilibrium magnetization. Now we con-
sider ways to evaluate $T_1$ (which means evaluating A and
B, too) in the most efficient way, without having to measure
B directly.

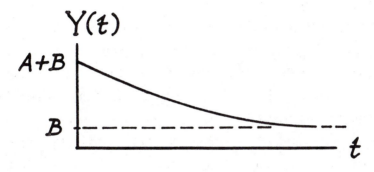

Obviously, one plan of attack would be to fit the data to

the expression above by a least squares algorithm in a computer (we need at least three data points to solve for the
three unknowns).   This is the brute force method which
works fine but requires some computer memory, both for the
program and the data (Sass and Ziessow, 1977; Kowalewski, et
al., 1977).

A graphical method for determining the decay time constant was mentioned some time ago (Mangelsdorf, 1959;
Livesey, 1979).   Mangelsdorf pointed out that a function linear
in $\exp(-t/T_1)$ has a linear relationship with any other
function linear in $\exp(-t/T_1)$.   In particular, eq. (1) and

$$Y(t+\Delta t) = A\exp(-t/T_1)\exp(-\Delta t/T_1) + B, \qquad (2)$$

which is obtained from eq. (1) by retarding the time by $\Delta t$,
are related in a linear way.   When these two functions are
plotted against each other, the slope is $\exp(-\Delta t/T_1)$.   Thus,
this method can be used to measure $T_1$, for example, by taking the data at equal delay intervals $\Delta t$ and plotting the data
against the same data table displaced by one data point.   The
$T_1$ value can then be obtained from the slope in one step on
a calculator (or even a slide rule!).   Another nice feature of
this method is that the plot is made on a regular graph paper
and not a semilog graph.

The main drawback of this method is that if $\Delta t/T_1$ <1 the
slope $\exp(-\Delta t/T_1)$ is not too different from 1 and it is quite
insensitive to $\Delta t/T_1$.   On the other hand, we cannot let
$\Delta t/T_1$ >1 because 95% of the signal will have recovered in $3T_1$
so not very many values of $Y(t+\Delta t)$ can be taken.

An easy way of overcoming this difficulty has been
known for half a century or more.   In fact, Mangelsdorf's
method is a special case of a method by Guggenheim (1926)

wherein the interval $\delta t$ is chosen to be many time-constants but the data points are taken at intervals much less than $\delta t$. Consider $Y_1$, $Y_2$,...$Y_n$ to be n readings taken at times $t_1$, $t_2$,...$t_n$ without any restrictions. Suppose n more readings are taken at times $t_1+\delta t$, $t_2+\delta t$,...$t_n+\delta t$, and they are called $Y_1'$, $Y_2'$,...$Y_n'$. Then from eqs. (1) and (2)

$$Y_i'-Y_i = A\exp(-t/T_1)[\exp(-\delta t/T_1)-1] \tag{3}$$

and, therefore,

$$\ln(Y_i'-Y_i) = -t/T_1 + \text{constant.} \tag{4}$$

So, we can plot $Y_i'-Y_i$ vs. t on semi-log paper to get $T_1$ from the slope. What this amounts to is to take data points $Y_i$ and compare them with points $Y_i'$ taken $\delta t$ (equal to one or two time constants) later instead of with $Y(\infty)$ taken five or more time-constants later.

When we were talking about fitting the data to eq. (1) by a computer, it should have occurred to us to consider solving three simultaneous equations to get the three unknowns. Considerations like this have led to several tries at developing a simple algebraic (but non-graphical) method of obtaining the decay rate of eq. (1).

Guggenheim (1926) suggested a method whereby N data points are divided into pairs N/2 apart and each pair solved for the decay time by the relation

$$T_1 = \delta t/[\ln Y(t + \delta t)-\ln Y(t)] \tag{5}$$

which is obtained from eq. (1) if B=0 and where $\delta t$ is the time delay between points N/2 apart in the data table. Moore and Yalcin (1973) have applied this method to NMR.  In their method, a trial t is calculated assuming B=0 and then B is estimated by a minimization of the variance of the $T_1$'s. Their error analysis showed that the optimum measurement time, $2\delta t$, is about $2T_1$.

Smith and Buckmaster (1975) solved eq. (1) directly for the baseline B and got

$$B = (Y_1Y_3-Y_2^2)/(Y_1+Y_3-2Y_2) \tag{6}$$

where

$$Y_i = A\exp\{-[t_o+(i-1)\delta t]/T_1\}+B. \tag{7}$$

Once B is known, $T_1$ can be obtained in the usual way. They then noticed that if the N data points were divided into blocks N/3 long, $\delta t$ could be set to N/3 times $\Delta t$ where the latter is the delay between consecutive data points.  There will now be N/3 equations for B and they may be averaged to yield B by eq. (4) where now

$$Y_i = (3/N)\sum Y(n\Delta t) \tag{8}$$

with the sum over the i-th third of the data table.  Notice the ease in data manipulation with this scheme since the first N/3 points are summed to get $Y_1$, the next N/3 for $Y_2$, and so on, and one simple calculation yields B from all the data points.  In fact, for a single line spectrum, the summing of eq. (8) can be performed as the data are being taken so the memory requirements are minimal.

McLachlan (1977) has extended this treatment by dividing the data table into four blocks each containing N/4 points and directly solving for an expression for the decay time $T_1$. In this case the decay time is

$$T_1 = [4/(N\Delta t)]\ell n[(Y_1-Y_3)/(Y_2-Y_4)] \tag{9}$$

where

$$Y_i = \sum_i Y(n\Delta t) \tag{10}$$

and now the sum is over the i-th quarter of the data table. He found that dividing the data table into four blocks is better than dividing it into three from noise considerations at least for $T_1$ because the noise is uncorrelated for the different data blocks and therefore for the numerator and the denominator of eq. (9).

McLachlan obtained an expression for a fractional error in $T_1$. One minimization criterion is to set $N\Delta t=5.2T_1$ and at this setting the error approaches the asymptotic limit of $N^{-1/2}$ (from below) as N becomes larger than about 30. McLachlan, therefore, recommends that $T_1$ measurements be made with N less than 30 and repeated rather than extending the measurement over larger N for the same time for optimum S/N.

Finally, we mention the $T_1$ determination method by variable nutation due to Christensen, et al., (1974) and Gupta (1977) which also does not require the of the equilibrium magnetization. The steady state magnetization as a function of the interval $\Delta t$ between repetitive $\theta$ degree pulses was given in III.D.2. as

$$M/M_o = \{[1-\exp(-\Delta t/T_1)]\sin\theta\}/[1-\exp(-\Delta t/T_1)\cos\theta] \tag{11}$$

which can be rearranged so that $M/\sin\theta$ can be plotted against

M/tanθ for various θ values to yield a straight line with a slope $\exp(-\Delta t/T_1)$ for magnetization obeying exponential recovery. In this case, the pulse interval $\Delta t$ is fixed and the adjustable parameter is the nutation angle which is most conveniently changed by the rf pulse width. This may make the method awkward for most existing apparatus.

One problem of this method when the FID is short is that since the time origin is in the center of the rf pulse, the FID moves with respect to the beginning of the pulse as the pulse length is changed. Another difficulty with solid samples requiring short transmitter pulses, is that the tip angle is usually not proportional to the pulse length because the pulse shape is not strictly a rectangle but is irregular. Even aside from these reasons, the method is best suited for high resolution lines which do not span a large spectral width because you do not want any dispersion of the tip angle across the spectrum due to finite $H_1$ as discussed in II.A.2.

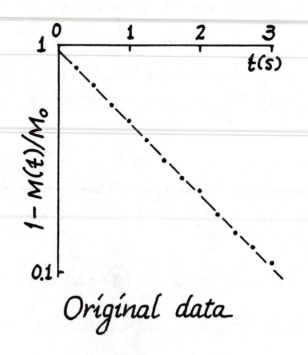

Original data

We now consider an example to illustrate the various methods for evaluating $T_1$ discussed above. The figure at left shows the usual decay curve for protons in $(CH_3)_4NB_3H_8$ at room temperature and 30.6 MHz. The straight line is a best fit by eye and leads to a $T_1$ of 1.4 seconds. Eight more data points were taken at different delays past the last point shown to try to determine the thermal equilibrium magnetization $M_o$. Since the points plotted are $1-M(t)/M_o$, $M_o$ should have been determined as accurately as $M(t)$ so we expect the uncertaintly in $M_o$ to be the major uncertainty in $T_1$.

If we assume that the total measurement time is the sum of the delay times in the experiment, it will have taken about 83 seconds for this run. (Of course the actual time is greater since it takes time to adjust experimental parameters and so on.) Given the fact that $M_o$ needed to be determined more accurately, a better strategy would have been to take fewer $M(t)$ points with more time spent in determining $M_o$.

The next figure shows a Mangelsdorf plot, i.e., our eq.

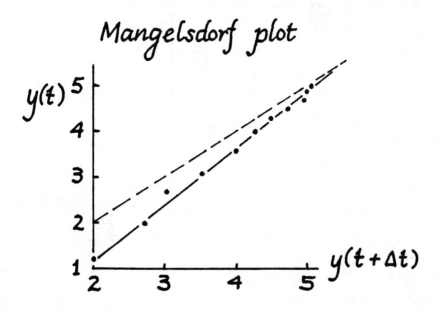

(1) plotted against eq. (3), with an "eyeballed" straight line through the points. The slope (and intercept) yields $T_1 = 1.2$ sec. From inspection of eqs. (1) and (2), we see that B from equations (1), (2), (6), and (7), (which is equal to $M_o$ in this case) is the value of $Y(t)$ when it equals $Y(t + \Delta t)$ and this is about 5.5 from the figure. The value of $M_o$ used in calculating the original points was 5.8 so this is a significant disagreement and explains the difference between the $T_1$'s obtained.

   B can be calculated from eqs. (6) and (8) and the result is 5.6 so the "eyeball" fit was quite good. Using this value of $M_o$, we get $T_1 = 1.23$ sec from eq. (3), whereas a least squares fit of the data with $M_o = 5.6$ yields 1.25 sec. Eqs. (9) and (10) yield $T_1 = 1.32$ sec while its analog with three (rather than four) data blocks yields 1.24 seconds.

   The third figure in this sequence shows a Guggenheim plot of the same data. There are six points because the original 12 point data set is considered in pairs, each of which are six data points apart. The least square fit of these points yields 1.32 sec for $T_1$.

   As a comparison to the original data shown in the first graph of this series, we present the next figure which is a decay curve of $M(t)/M_o$ with $M_o = 5.6$. The slope yields

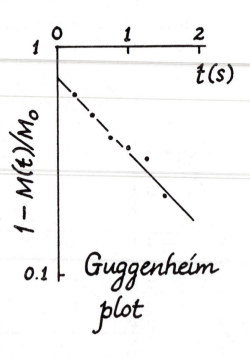

$T_1 = 1.25$ sec but more significant is the fact that the plot looks

straighter than the original plot (which seems to curve up
ever-so-slightly). Parenthetically, both of these plots have
intercepts of 0.95 because the rf pulse was not a good 90°
pulse (but this does not affect the $T_1$ determination).

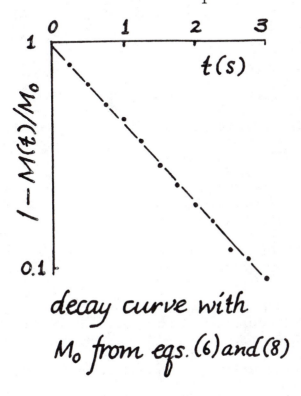

decay curve with
$M_0$ from eqs. (6) and (8)

We conclude that all these methods are quite good in
estimating relaxation times from data which do not include
sufficiently good data, if any, at long times. The total
delay in the 12 data points used here was about 20 seconds
so we spent about one-fourth as much time as the original
experiment and got a better determination of $T_1$. Even 80
seconds doesn't sound too bad for a $T_1$ determination but if
the $T_1$ were longer, say one hour, the difference between a
full recovery curve like that shown in the first figure of this
series and one of these other methods become significant,

indeed.   The time difference could become even longer, of course, if the signal has to be accumulated. Remember that these methods are limited strictly to exponential recoveries and this is a nontrivial point.


## REFERENCES

K. A. Christensen, D. M. Grant, E. M. Schulman, and C. Walling, "Optimal determination of relaxation times of Fourier transform nuclear magnetic resonance. Determination of spin-lattice relaxation times in chemically polarized species," J. Phys. Chem. 78, 1971-1977 (1974).

H. T. Edzes, "An analysis of the use of pulse multiplets in the single scan determination of spin-lattice relaxation rates," J. Magn. Resonance 17, 301-313 (1975).

E. A. Guggenheim, "XLVI. On the determination of the velocity constant of a unimolecular reaction," Phil. Mag. 2, 538-543 (1926).

R. K. Gupta, "A new look at the method of variable nutation angle for the measurement of spin-lattice relaxation times using Fourier transform NMR," J. Magn. Resonance 25, 231-235 (1977).

F. Heatley, "Measurement of nuclear magnetic spin-lattice relaxation times by the use of repetitive sweep techniques," Faraday Trans. II 69, 831-841 (1973).

J. Kowalewski, G. C. Levy, L. F. Johnson, and L. Palmer, "A three-parameter non-linear procedure for fitting inversion recovery measurements for spin-lattice relaxation times," J. Magn. Resonance 26, 533-536 (1977).

S. Linder, "Nuclear magnetic relaxation in 1,2-dichloroethane," J. Chem. Phys. 26, 900-905 (1957).

D. L. Livesey, "Determination of exponential growth factors by Mangelsdorf's method," Am. J. Phys. 47, 558-559 (1979).

D. C. Look and D. R. Locher, "Nuclear spin-lattice relaxation measurements by tone-burst modulation," Phys. Rev. Letters 20, 987-989 (1968); "Pulsed NMR by tone-burst generation," J. Chem. Phys. 50, 2269-2270 (1969); "Time saving in measurement of NMR and EPR relaxation times," Rev. Sci. Instrum. 41, 250-251 (1970).

L. A. McLachlan, "Algebraic analysis of noisy exponential decays," J. Magn. Resonance 26, 223-228 (1977).

P. C. Mangelsdorf, Jr., "Convenient plot for exponential functions with unknown asymptotes," J. Appl. Phys. 30, 442-443 (1959).

W. S. Moore and T. Yalcin, "The experimental measurement of exponential time constants in the presence of noise," J. Magn. Resonance 11, 50-57 (1973).

M. Sass and D. Ziessow, "Error analysis for optimized inversion recovery spin-lattice relaxation measurements," J. Magn. Resonance 25, 263-276 (1977).

M. R. Smith and H. A. Buckmaster, "The analysis of noisy exponential decays," J. Magn. Resonance 17, 29-33 (1975).

J. Stepisnik, J. Porok, and V. Erzen, "Spin-lattice relaxation measurements by improved signal decay method," J. Phys. E: Sci. Instrum. 3, 525-526 (1970).

## III.E.   MOLECULAR DIFFUSION IN A FIELD GRADIENT

The next three sections deal with steady and pulsed field gradient methods for measuring translational diffusion in liquids and plastic solids. There are other methods for measuring slower diffusion which we will not really touch upon. See, for example, Ailion, referenced in IV.C.1., and Burnett and Harmon (1972).

## REFERENCE

L. J. Burnett and J. F. Harmon, "Self-diffusion in viscous liquids: Pulse NMR measurements," J. Chem. Phys. 57, 1293-1297 (1972).

## III.E.1.  DIFFUSION IN A STEADY FIELD GRADIENT

In isotropic systems, the self-diffusion coefficient D can be related to the mean square jump distance $\ell^2$ and the correlation time of the jumping motion $\tau_c$ as (Manning, 1968)

$$D = \ell^2/(6\tau_c).$$

The NMR gradient diffusion experiments yield values of D directly.  To obtain the mean jump distance $\ell$, $\tau_c$ must be determined usually from the intermolecular contribution to $1/T_1$ or $1/T_{1\rho}$.

In many cases, the NMR method of measuring D is superior to tracer diffusion studies.  For liquids, at least, the diffusion coefficient of the tracer does not correspond to that of the major chemical species except for very small tracer concentrations (Ahn, et al., 1972).  Furthermore, the NMR techniques do not contaminate the sample, the measured value of D is not disturbed by isotope effects, and the method is not limited to the availability of suitable isotopes, although it is, of course, limited to suitable NMR nuclei and the existence of Hahn echoes.

In the ordinary spin echo experiment with a $\pi/2-\tau-\pi$ pulse sequence, an echo with amplitude

$$M = M_o \exp[-(2\tau/T_2 + 2\gamma^2 DG^2\tau^3/3)] \qquad (2)$$

appears at $2\tau$ (Carr and Purcell, 1954).  Note that the echo amplitude decays expontentially only if the field gradient G is small, or if diffusion is slow (small D).  Otherwise, the echo amplitude decreases very rapidly with increasing $\tau$ and $T_2$ must be measured with the Carr-Purcell pulse train

$$\pi/2-\tau-\pi-2\tau-\pi-2\tau-\pi-2\tau-\pi-$$

which induces echoes with maxima halfway between each pair of $\pi$ pulses (See I.C.2). The amplitude of the n-th echo is given by

$$M = M_o \exp(-2n\tau/T_2)\exp(-2\gamma^2 DG^2 \tau^3 n/3)$$

$$= M_o \exp(-t/T_2)\exp(-\gamma^2 DG^2 \tau^2 t/3) \qquad (3)$$

where t is the total time elapsed to the n-th echo. Here, the effect of diffusion in the field gradient is reduced by a factor of $4n^2$ for the echo observed after n $\pi$ pulses compared to the ordinary echo of eq. (2). Clearly, the Carr-Purcell sequence (in fact the CPMG sequence as described in I.C.2.) should be used to minimize the effects of diffusion on a $T_2$ measurement, and the simple echo sequence with a known gradient G should be used to measure diffusion, although if $G^2 D$ is sufficiently large the CPMG sequence may be more appropriate.

Diffusion in an applied field gradient serves to diminish the amount of magnetization recoverable by echo formation. When the sample experiences a distribution of applied fields, the magnetization disperses because different regions of the sample give rise to magnetization components (isochromats) with different Larmor frequencies. A $\pi$ pulse can refocus those various components only if the nuclei do not migrate to a region with a different field during the course of the experiment. This phenomenon can be exploited to measure diffusion by applying a large enough gradient to make this irreversible dephasing significant. If G and $T_2$ are known, then D can be calculated.

The field gradient is applied with a pair of coils. A

common scheme is to use Helmholz coils with current flowing through the two halves in opposite directions.    Another scheme is a quadrupole coil wherein the ends of the coil do not contribute to the field (Assink, 1976).    A calibration of G versus the current in the coils is needed in the experiment to evaluate D.    This is a nontrivial task and will be discussed in III.E.3.

Suppose that on exact resonance, the FID, and the echo, in the absence of an applied field gradient, appear as in the left half of the figure below.    Then, the application of a magnetic field gradient G will generate signals as shown on the right half for a cylindrical sample of radius R.    The FID

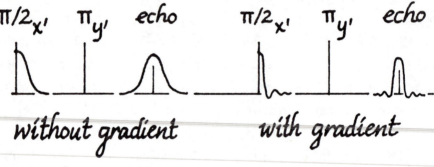

is considerably shorter in the presence of a field gradient due to the inhomogeneous field.    In the presence of a gradient, the FID has the form of a first order Bessel function for a cylindrical sample.    The echo, representing two FID's back-to-back, is similarly affected.

The fact that the FID becomes very short in the presence of large field gradients is equivalent to the linewidth becoming very large.    Unless a very short $\pi/2$ pulse is used, not all parts of the resonance line will experience the same rf power. Equivalently, we can consider the problem from the perspective of the rotating reference frame.    If the sample is centered on the gradient, only the nuclei at the center of the sample experience an effective field $H_e$ in the rotating frame equal to $H_1$. Nuclei in other parts of the sample precess about

$$H_e = [H_1^2 + (H_0 - \omega/\gamma)^2]^{\frac{1}{2}}.$$

Clearly this will produce magnetization components which do not focus properly in the x-y plane. In order to avoid these problems, the condition $H_1 > G \cdot 2R$ should be met. Typically, this will limit continuous gradient diffusion measurements to G of 30 gauss/cm or less. One can do an approximate evalua-tion of G from the echo duration and of $H_1$ from the duration $\tau$ of a $\pi/2$ pulse, where $\gamma H_1 \tau = \pi/2$, to check the above condition. The pulsed gradient method described in the following section alleviates this problem.

## REFERENCES

M.-K. Ahn, S. J. K. Jensen, and D. Kivelson, "Molecular theory of diffusion constants in liquids," J. Chem. Phys. 57, 2940-2951 (1972).

R. A. Assink, "A quadrupolar coil for nuclear magnetic reso-nance spin-echo diffusion measurements as a function of pres-sure," J. Magn. Resonance 22, 165-168 (1976).

H. Carr and E. M. Purcell, "Effects of diffusion on free pre-cession in NMR experiments," Phys. Rev. 94, 630-638 (1954).

J. R. Manning, Diffusion Kinetics for Atoms in Crystals, (D. Van Nostrand, Princeton, 1968), p. 45.

## III.E.2  THE PULSED FIELD GRADIENT METHOD

The gradient diffusion measurement can be extended to much lower D, which would normally require much higher gradients, by the use of pulsed field gradients. The continuous field gradient method is restricted to $D \gtrsim 10^{-7}$ $cm^2/sec$ and to systems with unrestricted diffusion processes. The pulse field gradient method extends the range to D less than $10^{-8}$ $cm^2/sec$ and can be used when the molecules are diffusing in a finite volume as described at the end of this section. Because the field gradient is off when the rf pulses are on, it is easier to cover the required spectral range with the frequency components of the pulse and, in addition, more filtering can be applied in order to get better S/N than in a static gradient experiment where the echo wiggles more rapidly.

Another advantage of the pulse gradient method is that a variety of diffusional times (times between pulse gradients) can be chosen without changing the time between the rf pulses. As a consequence, the echo amplitude for only one rf pulse-spacing without applied gradient need to be obtained for a series of experiments with different diffusional times. In contrast, in the continuous gradient method, the time during which diffusion is monitored is the time between rf pulses; therefore a new echo amplitude without gradient must be determined for each time interval.

Finally, although the pulse gradient experiment is more complex to perform than the cw gradient experiment, the average transmitter power required for a given G is much smaller that it is for the continuous gradient because of the lower duty cycle. In addition, much higher peak currents can be applied to the coils without burning them out or overheating them.

Stejskal and Tanner sequence

A common pulse field gradient sequence due to Stejskal and Tanner (1965) is shown below.

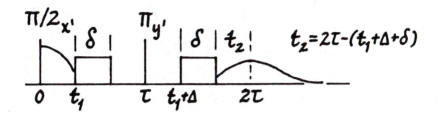

If there is no dc field gradient and identical width field gradient pulses of duration $\delta$ are used, then the appropriate relationship for the echo amplitude M is

$$\ell n[M(G)/M(0)] = -\gamma^2 D\delta^2(\Delta-\delta/3)G^2. \tag{1}$$

When a dc gradient $G_0$ exists in addition to the gradient pulses

$$\ell n[M(G)/M(0)] = -\gamma^2 D\{2G_0^2\tau^3/3 + G^2\delta^2(\Delta-\delta/3)$$

$$-\delta GG_0[(t_1^2+t_2^2)+\delta(t_1+t_2)+(2\delta^2/3)-2\tau^2]\}. \tag{2}$$

The $\pi/2$ pulse, with $H_1$ along x', rotates the magnetization to lie along the y' axis in the rotating frame and spin isochromats fan out symmetrically about the rotating y' axis. Upon application of the first gradient pulse, the isochromats precess rapidly to new positions in the x'-y' plane and after the gradient pulse, the isochromats resume their slow precession. The $\pi$ pulse along y' transposes these isochromats so that each isochromat is now precessing in the same sense as before but with its position transposed across the rotating y' axis. The second gradient pulse causes a repositioning of the isochromats which corresponds to undoing the effects of the first gradient pulse and is necessary if the isochromats are to focus properly along the y' axis. Those molecules which have moved from one area of the sample to another in the interval between the two gradient pulses will precess at the wrong rates and will not refocus properly. See the sketch for the $(\pi/2)_{x'}-(\pi)_{y'}$-echo sequence in I.C.2. as the trajectories for the spin isochromats are not altered by the pulsed gradients except for their angular velocities.

It should be noted that the duration over which the diffusion is being measured is well defined, which is one of the advantages of the pulse gradient method over the steady gradient method. This time is $\Delta-\delta/3$ for the condition $\Delta\geq\delta$ and becomes simply $\Delta$ when $\Delta>>\delta$. For an implementation of such a sequence for solids see, for example, the article by Gordon, et al. (1978).

## Potential problems with pulsed gradient methods

A variety of experimental considerations not necessary in the steady gradient experiments affect the pulse gradient experiment. The pulse gradients can be calibrated the same way as the cw gradients by adjusting the position and duration so that it overlaps the echo but not the $\pi$ pulse (see the following section). The gradient pulse must have relatively

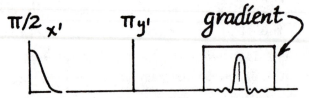

short rise and fall times so that a well defined gradient exists during the pulse. Such gradient pulses may induce eddy currents in the probe body and reflections in the magnet pole faces which may persist for as long as a millisecond. Therefore, one should not put the gradient pulses any closer to the $\pi$ pulse than the pulse fall time plus the time for decay of eddy currents. It may be desirable to use poor conductivity walls near the gradient coils. This may mean removal of aluminum side plates on the probe and substituting these with Faraday shield plates or simply plates of other material. See section VI.B.5. on spurious ringing for related discussions.

If the echo does not focus at $2\tau$ this may mean that the two gradient pulses may not have equal areas. Therefore, the instrument should permit the adjustment of one of the pulse amplitudes. You should use a sample with a very small D, such as glycerol, and adjust the gradient pulses for maximum echo amplitude because any mismatch results in a reduced echo.

The above effect as well as the echo position shifting back and forth due to small fluctuations in the gradient amplitudes or widths may be particularly troublesome when the gradients are very short and intense as they will be when D is small and $T_2$ is short. The suggested causes of these problems include eddy current associated gradients from nearby metallic surfaces, sample movement, field/frequency variations, and gradients stimulated by pickup on the sensing resistor of the current supply (Callaghan, et al., 1980). Whenever possible, attacking these problems at their sources yields the best solutions. The above authors, for example, claim that cleaning up these sources of noise and fluctuations has allowed them to measure D to a precision of 1% without any need for trimming the gradient pulse durations. In contrast, the more traditional way to cope with these problems is to apply a small constant gradient on top of the pulsed gradient.

Even though the heating effects for a given amount of gradient is much less for pulsed gradients than for cw gradients, you can still run into heating problems with the pulsed gradient method -- this is another possible source of trouble (von Meerwall, et al., 1979).

### Lowe and Karlicek method

The expression given by eq. (2) is not at all straight-forward to apply. It assumes the simple form of eq. (1) only when $G\delta \gg G_0\tau$. In many polycrystalline solids this condition is violated because of the volume susceptibility of each crystallite which can create large $G_0$'s which are, in addition, distributed in magnitude and direction. Lowe and Karlicek (1978) have proposed a modification to the pulsed gradient method to minimize the effects of such susceptibility distributions which eliminate the last term of eq. (2) and increases the importance of the second term with respect to the first. Eq. (2) then becomes

$$\ln[M(G)/M(0)] = -(\gamma^2 D\tau^3/3)(10G_0^2 + 128G^2) \qquad (3)$$

and the simplest form of their pulse sequence is sketched below. The new feature is the alternation of the sign of the field gradient G.

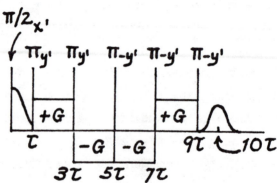

## Stimulated echo method

Situations can occur in which it is not possible to measure D by these methods owing to very slow diffusion or to a short $T_2$, or both. In systems where $T_1 \gg T_2$, the stimulated echo sequence (II.D.5.) can be used instead of the $\pi/2-\pi$ sequence because it is limited by $T_1$ rather than $T_2$. As pointed out by Hahn (1950), a three pulse sequence of $\pi/2$ pulses, all with the same phase, can yield five echoes.

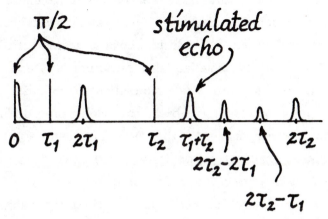

The echo occurring at $\tau_1 + \tau_2$ is called the stimulated echo. If a continuous gradient G is applied for the duration of the experiment,

$$\ln(M/M_o) = -[(\tau_2-\tau_1)/T_1 + 2\tau_1/T_2 + \ln 2 + \gamma^2 D G^2 \tau_1^2 (\tau_2 - \tau_1/3)].$$

If, on the other hand, a pulse gradient is applied just following the first and the third $\pi/2$ pulses in the presence of a constant gradient $G_0$, then the appropriate equation is (Tanner, 1970)

$$\ln(M/M_o) = -\gamma^2 D \{ G^2 \delta^2 (\Delta - \delta/3) + G_0^2 \tau_1^2 (\tau_2 - \tau_1/3)$$

$$-GG_0 \delta[t_1^2 + t_2^2 + \delta(t_1+t_2) + (2\delta^2/3) - 2\tau_1\tau_2]\}.$$

Note that if there is no dc field gradient, then this expression reduces to

$$\ln(M/M_o) = -\gamma^2 DG^2\delta^2(\Delta-\delta/3)$$

in agreement with the first equation in III.E.2.

### Restricted diffusion

In many systems, the diffusion of small molecules is not free but is restricted.   Examples of systems with restricted diffusion are molecules in cellular compartments, fluids between long flat plates, lamellar and vesicular systems, water-filled pores in rocks, and small molecules in colloidal suspensions.   If the time during which the molecular diffusion is monitored in the experiment (the time between rf pulses in the continuous gradient experiment and $\Delta-\delta/3$ in a pulse gradient diffusion experiment) is much longer than the time for a molecule to travel to a boundary, then the calculated D will be too small (Woessner, 1963).   In such cases of restricted diffusion, appropriate equations must be derived for the particular geometry of the constraint (Stejskal, 1965; Wayne and Cotts, 1966; Robertson, 1966; Tanner and Stejskal, 1968; Boss and Stejskal, 1968).   In favorable cases, the diffusion measurement can yield not only D but an estimate of the restraining dimension as well.

## REFERENCES

B. D. Boss and E. O. Stejskal, "Restricted, anisotropic diffusion and anisotropic nuclear spin relaxation of protons in hydrated vermiculite crystals," J. Colloid Interface Sci. 26, 271-278 (1968).

P. T. Callaghan, K. W. Jolley, and C. M. Trotter, "Stable and accurate spin echoes in pulsed field gradient NMR," J. Magn. Resonance 39, 525-527 (1980).

R. E. Gordon, J. H. Strange, and J. B. W. Webber, "An NMR apparatus for the measurement of self-diffusion in solids," J. Phys. E: Sci. Instrum. 11, 1051-1055 (1978).

E. L. Hahn, "Spin echoes," Phys. Rev. 80, 580-594 (1950).

I. J. Lowe and R. F. Karlicek, Jr., "A modified pulsed gradient technique for measuring diffusion in the presence of large background gradients," J. Magn. Resonance 37, 75-91 (1980).

B. Robertson, "Spin-echo decay of spins diffusing in a bounded region," Phys. Rev. 151, 273-277 (1966).

E. O. Stejskal, "Use of spin echoes in a pulsed magnetic-field gradient to study anisotropic restricted diffusion and flow," J. Chem. Phys. 43, 3597-3603 (1965).

E. O. Stejskal and J. E. Tanner, "Spin diffusion measurements: spin echoes in the presence of a time-dependent field gradient," J. Chem. Phys. 42, 288-292 (1965).

J. E. Tanner, "Use of the stimulated echo in NMR diffusion studies," J. Chem. Phys. 52, 2523-2526 (1970).

J. E. Tanner and E. O. Stejskal, "Restricted self-diffusion of protons in colloidal system by the pulsed-gradient, spin-echo method," J. Chem. Phys. 49, 1768-1777 (1968).

E. von Meerwall, R. D. Burgan, and R. D. Ferguson, "Pulsed gradient NMR diffusion measurements with a microcomputer," J. Magn. Resonance 34, 339-347 (1979).

R. C. Wayne and R. M. Cotts, "Nuclear-magnetic-resonance study of self-diffusion in a bounded medium," Phys. Rev. 151, 264-272 (1966).

D. E. Woessner, "N.M.R. spin-echo self-diffusion measurements on fluids undergoing restricted diffusion," J. Phys.
Chem. 67, 1365-1367 (1963).

## III.E.3.   GRADIENT CALIBRATION

We have utilized in the last two sections the fact that the
bulk diffusion coefficient D is related to NMR relaxation times
through the magnetic field gradient.   The precision in the
determination of D is directly related to the precision of the
relaxation time measurement and of the gradient G.   Therefore,
it is important to determine G as accurately as possible.   Very
often, this is the weak link in the experiment, a fact which is
not always recognized.

Since NMR is probably the most accurate method for the
determination of G (provided that D is known!), one obvious
method for determining G is to examine a sample of known D.
This is done quite often and it works provided that it is done
carefully.   The D value for the calibration sample had to be
determined sometime on an absolute scale and you have to make
sure that this is a good value.

Another method is to calculate G from the geometry of the
coil and the current through the coil.   As you will see from
the example later in this section, a small geometric effect can
have a large effect on G so that this is not a good way to get
G except as a check for reasonableness.

We need a method to determine G which is independent of
D.   This is possible because NMR measures the distribution of
fields seen by the nuclei in the sample so that, in addition,
only the sample size needs to be known to calculate the gradient.

The majority of samples used in NMR are cylindrical. The echo decay of a cylindrical sample in the presence of a uniform field gradient perpendicular to the cylinder axis represents a Fourier transform of a distribution of chords of a circle which, in turn, is a semi-ellipse. Thus, the echo can be written in terms of a Bessel function as $J_1(t)/t$, where t is the time coordinate, because this represents a Fourier transform of a semi-circle which differs from a semi-ellipse only by a multiplicative constant. (See, for

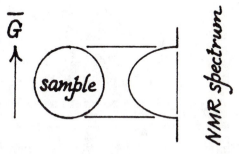

example, Chapter 19 of Bracewell, Appendix A.) Because of the way in which the two domains are related (as discussed in sections I.D.1. and I.D.2.), there is a reciprocal relationship between the spread of the Larmor frequencies across the cylindrical sample and the extent of the signal with the shape $J_1(t)/t$ in the time domain. Thus the information on the field gradient experienced by the sample is contained in the echo if the sample dimension is known.

The most common method for static field gradient determination is to use the zeros of the oscillatory NMR echo signal and compute the field spread of the Larmor frequencies by the known characteristics of the zeros in the function $J_1(t)/t$. For this method, sufficient rf pulse power must be applied to cover the absorption line, which is greatly broadened by the gradient. The symptom of inadequate coverage of the line is a calibration graph of G vs. I which is linear at low currents but which shows successively smaller increase in G for equal increases in I at high currents. The rule of thumb is that $H_1$ must exceed $G \cdot 2R$ where R is the radius of a cylindrical sample with its axis perpendicular to the gradient. If this condition cannot be met, the rf pulse lengths can be reduced to values much below that required for the conventional $\pi/2$-$\tau$-$\pi$ sequence. You then get a reduced amplitude Hahn echo but with more adequate linewidth coverage. For calibration of G vs. I, we are only interested in the time between minima in the echo, so the reduction in echo amplitude is of no consequence other than the less than optimal S/N.

In calibrating the pulsed gradient, we can position it over the echo and the width between the first minima of the resulting echo can be determined without applying any gradient when the rf pulses are on. However, one must be sure to use adequate receiver bandwidth for observing the echo properly, unlike the case when the actual diffusion constant measurement is being made without the gradient during the echo. In addition, there is a difficulty with the position of the echo shifting in time when the gradient is applied. The echo must lie entirely within the gradient pulse's duration. A scope display of the current pulse fed to the gradient coils is useful for this purpose.

The above should be a foolproof method if the echo signal, indeed, has the shape given by $J_1(t)/t$. Unfortunately, this is very often not the case for reasons discussed, for example, by Murday (1973) and Vold, et al. (1973). It turns out that the zeros of the echo signal change in a sensitive way when the field distribution seen by the cylindrical sample changes, thus leading to wrong values for the field gradient and throwing off the calculation of D.

Clearly, the best solution is to utilize the entire echo signal and Fourier transform it. Then, you would have the field distribution itself and it can be examined for any distortions due to non-uniform field gradient, sample tilting (except for tilting parallel to the gradient axis), being off resonance, etc., in addition to being able to measure exactly what the field difference across the sample is so that an accurate determination of the field gradient can be made.

As an example, consider the echo shown at right. It is a carbon-13 echo in liquid CO (Fukushima, et al., 1979) and it is not at all clear from just looking at this signal what

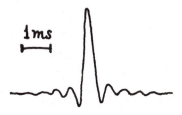

the departure from the predicted $J_1(t)/t$, if any, might be. Its Fourier transform, on the other hand, shows exactly what is going on and the sketch below shows it broken down into different contributions. In order to understand it, you have

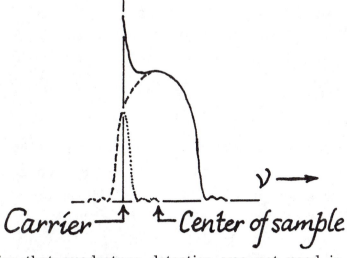

Carrier → ↑ ↑ ← Center of sample

to realize that quadrature detection was not used in this experiment, so the frequency spectrum below the origin is folded back and added. The frequency origin is at the carrier which, in our case, is not at the center of the sample. When you think about it for a while, it becomes clear that having the nuclei at the center of the sample on resonance is not so easy to arrange.  Even if the entire sample were on resonance in the absence of the field gradient, the field gradient will most likely have an offset so that the additional field will not be exactly zero at the center of the sample.  As we have mentioned, just this fact alone can throw off the gradient determination by changing the shape of the echo.

In the above sketch, the dashed line indicates the true distribution of the Larmor frequencies and the dotted line the reflection of the dashed line across the origin.  So, if the dotted line is added to the dashed line, you get the experimental distribution which is the solid line.  Of course, in the actual experiment, we turn the problem around and optimize the fit between such a constructed curve and the real signal obtained by Fourier transforming the echo.  The field offset parameter is adjusted in the computer for the best fit.  The gradient is the measured difference in the field intensity across the sample divided by the dimension of the sample.  In this example, the misalignment between the center of the cylinder and the zero offset point of the applied gradient was 3 mm out of an overall coil dimension of 90 mm which is entirely reasonable.

Here is another case in which quadrature detection would have been a big help.  What threw the zeros off in the echo was the contribution from the spike at the origin which contributes a broad hump to the spectrum.  Such a broad hump can change the zeros because just adding a constant to $J_1(t)/t$ will alter the zeros.  QD would have simply removed the reflection about the origin which caused the spike.  Therefore, the standard method of deducing the gradient from the zeros of the echo should work more reliably with QD.  At any rate, Fourier transforming the echo still is the most reliable method since other factors such as the gradient uniformity can be checked, as mentioned above.

If you do not have a computer nor QD, your best bet is to deliberately go off resonance enough so that all the nuclei are known to be offset to one side. Then $J_1(t)/t$ will be well defined as the envelope to the higher frequency off-resonance beats.

## REFERENCES

E. Fukushima, A. A. V. Gibson, and T. A. Scott, "Carbon-13 NMR of carbon monoxide. II. Molecular diffusion and spin-rotation interaction in liquid CO," J. Chem. Phys. $\underline{71}$, 1531-1536 (1979).

J. S. Murday, "Measurement of magnetic field gradient by its effect on the NMR free induction decay," J. Magn. Resonance $\underline{10}$, 111-120 (1973).

R. L. Vold, R. R. Vold, and H. E. Simon, "Errors in measurements of transverse relaxation rates," J. Magn. Resonance $\underline{11}$, 283-298 (1973).

CHAPTER IV

## NMR OF SOLIDS

## IV.A. LINESHAPES IN SOLIDS

In the first section, we give a qualitative description of the different kinds of lineshapes possible for solids. Although, historically most solid state line shapes were studied by cw NMR, lineshape studies are getting very popular in pulse FT NMR because of the increasing popularity of experiments in which the dipole-dipole interaction can be suppressed to yield informative lineshapes, especially in non-spinning powders. It is also possible to perform moments analyses from the FID's without Fourier transformation and this is covered in the second section.

Eiichi Fukushima and Stephen B. W. Roeder, Experimental Pulse NMR: A Nuts and Bolts Approach

## IV.A.1.   HOMOGENEOUS AND INHOMOGENEOUS LINES

All contributions to an NMR line can be classified as homogeneous or inhomogeneous.  The purest form of a homogeneous line is one without separate contributions to different parts of the line, i.e., all the nuclei in that line make identical contributions with their intrinsic linewidths the same as the composite.  High resolution lines in solution are homogeneous, in principle.  They are usually Lorentzian in shape and their linewidths are determined by their lifetimes so that their half-widths at half height are equal to $1/T_2 \sim 1/T_1$, provided that the magnet homogeneity is high enough.

Dipolar coupled lines in solids, especially in single crystals, are usually homogeneous,. too, but in a different way. In this case, there are separate contributions to different parts of the line but they cannot be so identified because the coupling between the separate parts of the line is so good. The basic lineshape approximates a Gaussian and its width is totally unrelated to $1/T_1$ except that it must equal or exceed it.

An inhomogeneous line is one in which separate parts of the line can be identified as due to separate contributions. The most common example is a powder pattern due to anisotropies such as quadrupolar or chemical shifts in which crystallites with different orientations with the static magnetic field contribute to different parts of the line.

One way of determining whether the line is broadened homogeneously or inhomogeneously, is to irradiate and saturate a specific spot in the line with a large, coherent $H_1$ field. Immediately after the irradiation, the line is detected either by cw or pulse FT NMR and an inhomogeneous line will have a "hole" burned into it (Bloembergen, et al., 1948).

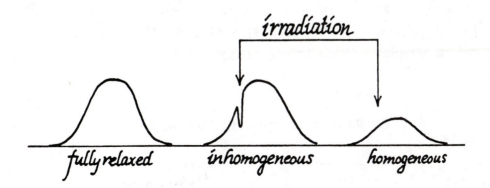

As the shape of the "hole" will approximate the shape of a line in a homogeneous part of the magnetic field, it can be used to study natural lineshapes under conditions of poor $H_0$ homogeneity (Schaefer, 1974). If, on the other hand, the line is homogeneously broadened, the attempt to burn a hole in the line will result in a collapse of the entire line.

How about some other examples of homogeneously broadened lines? The application of high radio frequency power to a transition not only causes saturation (if $T_1$ is sufficiently long) but also causes power broadening for the entire system. When a strongly coupled system, coupled by dipolar interactions, permits spin diffusion and the establishment of a common spin temperature, the attempt to burn a hole in the line will also result in a collapse of the entire line. Dipolar interactions, spin diffusion, and spin temperatures are discussed later in this chapter, as well as in III.C.1. and in IV.D.

A simple example of an inhomogeneously broadened line is a resonance broadened by inhomogeneity of the applied dc magnetic field. If the applied field has a spatial inhomogeneity, otherwise identical spins in different parts of the sample will have slightly different resonance frequencies. As a result, the FID will be much shorter than that determined by the natural $T_2$ of the sample and the NMR line will be much broader because it is actually a superposition of lines at various frequencies distributed according to the distribution of the dc magnetic fields in the sample.

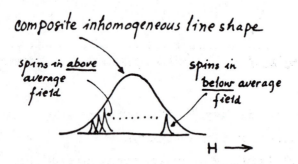

A dramatic example of such an inhomogeneous line is that due to a cylindrical sample in a linear magnetic field gradient. This situation exists during the standard gradient diffusion measurement as described in Section III.E. and it was pointed out that the lineshape obtained is a "profile" of the sample so that a cylindrical sample would yield a semi-ellipse. Another interesting sample cross section to consider in a linear gradient is a square and we leave to the reader to consider how it can give rise to the two spectra shown. In all of these cases,

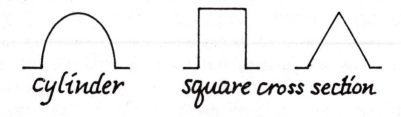

it is clear that the signal at a specific point in the spectra arises from nuclei at special locations in the sample and that there is no crosstalk between those different nuclei. [We parenthetically remark that the ability to map nuclei in physically different locations within the sample to different parts of the spectrum is utilized in the growing field of NMR imaging, also called zeugmatography.] In real samples, especially solids, there are interactions between nuclei which are close to each other and they lead to a broadening. Therefore, in an actual experiment, the sharp features shown in the figures above will appear rounded.

Additional examples of inhomogeneously broadened lines are: (a) Spatial distribution of electric field gradients in amorphous or stressed solids yielding inhomogeneously broadened NMR and NQR lines for nuclei with $I > \frac{1}{2}$ in complete analogy with the spatial distribution of the magnetic field. This occurs most frequently for large Z nuclei because, for a given nuclear quadrupole moment, they are much more susceptible to interaction with an EFG. This, in turn, is due to the inner electrons being much closer to the nucleus so any deformation of the electron cloud by the external field gradient produces a larger effect at the nucleus. [This effect is usually referred to in terms of the Sternheimer anti-shielding factor. See, for example, Slichter (Appendix A) and references therein as well as a brief discussion in III.C.3.]

(b) In a polycrystalline sample, $I > \frac{1}{2}$ nuclei usually will exhibit inhomogeneously broadened lines because the resonance frequency depends on the angle between the EFG axis (which is related to the crystallite axis) and the magnetic field. (c) Similarly, any other anisotropic broadening effect can yield an inhomogeneously broadened line in a polycrystalline sample. For example, in $CaSO_4 \cdot 2H_2O$ the dipolar interaction between

the two protons in each water molecule will be angle dependent and the spectrum of the polycrystalline material will be
the superposition of a distribution of pairs of lines.  The
reason the dipolar interaction leads to inhomogeneous rather
than homogeneous broadening in this case is the isolation of
each pair of protons from others which prevents spin diffusion between protons in different crystallites having different
orientation with the magnetic field.

(d) In addition to the dipolar and the quadrupolar interactions, the chemical shift anisotropy will be angle dependent
and the observed line will be the superposition of lines resulting from crystallites lying at different angles with respect to
the magnetic field.  For example, in a bulk sample of a lamellar phase of a soap or lipid, the random distribution of lamellae will give rise to a chemical shift anisotropy pattern with
uniaxial symmetry if the molecules are rotating rapidly about
their long axes.  (e) Finally, distributions of bulk susceptibility can produce magnetic field distortions in the sample resulting in an appropriately broadened line.

All of these spectra are inhomogeneously broadened and
composed of superposition of individual Gaussian-like lines.
If the individual linewidths are significant in comparison to
the features of the powder patterns, the powder patterns are
not as sharp as represented here, except in line narrowing
experiments as described later in this chapter.  In the following paragraphs, we discuss how these patterns come about.

Suppose we have a molecule containing an axis of symmetry 3-fold or higher or is rotating rapidly about one axis
in the solid.  $CO_2$ would be a good example.  As the shielding experienced by the $^{13}C$ nucleus from the applied magnetic
field would be different if the molecule had its axis parallel to
the applied field rather than perpendicular to the field, we

expect to find a chemical shift anisotropy.  Molecules in the powder oriented parallel to the applied field exhibit one $^{13}C$ shift, those perpendicular to the applied field a different chemical shift, and those oriented at angles in between might have chemical shifts between these two extremes.

Statistically we expect to find many more molecules perpendicular than parallel to the applied field because at a given value of the polar angle all values of the azimuthal angle give rise to the same chemical shift and the number of molecules at any given polar angle should decrease monotonically as the polar angle decreases.  The expected powder pattern is sketched below.  The extremes of the powder pattern can be immediately associated with those components of the chemical shift  parallel  to  and  perpendicular  to  the  magnetic  field.

Without  axial  symmetry,  the  chemical  shift  anisotropy  will
look  like  the  pattern  shown.    Here,  one  cannot  as  easily  de-

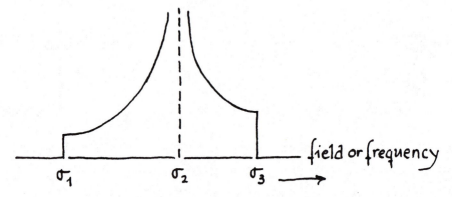

termine  which  of  the  principle  values  of  the  chemical  shift
tensor  is  associated  with  which  direction  in  the  molecule.

Now  consider  a  dipolar  coupled  pair  of  isolated  spin-1/2
particles  in  a  powder.    For  each  angle  $\theta$  between  the  internuclear  vector  and  the  applied  field,  there  is  a  corresponding
dipolar  coupling  containing  the  term  $3\cos^2\theta-1$.    Each  spin  in
the  presence  of  the  local  field  of  the  other  gives  rise  to  a  distribution  of  lines  with  the  same  qualitative  shape  as  that  of  an
axially  symmetric  molecule.    The  resulting  spectrum  with  its
appropriate  limits  indicated  is  shown  in  the  following  diagram.

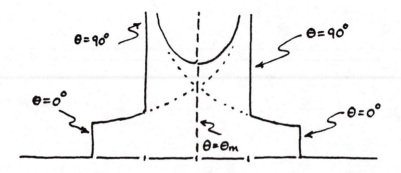

The angle $\theta_m$ corresponding to the point where the two curves cross is called the "magic angle," and is equal to 54.7° as required by the condition $3\cos^2\theta_m - 1 = 0$. We will explain the significance of the magic angle in IV.E.

Next, consider the energy levels of a spin-3/2 nucleus in a single crystal in the presence of an applied magnetic field and an electric field gradient so that the perturbation provided by the quadrupole interaction with the applied EFG is small; a first order perturbation. We already considered this case in II.D.1. where we saw that the quadrupole interaction split the single line into three by shifting the $3/2 \leftrightarrow 1/2$ and the $-3/2 \leftrightarrow -1/2$ lines in opposite directions. We state without proof that the energy of $\pm 3/2$ to $\pm 1/2$ transitions are proportional to $3\cos^2\theta - 1$ where $\theta$ is the angle between the EFG and the applied magnetic field. Therefore, in a polycrystalline sample, there will be a distribution of $\theta$ so that it results in a powder pattern entirely analogous to that of a chemical shift anisotropy in an axially symmetric molecule for each transition.

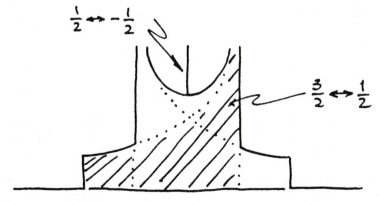

The two of them together form a pattern identical to that from a pair of interacting magnetic dipoles. Finally, the picture is completed by adding the 1/2 to -1/2 transition. This transition is not displaced nor broadened to first order by the variation in θ and therefore is a sharp peak. Such a pattern is called the first order quadrupolar powder pattern for spin 3/2 and is sketched above. As already mentioned, there is nearly always additional dipolar broadening of the lineshape.

These anisotropies, despite their obvious sensitivity to molecular structure and environment will not be discussed further in this book. The chemical shift anisotropies, at this writing, are already proving to be extremely valuable parameters, as obtained by line narrowing experiments in solids (Section IV.E.) and good reviews of the resulting powder patterns are given by Haeberlen (1976), Mehring (1976), and Spiess (1978). For quadrupolar powder patterns, see Creyghton, et al. (1973) and references therein.

We close this section by pointing out that it is not unusual to have a combination of these effects in a real solid. For example, the combination of chemical shift and quadrupolar anisotropies in a polycrystalline solid has been treated by Jones, et al. (1963) whereas an exmple of chemical shift and dipolar anisotropies was given by O'Reilly, et al. (1971).

## REFERENCES

N. Bloembergen, E. M. Purcell, R. V. Pound, "Relaxation effects in nuclear magnetic resonance absorption," Phys. Rev. 73, 679-712 (1948).

J. H. N. Creyghton, P. R. Locher, and K. H. J. Buschow, "Nuclear magnetic resonance of $^{11}B$ at the three boron sites in rare-earth tetraborides," Phys. Rev. B7, 4829-4843 (1973).

W. H. Jones, Jr., T. P. Graham, and R G. Barnes, "Nuclear magnetic resonance line shapes resulting from the combined effects of nuclear quadrupole and anisotropic shift interactions," Phys. Rev. 132, 1898-1909, (1963).

U. Haeberlen, High Resolution NMR in Solids (Academic Press,

New York, 1976).

M. Mehring, High Resolution NMR Spectroscopy in Solids, (Springer-Verlag, Berlin, 1976).

D. E. O'Reilly, M. E. Peterson, Z. M. El Saffar, and C. E. Scheie, "Anisotropy of the chemical shift tensor in solid fluorine," Phys. Letters 8, 470-472 (1971).

J. Schaefer, "The Carbon-13 NMR analysis of synthetic high polymers," in Topics in Carbon-13NMR Spectroscopy, edited by G. C. Levy (Wiley-Interscience, New York, 1974), vol. 1.

H. W. Spiess, "Rotation of molecules and nuclear spin relaxation," in Dynamic NMR Spectroscopy, edited by P. Diehl, E. Fluck, and R. Kosfeld (Springer-Verlag, Berlin, 1978).

## IV.A.2 MOMENTS OF ABSORPTION LINES

Dipolar interactions between more than two or three spins can produce a nearly featureless line which could not be analyzed in terms of discrete splittings but nevertheless can be analyzed to yield much information. Because dipolar coupling is a "through space" interaction, it is directly affected by structural parameters of the system. Unfortunately, it is difficult to extract the structural parameters from the spectrum because of the difficulty of calculating the spectrum in such a many-body problem and also because of the relatively subtle changes in the spectrum caused by changes in the structural parameters. Although the increasingly popular field of high resolution NMR in solids (Section IV.E.) and 2-D spectroscopy (II.D.3.) are well suited to studying the dipolar interaction, the traditional method of studying this interaction has been that of moments (Van Vleck, 1948; Abragam, Appendix A). In general, the n-th moment of the lineshape $f(x)$ about $x_o$ is where x is a spectral coordinate which is either the frequency

$$S_n = \int_0^\infty (x-x_0)^n f(x)\,dx \Big/ \int_0^\infty f(x)\,dx$$

or the field in NMR. In principle, the description of a complex line requires specification of all its moments.

The second moment of an NMR line is completely analogous to the moment of inertia of an object with the same shape as the line. In particular, the second moment is where the

$$S_2 = [\int x^2 f(x)\,dx]/[\int f(x)\,dx]$$

independent variable x can be the frequency or the field and $f(x)$ is the NMR lineshape. We now consider the question "where is the origin of x?"

Let us start by taking a simple shape and choosing the origin of the independent variable arbitrarily. We will consider a rectanglular lineshape with its center at distance h from the origin. Let the width 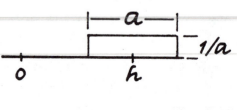 of the rectangle be 'a' and normalize the area to unity so that the height is 1/a. We can now use the parallel-axis theorem from elementary physics and see that the second moment about the origin is the moment about the center of the rectangle $a^2/12$ plus $h^2$. For a single line we usually want just the former contribution so we want a rule for deciding on the choice of origin for x.

The most intuitively obvious point to choose, in general, is the center of gravity. For a symmetric line like our rectangle, the center of gravity is in the middle. Now the first moment vanishes about he center of gravity, that is,

$$\int x f(x) dx = 0.$$

So, the rule for calculating the second moment of a general line is to calculate the first moment and then calculate the second moment about the point about which the first moment is zero. Starting with an arbitrary origin, the second moment expression, then, is

$$S_2 = \frac{\int x^2 f(x) dx}{\int f(x) dx} - [\frac{\int x f(x) dx}{\int f(x) dx}]^2.$$

What if there are two NMR lines? If they are well separated, just calculate the second moment of each. If the two lines are not well separated, then only one second moment for the whole signal is accessible and we need to ask what this total second moment means.

   Let us do this problem backwards. We will concoct our own two line spectrum as shown. About their own centers of gravity, the two "peaks" at $h_1$ and $h_2$ have second moments of $a^2/3$ and $a^2/12$, respectively (which goes to show the importance of the areas away from the centers of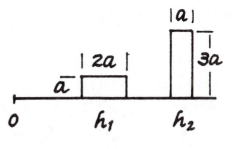
gravity since the peak at $h_2$ has a second moment only 1/4 as large as that at $h_1$ even though its area is 3/2 times larger).

   Now, what is the "total" second moment of these two lines? The prescription given above was to find the overall center of gravity and then calculate the second moment about it. The center of gravity is at $2h_1/5 + 3h_2/5$ and the total moment is $a^2/3 + a^2/12 + 13(h_2-h_1)^2/25$. All of this is valid even when the two lines overlap, i.e., when $(h_2-h_1)<3a/2$, so we now know what the contributions to the total composite line second moment are. The first two terms, of course, are the second moments of the separate lines about their respective centers of gravity while the third term represents the separation of the two lines and the relative areas.

Several things should be noted: First, the second moment of a line is simply the mean square linewidth. Next, if the absorption line is symmetric, all odd moments about the center of the line vanish. This is the common case in liquids but asymmetric lines can exist in the presence of chemical shifts, Knight shifts, or quadrupolar interaction in polycrystalline samples (Cohen and Reif, 1957; Haeberlen, 1976). Third, the limits of integration should be set to exclude the satellite lines which occur at 0, 2x, 3x, etc. (Van Vleck, 1948; Sections 3.2 and 3.3 of Slichter, Appendix A). These lines, resulting from the zero quantum, two quantum, three quantum, etc. transitions are "forbidden" transitions and are greatly reduced in intensity compared to the line at $x_o$ but, because of the magnitude of $(x-x_o)^2$, can give considerable contributions to $S_2$ and even more so to the higher moments. When working at high fields it is easy to exclude these contributions which are normally far away but at low fields it may be difficult to eliminate their influence. [The second moment including the effect of the satellites is 10/3 times larger than that without the satellites (Van Vleck, 1948).]

Two lineshapes commonly used in NMR are the Gaussian

$$f(x)=(1/\sigma\sqrt{2\pi})exp[-(x-x_o)^2/2\sigma^2]$$

and the Lorentzian

$$f(x) = \frac{1}{\pi T_2} \frac{1}{(1/T_2)^2 + (x-x_o)^2} \, .$$

The Gaussian lineshape is completely characterized by its second moment and from it all other moments can be calculated:

$$S_2 = \sigma^2; \ S_4 = 3\sigma^4; \ S_{2n} = 1 \cdot 3 \cdot (2n-1)\sigma^{2n} \, .$$

Verification of the above relationship between the moments of the Gaussian line is easy if one notes that the Fourier transform of a Gaussian is another Gaussian. The Gaussian in the time domain can then be expanded and this expansion equated with the moment expansion for the FID, introduced later in this section.

Let us account for the Gaussian lineshape, commonly found in solids. Stop for a moment and consider an analogy. Suppose that you shoot at a target which has a bull's-eye and concentric circles drawn around it. Without systematic errors, the resulting pattern of bullet holes will be a circularly sym-

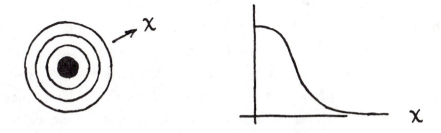

metric with the highest density at the center. The plot of the number of points lying at a distance x from the center versus that distance x is a bell-shaped curve. Such a curve,

called a Gaussian, gives the probability p that a hole is a
distance x away from the center of the target

$$p=(1/\sigma\sqrt{2\pi})\exp[-(x-x_o)^2/2\sigma]$$

where $x_o$ is the mean value of x and $\sigma$ is the standard deviation.  The width of the Gaussian curve is related to $\sigma$ which,
in turn, is defined by

$$\sigma^2 = N^{-1} \sum_{i=1}^{N} (x_i-x_o)^2 \quad .$$

The full width at half maximum is a measure of how bad a
shot you are and turns out to be $2.36\sigma$ as stated in III.B..

Now consider the NMR experiment.  In a rigid solid in a
static magnetic field $\overline{H}_0$ at room temperature, about one-half
of the nuclear spins will be essentially parallel to $\overline{H}_0$ and the
remainder will be essentially antiparallel.  A local field due to
the neighboring magnetic moments which depends on the orientation of these moments exists at each nucleus.  The local
fields at various other nuclear sites in the molecule are different depending upon the number of neighboring moments
(assumed to be more than one or two) with their spins up and
down and their distances from the nucleus.  Thus, the distribution of the local fields will also be a Gaussian.  As each
magnetic moment precesses at a frequency which is determined
by the field at that location which, in turn, will be the vector
sum of the applied field and the local field, the NMR lineshape
is approximately a Gaussian with the width proportional to the
standard deviation of the local field distribution.

The Lorentzian line, observed when there is considerable

motional narrowing as in the usual high resolution solution NMR, does not yield finite values for the moments above the first moment because the integrals diverge due to the wings of the line not dropping off fast enough. If a finite second moment is predicted from a calculation for a line which is apparently Lorentzian, the wings of the actual line must decay faster than Lorentzian out in the region where the line is too small to be detected.

In general, if the NMR line has any Lorentzian character to it, the second moment is not an ideal parameter for study because so much of the crucial information is lost in the wings of the spectrum. A line can be considered to be a Gaussian when $S_4/(S_2)^2 \cong 3$. When such difficulties exist, it may be easier to get the true second and fourth moments from the FID as will be described shortly. For discussions of moments of Lorentzian-like lines, the relationship between the true moment and the measurable moment, and which component of the second moment is affected by molecular motion, see Abragam (Appendix A), Sections IV.II.B, IV.IV.A, X.V.A.

In a homogeneously broadened system, each magnetic resonance line is symmetric (Abragam, Appendix A, p.108) and the FID can be expanded in a series

$$f(t) = M_0(1 - S_2 t^2/2! + S_4 t^4/4! - S_6 t^6/6! + . . .) .$$

It is clear that the second moment describes the short time behavior of the system. In particular, $(1 - S_2 t^2/2!)$ describes the FID shape to within 1% when it has decayed to 89% of its initial height and $(1 - S_2 t^2/2! + S_4 t^4/4!)$ describes the FID down to 70% of its initial height with the same accuracy (Powles and Carazza, 1970). The 2n-th moment can be determined by taking the appropriate derivative of the FID shape f(t):

$$S_{2n} = (-1)^n \left. \frac{d^{2n}f(t)}{dt^{2n}} \right|_{t=0} \div f(0) .$$

Now we mention several points which are important in the actual determination of the second (or higher) moment from the FID. In order to obtain the baseline just before the pulse and to observe the pulse itself, it may be necessary to pretrigger the transient recorder or computer. In typical cases, more rapid digitization than is commonly used in FT NMR is needed. The appropriate choice of time origin for fitting the start of the FID is the middle of the rf pulse. (see Section VI.D.4.). The $S_2$ obtained from the fit has units of $rad^2/sec^2$. The method is limited by the recovery of the receiver and ring-down of the probe. It also does not work when the region linear in $t^2$ is past by the time the receiver has recovered. Since the major obstacle to an accurate determination of the

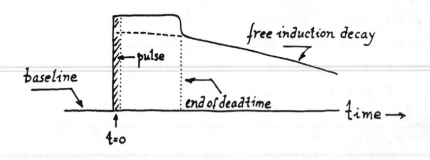

moment from the FID is due to the receiver deadtime, the reader is referred to sections IV.B.3., IV.B.4., and VI.D.4. for alternative methods.

The contribution to the second moment from the interaction of like spins is

$$(S_2)_{II} = \frac{3}{4} \frac{\gamma^4 \hbar^2 I(I+1)}{N} \sum_{k,j} \frac{(1-3\cos^2\theta_{kj})^2}{R_{kj}^6}.$$

If the nucleus under observation by NMR has unlike, rather than like neighbors, everything else like crystal geometries being the same, the second moment is different by the factor

$$(4/9)(\gamma_S/\gamma_I)^2[S(S+1)/I(I+1)]$$

and becomes

$$(S_2)_{IS} = \frac{1}{3} \frac{\gamma_I^2 \gamma_S^2 \hbar^2 S(S+1)}{N} \sum_{k,j} \frac{(1-3\cos^2\theta_{kj})^2}{R_{kj}^6}.$$

This means that even if the two kinds of nuclei are only moderately different in $\gamma$ and spin, second moment is different by about 4/9. The difference in $S_2$ between like and unlike spins arises because the dipole coupling between unlike spins cannot lead to an energy conserving mutual spin flip, namely $(\uparrow\downarrow)\leftrightarrow(\downarrow\uparrow)$. Thus, there is a qualitative difference between the "unlike" spins and, in fact, the presence of the mutual spin flip is a good operating definition of "like" spins. Conversely "different" lines are those separated sufficiently from each other so that the difference is large compared to the dipolar coupling regardless of whether the nuclear species are the same or different.

The usual notation for $S_2$ in units of field is $\langle \Delta H^2 \rangle$ while $S_2$ in terms of frequency is $\langle \Delta\omega^2 \rangle$. Since $\omega$ and H are related by the Larmor theorem, the two second moments are related by $\langle \Delta\omega^2 \rangle = \gamma^2 \langle \Delta H^2 \rangle$. The second moment $\langle \Delta H^2 \rangle$ is a measure of the mean square local dipolar field, i.e.,

$$S_2 = \langle \Delta H^2 \rangle \sim H^2_{loc}/3.$$

The second moment contains line-broadening contributions only from pairwise interactions between spins. The effects of multi-spin correlations, if any, are contained in the higher moments.

The relaxation rates for spins in a molecule or a molecular unit undergoing isotropic motion with motional correlation time $\tau_c$ in the presence of other like spins [see III.C.1. for discussion] can be written in terms of the rigid second moment for the uniformly distributed spins as

$$\frac{1}{T_1} = \frac{2}{3}\langle \Delta\omega^2 \rangle \left[ \frac{\tau_c}{1 + \omega^2\tau_c^2} + \frac{4\tau_c}{1 + 4\omega^2\tau_c^2} \right]$$

and

$$\frac{1}{T_2} = \frac{1}{3}\langle \Delta\omega^2 \rangle \left[ 3\tau_c + \frac{5\tau_c}{1 + \omega^2\tau_c^2} + \frac{2\tau_c}{1 + \omega^2\tau_c^2} \right].$$

In many solids, new degrees of molecular motion become possible as the kinetic energy of the molecule increases. There will be a drop in the second moment as part of the dipolar interaction is averaged out by an onset of a specific motion. [Strictly speaking, it is only the observable part of $S_2$ which decreases due to the motion, because the complete second moment is invarient to such motions. See Section X.V.A. of Abragam (Appendix A) for details.] A typical example is solid benzene (Andrew and Eades, 1953). The Van Vleck second moment expression leads to quite a large second moment when the system is very rigid, i.e., when the internuclear vectors are not averaged over time, and the experimental value is in good agreement with the calculated rigid body second moment below 90K. The second moment drops

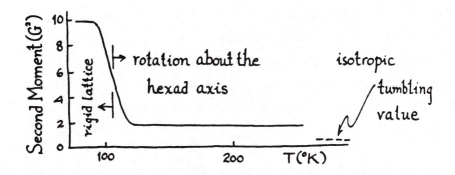

precipitously to a much lower value at temperatures above 100K and we can determine that the rotation about the hexad axis is rapid enough to average the intramolecular dipole interaction to give the reduced second moment. An isotropic tumbling motion of the molecule would lead to an even smaller second moment. There have been many such studies of the second moment to determine the nature of molecular motion in solids.

The process of motional narrowing can be described by three regimes: The rigid lattice regime $(S_2)^{\frac{1}{2}}\tau_c \gg 1$, the motionally narrowed regime $(S_2)^{\frac{1}{2}}\tau_c \ll 1$, and the in-between region where $(S_2)^{\frac{1}{2}}\tau_c \sim 1$. For a particular contribution of the second moment to the linewidth, there is a region over which it drops from a "rigid lattice" value to a motionally narrowed regime as $\tau_c$ gets shorter with increasing temperature. Significant information concerning the nature of molecular motions in solids can be obtained, in principle, by examining this process in detail, but the actual shape of the linewidth and second moment transition is usually not sensitive enough to even yield a reliable value for the activation enthalpy. It has been found (Waugh and Fedin, 1963), in fact, that the temperature of the transition is a much more reliable measure of the activation enthalpy than the shape of the transition. Their semi-empirical criterion is $E_a \sim 37T_o$ where $E_a$ is the activation enthalpy in calories/mole and $T_o$ is the center of the transition in K. This leads to a very useful rule of thumb that a molecule which undergoes a line-narrowing at room temperature is likely to have an activation enthalpy of about 11 kcal/mole = 46 kJ/mole, and this number scales linearly with the transition temperature.

Let us now consider how we might treat an observed NMR line with respect to the question "Is it a single line or an unresolved pair with a small chemical shift between them?" The key to answering this question is the fact that the chemical shift is proportional to the applied field. Therefore, a field dependence study of the second moment should yield the field independent dipole-dipole contribution as the intercept at zero field. The remaining component which is field dependent will represent the chemical shift of the two lines.

Alternatively, this effect can be used to monitor a collapse of incompletely resolved lines into one line due to, for example, thermally activated molecular motion, because the term in the second moment having to do with the relative displacement of the two lines goes away when the two sites become one. This is shown in the figure of the reciprocal rms second moment of $^{19}F$ in

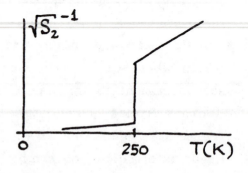

$Na_3UF_8$. The jog at 250 K represents such a collapse (Fukushima and Hecht, 1971).

Now, let us consider the chemical shift anisotropy as it pertains to the second moment. Here we have a familiar line-shape. Since the different parts of this line arise from different nuclei (chemically shifted from others), the  separation between the different parts of the line (and there-fore the width of the pattern) is proportional to the magnetic field just like any other features which are chemically shifted. Therefore, if there is an "inherent" linewidth to each line, i.e., the homogeneous contribution, we can separate it from the inhomogeneous contribution by measuring the line as a function of the magnetic field intensity just like we suggested for seeing if there are two unresolvable homogeneously broad-ened lines. If we take, for simplicity, an axially symmetric case, the second moment turns out to be $S_2 = S_{2d} + kH^2(\Delta\sigma)^2$ where $S_{2d}$ is the field independent homogeneous contribution (usually due to the dipole-dipole interaction), k is a constant, and $\Delta\sigma$ is the shift anisotropy. The canonical procedure (Andrew and Tunstall, 1963; Blinc, et al., 1966) is to plot the total second moment $S_2$ against the square of the field $H^2$ which gives rise to a straight line with intercept $S_{2d}$ and slope $k(\Delta\sigma)^2$.

Further details of how the second moment depends on the structural parameters, how it can be used to separate kinds of molecular motion, e.g., isotropic vs. anisotropic, etc. are not within the domain of this book and we refer you to some of the references (Abragam, Appendix A; Hughes and MacDonald, 1961; Slichter, Appendix A; VanderHart and

Gutowsky, 1968). For more on how to obtain $S_2$ from the FID, see sections IV.B.3., IV.B.4, and VI.D.4.

# REFERENCES

E. R. Andrew and R. G. Eades, "A nuclear magnetic resonance investigation of three solid benzenes," Proc. Roy. Soc. A 218, 537-552 (1953).

E. R. Andrew and D. P. Tunstall, "Anisotropy of chemical shift for some fluorine compounds," Proc. Phys. Soc. (London) 81, 986-995 (1963).

R. Blinc, E. Pirkmajer, J. Slivnik, and I. Zupancic, "Nuclear magnetic resonance and relaxation of hexafluoride molecules in the solid," J. Chem. Phys. 45, 1488-1495 (1966).

M. H. Cohen and F. Reif, "Quadrupole effects in nuclear magnetic resonance studies of solids" in Solid State Physics, vol. 5, edited by F. Seitz and D. Turnfull (Academic Press, New York, 1957), pp. 321-438.

E. Fukushima and H. G. Hecht, "NMR study of $Na_3UF_8$ structure and hyperfine effects," J. Chem. Phys. 54, 4341-4344 (1971).

U. Haeberlen High Resolution NMR in Solids: Selective Averaging (Academic Press, New York, 1976).

D. G. Hughes and D. K. C. MacDonald, "Some properties of resonance lineshape functions," Proc. Phys. Soc. (London) 78, 75-80 (1961).

J. G. Powles and B. Carazza, "An information theory of line shape in nuclear magnetic resonance," in Magnetic Resonance, edited by C. K. Coogan, N. S. Ham, S. N. Stuart, J. R. Pilbrow, and G. V. H. Wilson (Plenum Press, New York, 1970) pp. 133-161.

J. H. Van Vleck, "The dipolar broadening of magnetic resonance lines in crystals," Phys. Rev. 74, 1168-1183 (1948).

D. L. VanderHart and H. S. Gutowsky, "Rigid-lattice NMR moments and line shapes with chemical-shift anisotropy," J. Chem. Phys, 49, 261-271 (1968).

J. S. Waugh and E. I. Fedin, "Determination of hindered rotation barriers in solids," Sov. Phys.-Solid State 4, 1633-1636 (1963). [Translation of Fizika Tverdogo Tela 4, 2233-2237 (1962).]

## IV.B.    DIPOLAR AND ZEEMAN ORDER

We now describe the concept of dipolar and Zeeman order and then discuss some experimental techniques which utilize the concepts. Dipolar order is a very useful concept to know when you are dealing with solids.

### IV.B.1.  THE MEANING OF DIPOLAR AND ZEEMAN ORDER

The concept of Zeeman order has already been discussed in terms of relaxation in III.A. If a sample containing nuclear spins is introduced into an external magnetic field $\overline{H}_0$, no magnetization appears immediately. However, as the sample is allowed to soak in the field, a magnetization parallel to the applied field develops. This happens because the spins, originally equally divided between the two energy states, i.e.,

orientations, in the field redistribute themselves so that a
Boltzmann population difference is set up. This is illustrated
in the figure above. The spin-lattice relaxation time $T_1$ is the

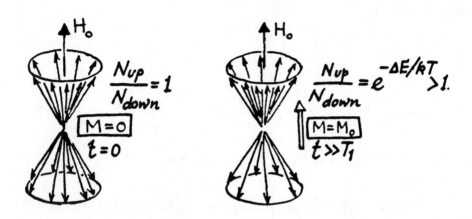

time at which the magnetization M has reached a value which is
$1-1/e$ of the equilibrium magnetization, $M_o$.

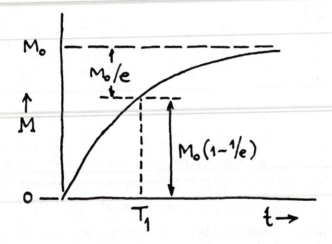

The concept of dipolar order is less obvious. Consider
the following experiment: A sample with a long $T_1$ which has
been polarized in a high field as described in the last para-

graph is slowly removed from the field. When the sample is outside the large field, there can be no macroscopic magnetization (we assume a solid sample in which there is good dipolar coupling and, therefore, a short $T_2$). If the sample is now reintroduced into the magnetic field at the same rate as it was removed and the total duration of the experiment is short compared to $T_1$, almost all of the magnetization will reappear in a time short compared to $T_1$.

An explanation of this phenomenon is the following. The sample was removed adiabatically from the field. An adiabatic and reversible change is isentropic and so the entropy did not change. Therefore, the initial ordering along the applied field was preserved in another form when it was in low field. In fact, in each small region of the sample the magnetization becomes aligned along the local field at that point in the sample. Since the local fields are oriented randomly throughout the solid, there is no net macroscopic magnetization to observe. When the sample is reintroduced to the field isentropically, the ordering again appears as Zeeman order. The spin order when the system consisted of the magnetizations aligned along the local field is called dipolar order.

The magnitude of the original magnetization in equilibrium

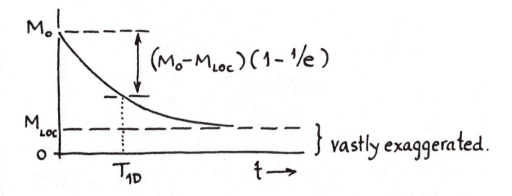

with the applied field is determined by Curie's law to be $M_0 = CH_0/T$ where C is the Curie constant. Upon transformation to the dipolar ordered state, the magnetization aligning along the local fields is much larger than Curie's law would require for equilibrium, i.e., $M_0 \gg M_{loc} = CH_{loc}/T$, because the local field $H_{loc}$ is much smaller than the laboratory field $H_0$. Therefore, the magnetization will decrease while it is in the dipolar ordered state. The time constant for the exponential decay along each of the local field vectors is $T_{1D}$, the dipolar relaxation time.

While $T_1$, the Zeeman relaxation time, is sensitive to motions occurring at the Larmor speed in the large applied field, the dipolar relaxation time is sensitive to fluctuations occurring at the Larmor speed in the local field which is three to five orders of magnitude smaller. The former might be in the range of 5-300 MHz while the latter will be in the few kHz range.

Provided that the system is in a high field $H_0$, and that no rf fields are applied on resonance, the dipolar ordered system and the Zeeman system are isolated from each other, the only connection being through the lattice. The dipolar

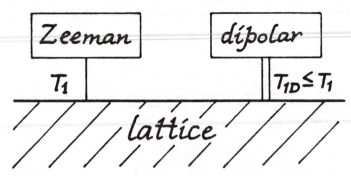

and Zeeman spin systems will achieve a common spin temperature in a time of the order of $T_1$ which is usually comparable to or greater than $T_{1D}$.

There are other ways in which dipolar order can be envisioned. Consider the possible Zeeman energy levels for an N spin system, consisting of all spins parallel to the field, antiparallel to the field, and all combinations in between. If there are no spin-spin interactions, the result would be N equally spaced levels as shown on the left side of the figure below. If, on the other hand, dipole-dipole interactions occur, each of these levels will be split into many fine levels owing to the energies of dipole-dipole interactions as shown on the right side. Ordering among the main levels, giving rise to a net polarization, is Zeeman order. Ordering among the fine levels

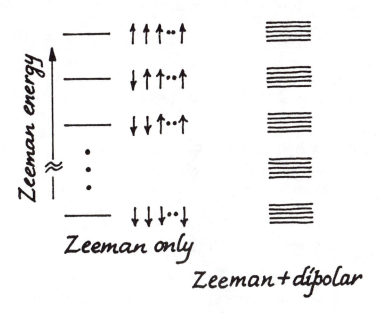

within each major level is dipolar order.

Note that one could have transitions of a nucleus up one Zeeman level and another nucleus down one Zeeman level to maintain the magnetization while altering the dipolar order. Because the dipolar reservoir is usually isolated from the Zeeman reservoir, a system can be prepared so that its Zeeman spin temperature $T_Z$ is different from its dipolar temperature

$T_D$, and, in turn, even different from the lattice temperature $T_L$. In the figure below, a curve superimposed on the energy levels indicates the relative populations of the states.

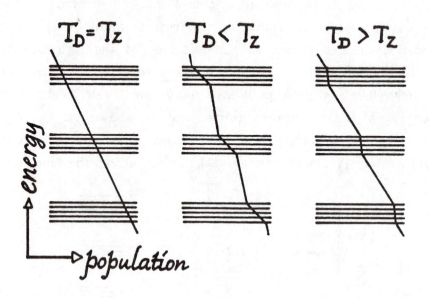

For further insights and examples of dipolar order, see, for example, Anderson and Hartmann, 1962, sections IV.B.2. and IV.B.4. of this book, and references therein.

REFERENCE

A. G. Anderson and S. R. Hartmann, "Nuclear magnetic resonance in the demagnetized state," Phys. Rev. 128, 2023-2041 (1962).

## IV.B.2.  ROTATING FRAME ADIABATIC DEMAGNETIZATION

There are two major means of transforming Zeeman order into dipolar order.  The technique, mentioned in the previous section, of physically removing the sample from the magnet and then replacing it is quite slow and cumbersome to be practical in most cases although it was used in at least one interesting set of experiments (Jones and Daycock, 1967; Jones, et al., 1969).

The two practical techniques are a̲diabatic d̲emagnetization in the r̲otating f̲rame (ADRF) (Anderson and Hartmann, 1962) and the Jeener echo (or dipolar echo) (Jeener and Broekaert, 1967).  The ADRF technique can be performed in several different ways, each of which involve aligning the magnetization with a rotating rf field and then adiabatically reducing the rf field amplitude to zero.  One technique involves using the spin-locking sequence $(\pi/2)_0$-$(\text{long})_{90}$.  The rf field, after the magnetization is locked on it, is adiabatically reduced to zero. The requirement for adiabatic reduction is that

$$(dH_e/dt) \ll \gamma H_e^2,$$

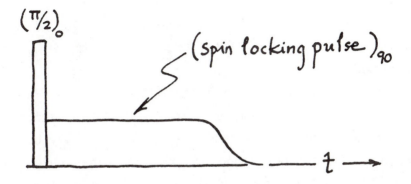

i.e., that the change in the precession frequency about the
effective field is smaller than the precession frequency itself
about the effective field.

Another technique for ADRF is identical to the first
except that an adiabatic fast passage (section I.B.) to the
center of the line is used to spin-lock rather than a phase
manipulated pulse sequence. First the applied field is
changed from a position off resonance to exact resonance. If
done adiabatically, the magnetization follows the effective field
which is being tipped away from the z direction. The result
is the same as that of the spin-locking sequence, and the
magnetization ends up aligned along the rotating rf field.

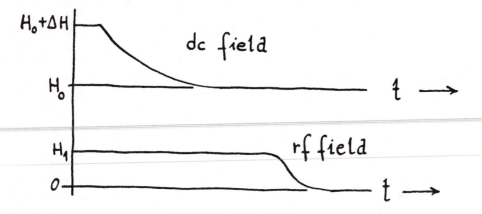

Now the rf field can be reduced adiabatically to zero as before. Any other spin-locking scheme may be used to accomplish the preparation for ADRF. See IV.C.1. for other spin-
locking schemes.

The next technique for ADRF differs from the previous
in the way the rf is turned off. In the first two methods the
rf amplitude is reduced by applying an appropriately changing
dc level to the screen or control grid of a tube in the transmitter. (The actual shape of the bias voltage applied to the
grid must be chosen to reflect both the desired envelope of

the rf and the characteristics of the tube.) In this technique, a sequence of rf pulses which have the same amplitudes but different durations is used. The sequence consists of spin locking followed by a train of pulses (of the same phase) in which the "on" times are progressively reduced. If the off

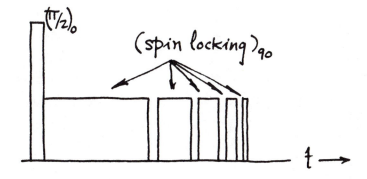

times between the pulses is made small in comparison to $T_2$, the time average effect on the spins of this pulse sequence is equivalent to reducing the rf field adiabatically by reducing its amplitude. This sequence can be implemented with ease when the rf pulses are controlled by a digital pulse programmer.

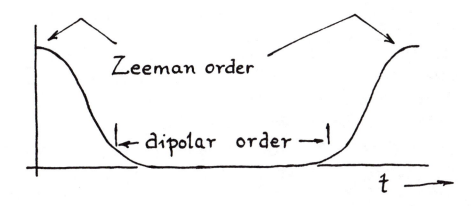

A final method, not involving spin-locking, is simply to reduce the applied magnetic field to zero adiabatically, which is equivalent to removing the sample (Strombotne and Hahn, 1964). This is sketched at the bottom of last page.

Any of the above ADRF methods, in principle, can convert Zeeman order into dipolar order with an efficiency approaching 100%. In contrast, the Jeener echo technique (to be discussed in IV.B.4.) can only achieve approximately 50% conversion.

Once the system has been converted into predominantly dipolar order, we can measure the dipolar relaxation time $T_{1D}$ by monitoring the magnetization as a function of time spent in the demagnetized state. One method of doing this is by sweeping through resonance after ADRF (Anderson and Hartmann, 1962). For this technique to work, it is essential that the system be prepared in a state of dipolar order while in a high dc field. The techniques of reducing the field to zero or removing the sample from the dc field would not be suitable. The lineshape of the absorption spectrum after sweeping through resonance with a wide-line spectrometer appears similar to the derivative of the absorption line measured under conventional conditions. Alternatively, one could remagnetize the system by isentropically turning on an rf field, the result of which would be the magnetization aligned with the rf field. If, after remagnetization, the rf field were abruptly terminated, an FID would appear. In common with the usual pulse methods, this method of observing the magnetization after the system has been reconstituted into Zeeman order, suffers from the disadvantage of having the beginning of the FID obscured by instrumental problems but it has the advantage that the measurement can be performed very quickly. The Jeener echo technique, described a little later, does not suffer from this limitation.


## REFERENCES

A. G. Anderson and S. R. Hartmann, "Nuclear magnetic resonance in the demagnetized state." Phys. Rev. $\underline{128}$, 2023-2041 (1962).

J. Jeener and P. Broekaert, "Nuclear magnetic resonance in solids: Thermodynamic effects of a pair of rf pulses." Phys. Rev. $\underline{157}$, 232-240 (1967).

G. P. Jones and J. T. Daycock, "Subsidiary proton resonances in N.M.R.," Phys. Letters $\underline{24A}$, 302-303 (1967).

G. P. Jones, J. T. Daycock and T. T. Roberts, "A sample moving system for nuclear magnetic resonance adiabatic demagnetization experiments." J. Phys. E: Sci. Instrum. $\underline{2}$, 630-631 (1969).

R. L. Strombotne and E. L. Hahn, "Longitudinal nuclear spin-spin relaxation." Phys. Rev. $\underline{133}$, A1616-A1629 (1964).


## IV.B.3.   THE SOLID ECHO AND
## OTHER TWO PULSE ECHOES

We discussed earlier (Section I.C.2.) that a magnetization which has decayed in the rotating x'-y' plane due to external field inhomogeneity can be refocussed into an echo by an appropriate pulse. Such an echo is called the ordinary spin echo or a Hahn echo. The usual sequence for a Hahn echo is $(\pi/2)_0$-$\tau$-$(\pi)_{90}$-$\tau$-echo where the subscript indicates the relative phase of the carrier. Hahn echoes in this sense cannot be formed in dipolar coupled solids where the local field is not

static.  It is possible to form echoes in solids, however, and
we will describe two of them, the Jeener echo in the following
section and the solid echo in this section.

Unfortunately, the solid echoes do not have nice physical
explanations as does the Hahn echo.  They are extremely useful, nevertheless, because they can overcome the deadtime
problem and, in addition, can yield information which is otherwise difficult or impossible to obtain.  The following discussion
describes the properties and uses of echoes without detailed
explanations.  Original references (Lowe, 1957; Powles and
Mansfield, 1962; Powles and Strange, 1963; Mansfield, 1965;
Warren and Norberg, 1967), reviews (Boden, 1971), and applications (Boden, et al., 1975 a, b; Moskvich, et al., 1975)
should be consulted.

Consider again the Hahn echo sequence $(\pi/2)_0$-$\tau$-$(\pi)_{90}$-
$\tau$-echo.  It should not give rise to any echoes in a dipolar
system of one kind of spins in solids because the second pulse
simply inverts all spins and, consequently, the local fields as
well.  A sequence which does result in an echo, however, is
$(\pi/2)_0$-$\tau$-$(\pi/2)_{90}$-$\tau$-echo,  $\tau < T_2$.  This is the solid echo and,
under the condition $\tau \ll T_2$, the trailing half of the echo is
equal to the FID (Powles and Strange, 1963).  Therefore, the

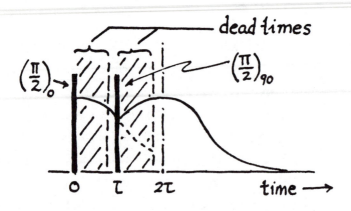

solid echo is a good technique for overcoming the effects of the deadtime in determining the FID shape. When $\tau$ is not small enough for the trailing half of the echo to equal the FID, the echo maximum tends to form too early.

The condition $\tau < T_2$ will be a rigid requirement for solid echo formation in the majority of cases. This is because this pulse sequence refocusses only what is called two-spin correlations and not the effects of many spins dipolar-coupled to each other as is usually the case in a typical solid. Thus, the dependence of echo formation on $\tau$ can yield information on the relative importance of correlations between pairs of spins and groups (larger than two) of spins.

Alternatively, this information can be gathered by the study of second and fourth moments as described in the section on moments. Because the FID can be expanded in powers of time with the even moments as coefficients, as already mentioned in section IV.A.2.,

$$f(t) = M_0(1 - S_2 t^2/2! + S_4 t^4/4! - \ldots)$$

and $S_2$ is determined by the pair correlations and the higher moments by correlations between larger numbers of spins, the importance of the pair correlations will be reflected by how good the fit to the FID can be made with the expansion including only the $t^2$ term.

Because the solid echo can reduce the deadtime problem, it can be used in conjunction with other pulse sequences to detect signals which are otherwise difficult to see. In particular, the solid echo sequence can be used in place of the monitoring pulse in any of the $T_1$ sequences mentioned in Section III.D.2.

In the $(\pi/2)_0 - \tau - (\pi/2)_{90} - \tau$-echo solid echo experiment,

the receiver reference should be set to 0° so that the first
FID should be a maximum at resonance and the second pulse
by itself should not produce any FID.  A typical solid echo
experiment uses pulse spacings between 10 and 100 µs depending on $T_2$ and the deadtime.  A good tuning sample for the
solid echo should have moderately long $T_2$ and short $T_1$ (but
still with $T_1 \gg T_2$) such as Teflon.

Hahn echo sequences work regardless of the relative
phases of the two pulses so that the solid echo sequence
$(\pi/2)_0 - \tau - (\pi/2)_{90}$ can induce Hahn echoes in an isotropic non-
viscous liquid as well as the echoes from two identical $\pi/2$
pulses.  Therefore, two different contributions to an echo can
exist for the solid echo in a sample like a liquid crystal which
has both a liquid and solid character (Cohen-Addad, 1974;
Cohen-Addad, and Vogin, 1974).

Now, in isotropic and non-viscous liquids, the Hahn
echoes from a $\pi/2 - \tau - \pi$ sequence will be about twice as large as
that from a $\pi/2 - \tau - \pi/2$ sequence if there are no instrumental
problems such as $H_1$ inhomogeneity (see II.D.5.).  On the
other hand, solid echoes, in principle, can recover the full
height of the FID.  Therefore, the maximum echo height from
a solid echo sequence in a sample with both liquid and solid
character is sensitive to the ratio of the Hahn and the solid
echo components.  In particular, if it is close to the full
height the solid echo component is dominant whereas if the
height is close to half of the full value the Hahn echo component dominates.

The possibility of echo formation is improved if there is a
second kind of spins present in the solid.  In fact, even the
Hahn echo sequence can produce an echo in the solid in that
case (Warren and Norberg, 1967).  Basically, the $\pi$ pulse
inverts only one kind of nuclei in the presence of the local

fields of the other kind of nuclei so that the effect of the other nuclei are used to rephase the spins of interest. Thus, from the fraction of the magnetization not refocussed in the echo, it is possible to measure the dephasing rate due only to similar nuclei in a multi-nuclear system by this method.

Other pulse sequences leading to echo formation in solids containing two different spin systems are the two with a pair of $\pi/2$ pulses including the solid echo sequence for one kind of spin, i.e., $(\pi/2)_0$-$\tau$-$(\pi/2)_0$-$\tau$-echo and $(\pi/2)_0$-$\tau$- $(\pi/2)_{90}$-$\tau$-echo (Mansfield, 1965). The first of these is sensitive to the dephasing due only to dissimilar nuclei. Thus, by a judicious choice of pulse sequences, is is possible to distinguish the separate contributions to the dephasing in a two-spin system.

## REFERENCES

N. Boden, "Pulsed NMR methods," in Determination of Organic Structures by Physical Methods, edited by F. C. Nachod and J. J. Zuckerman (Academic Press, New York, 1971), volume 4, chapter 2, pp. 51-137.

N. Boden, Y. K. Levine, D. Lightowlers, and R. T. Squires, "NMR dipolar echoes in liquid crystals," Chem. Phys. Letters 31, 511 (1975); "Internal molecular disorder in the nematic and smectic phases of thermotropic liquid crystals studied by NMR SPDE experiments," ibid 34, 63-68 (1975).

J. P. Cohen-Addad, "Effect of the anisotropic chain motion in molten polymers: The solid-like contribution of the nonzero average dipolar coupling to NMR signals. Theoretical description," J. Chem. Phys. 60, 2440-2453 (1974).

J. P. Cohen-Addad and R. Vogin, "Molecular motion anisotropy as reflected by a 'pseudo solid' nuclear spin echo: Observation of chain constraints in molten cis-1,4-polybutadiene," Phys. Rev. Letters 33, 940-943 (1974).

I. J. Lowe, "Double pulse nuclear resonance in solids," Bull. Am. Phys. Soc. 2, 344, (1957).

P. Mansfield, "Multiple-pulse nuclear magnetic resonance transients in solids", Phys. Rev. 137, A961-A974 (1965).

Yu. N. Moskvich, N. A. Sergeev, and G. I. Dotsenko, "Two-
pulse echo in solids containing isolated three-spin systems,"
Phys. Stat. Solidi (a) 30, 409-418 (1975).

J. G. Powles and P. Mansfield, "Double pulse nuclear-reso-
nance transients in Solids", Phys. Letters 2, 58-59 (1962).

J. G. Powles and J. H. Strange, "Zero time resolution nuclear
magnetic resonance transients in solids", Proc. Phys. Soc.
(London) 82, 6-15 (1963).

W. W. Warren, Jr. and R. E. Norberg, "Multiple-pulse nuclear-
magnetic-resonance transients of $Xe^{129}$ and $Xe^{131}$ in solid
xenon", Phys. Rev. 154, 277-286 (1967).

## IV.B.4.  THE JEENER ECHO

The Jeener echo sequence (Jeener and Broekaert, 1967;
Goldman, 1970) is

$$(\pi/2)_0 - \tau_1 - (\theta_1)_{90} - \tau_2 - (\theta_2)_{90} - \text{ECHO}$$

where the echo begins to form immediately following the third
pulse. The sequence and resulting NMR signals are demonstrated in the following figure. The first pulse initiates a
precessing magnetization in the x-y plane. The second pulse,
which is 90° phase shifted from the first is a rotating $H_1$ field
along the magnetization and performs a transformation from a
state of Zeeman order into a state of dipolar order. This
state of dipolar order is identical to that which would result
from an ADRF sequence except that, while the ADRF sequence
can perform the transformation with a theoretical 100% efficien-

cy, the first two pulses have a maximum efficiency, which occurs for $\theta_1=\theta_2=\pi/4$, of 56%. The system then sits in a state of dipolar order and relaxes at a rate dictated by its dipolar spin lattice relaxation time $T_{1D}$. The third pulse transforms the remaining dipolar polarization back into Zeeman order and a signal can be observed.

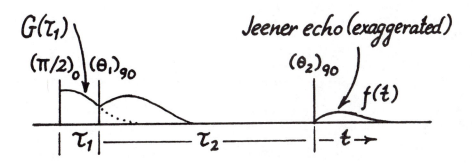

The requirements on the phases of the pulses is some-what less stringent than implied by the sequence above. The second pulse must be in quadrature with the first but the third pulse can have any arbitrary phase because the dipolar order is not referenced by any external directions. To ob-serve the Jeener echo, the reference phase of the receiver must be 90° out of phase with the third pulse. When the detector reference phase is identical with that of the third pulse, a Zeeman signal can be observed which is simply the FID resulting from the magnetization which has relaxed back to the z direction in the time since the first pulse.

Thus, the sequence stated at left and illustrated above is the most convenient one to set up. The reference phase for the receiver should be set to maximize the FID following the first pulse, in this case. In searching for it, it is good to know that the Jeener echo is typically an order of magni-tude smaller than the FID.

When $\tau_1 < T_2$ and $\tau_2 < 2T_{1D}$, the echo shape is given by

$$f(t) = (A/S_2)[dG(\tau_1)/d\tau_1][dG(t)/dt],$$

where $G(t)$ is the (normalized) FID shape, A is $(3/4)\sin(2\theta_1)\cdot \sin(2\theta_2)M_o\exp(-\tau_2/T_{1D})$, $S_2$ is the second moment, and $M_o$ is the initial magnetization. Because the echo is proportional to $dG/d\tau_1$, it is a maximum when the second pulse occurs at the steepest part of the FID. No echo is observed if $\tau_1 > T_2$.

The appearance of a Jeener echo depends upon the existence of a dipolar reservoir so that some dipolar interaction not averaged out by molecular motions is necessary. This does not nessarily restrict the application to rigid solids. For example, liquid crystals, including soaps and lipids in certain phases yield valuable information when studied by the Jeener echo sequence (Bloom, et al., 1977).

The most obvious application of the Jeener echo is to measure $T_{1D}$ by plotting echo amplitude vs. time spent in a state of dipolar order. If the system under study is homogeneously broadened, then the location on the echo where the amplitude is measured, provided it is chosen consistently, is immaterial for the measurement of $T_{1D}$. In an inhomogeneously broadened system, such as a polycrystalline material, the Jeener echo may be a superposition of Jeener echoes with different shapes and different $T_{1D}$'s. If so, it may matter where the echo amplitude is sampled. The most obvious test is to plot relaxation curves from different parts of the echo to determine whether the echo is relaxing uniformly. For a crystalline material, one expects that the dipolar relaxation time $T_{1D}$ will be independent of the crystal orientation in the external field because the state of dipolar order is independent of this external field.

A second application of the Jeener echo is to determine lineshape parameters such as the second moment (Bloom, et al. 1977). As the echo starts with zero magnitude at the third rf pulse and evolves in time from that point, it is less susceptible to problems of receiver deadtime than the FID is. From the equation for the expansion of the FID in terms of its moments

$$dG(\tau_1)/d\tau_1 \cong -2G(0)S_2\tau_1,$$

$$dG(t)/dt \cong -2G(0)S_2t,$$

and

$$f(t) \cong 3\sin(2\theta_1)\sin(2\theta_2)G^2(0)S_2\tau_1t.$$

This method requires evaluation of the initial slope of the Jeener echo which is linear in time. This linearity can be assumed for times shorter than $\tau_1$. Therefore it is less susceptible to errors from the recovery of the system after the third pulse. One must also obtain the initial height of the FID following the first pulse by extrapolating back to time zero. However, the extrapolation to obtain the magnitude of the FID is far less susceptible to errors than the estimate of the FID slope at time zero which is what most other methods require. Note also that this determination requires accurate knowledge of the rotation angles of the last two pulses.

There are two more ways to analyze the Jeener echo. From the relation

$$\int_t^\infty f(t)dt = -2AG(0)\tau_1 G(t), \quad \text{when} \quad \frac{dG(\tau_1)}{d\tau_1} \cong -2G(0)S_2\tau_1 ,$$

we see that plotting $\int_t^\infty f(t)dt$ vs. $\tau_1$ gives a representation of the FID from which the analysis described under the first method can be used.

A more sophisticated analysis indicates that we can evaluate $S_2$ from the Jeener echo without requiring knowledge of A or of G(0). This is because, for a homogeneous system, the shape $f(t)$ of the Jeener echo as opposed to its magnitude, is independent of $\tau_1$ and $\tau_2$. In the linear region

$$f(t) \cong 4AG^2(0)S_2\tau_1 t.$$

Now

$$\int_0^\infty f(t)dt = \frac{A}{S_2} \frac{dG(\tau_1)}{d\tau_1}[G(\infty)-G(0)]$$

$$= 2G(0)A\tau_1 G(0) = 2AG^2(0)\tau_1$$

so that

$$\frac{f(t)}{\int_0^\infty f(t)dt} = 2S_2 t.$$

In an inhomogeneously broadened system, such as the bulk sample of a lamellar phase, there will be a superposition of FID's with different $T_2$'s depending on the angle the lamellae's normal makes with $H_0$. As the effectiveness of the transformation from Zeeman to dipolar order depends upon the slope of the FID at the point of application of the second pulse, at short times one preferentially selects those lamellae with short FID's and at long $\tau_1$'s those lamellae with long decays, namely those with orientations close to the magic angle.

We add a final comment that an alternative method for

measuring $T_{1D}$ was proposed by Stepisnik and Slak (1973). It differs from the Jeener echo in that the dipolar state is prepared by an off-resonance saturation radiation rather than the first two pulses in the Jeener sequence. The application of the saturation method to short relaxation times comparable to the Jeener echo is discussed by Emid, et al. (1980).

## REFERENCES

M. Bloom, E. E. Burnell, S. B. W. Roeder, and M. I. Valic, "Nuclear magnetic resonance lineshapes in lyotropic liquid crystals and related systems," J. Chem. Phys. 66, 3012-3020 (1977).

S. Emid, J. Konijnendijk, and J. Smidt, "Measurement of short dipolar relaxation times by the saturation method," J. Magn. Resonance 37, 509-513 (1980).

M. Goldman,    Spin Temperature and Nuclear Magnetic Resonance in Solids,  (Oxford University Press, U.K 1970), pp. 89-91.

J. Jeener and P. Broekaert, "Nuclear magnetic resonance in solids: Thermodynamic effects of a pair of rf pulses", Phys. Rev. 157, 232-240 (1967).

J. Stepisnik and J. Slak, "Dipolar spin-lattice relaxation measurement by saturation," J. Magn. Resonnce 12, 148-151 (1973).

## IV.C.    ROTATING FRAME RELAXATION

In the following two sections, we briefly consider the concepts leading up to spin-lattice relaxation processes in the rotating frame which are reflected in the relaxation time $T_{1\rho}$.

## IV.C.1.  ROTATING FRAME SPIN-LATTICE RELAXATION
### MEASUREMENTS

We can understand spin-lattice relaxation in the rotating frame by first understanding the experiment to observe this process.  It is done by the following technique or some variation of it.  The sample is placed in a large applied field $\overline{H}_0$

and, in a time of several $T_1$'s, a magnetization $\overline{M}_0$ will appear along $\overline{H}_0$.  Next, a 90° pulse is applied along the rotating x' (refer to section I.B. for a discussion of the rotating frame) direction which puts the magnetization into the x'-y' plane. In the laboratory frame, $\overline{M}_o$ precesses in this plane with the Larmor speed; in the rotating frame (rotating with the Larmor speed) the magnetization is fixed along the rotating y' axis. Instead of turning off the rf, as in a conventional relaxation measurement, the rf is phase shifted 90° in the laboratory frame so that the $H_1$ field is rotated to lie along the magnetiza- tion in the rotating frame. In the laboratory frame the mag- netization and the $H_1$ field rotate together in the x-y plane.

In the rotating reference frame, the magnetization ex- periences only the $H_1$ field and not the $H_0$ field when the rf is applied exactly on resonance. Therefore, the $H_1$ field in the rotating frame now becomes analogous to the $H_0$ field in the laboratory frame and the magnetization is said to be spin- locked to $\overline{H}_1$. The magnetization, just prior to the application of the rf pulse is given by $M_o = CH_0/T_L$, where C is the Curie constant and $T_L$ is the lattice temperature. After the phase shift, the magnetization $\overline{M}_o$ is polarized along a much smaller field $\overline{H}_1$ in the rotating frame and, therefore, it will relax to a much smaller equilibrium value M given by $M = CH_1/T_L$ with an exponential time constant $T_{1\rho}$.

Two things should be noted: First, the relaxation rate in the rotating frame will be most strongly affected by motions occurring at the Larmor frequency in the rotating frame, i.e., $\omega_1 = \gamma H_1$. This makes this technique sensitive to ultra slow motion typically in the kHz region rather than in the greater than MHz region of laboratory frame relaxation measurements. Secondly, in principle, one could try a conventional relaxation measurement in a smaller $H_0$ field, to sample this slow regime of molecular motions, but the sensitivity will be very poor because S/N is proportional to $H_0^n$ where 1<n<2. The $T_{1\rho}$

technique maintains the sensitivity of the high field measurement while probing relaxation in a low field regime.

The measurement is made by turning off the rf field after
a suitable spin-lock period and monitoring the height of the
resulting FID as shown in the figure below.  The initial height
of the signal is recorded as a function of the length of the
spin-locking pulse to extract the relaxation time $T_{1\rho}$ in the
rotating frame.

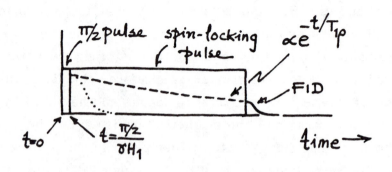

Several other techniques have been developed to prepare
the spin system in the spin-locked condition.  One technique
does not require phase-shifting the rf.  An rf pulse exactly

on resonance is applied along the y' axis while a dc magnetic field pulse of magnitude $H_1$ is applied along the z' axis. The effective field in the rotating frame then has magnitude $\sqrt{2}H_1$ and points 45° from the z' axis in the y'-z' plane. The magnetization precesses about this effective field and the field pulse is turned off when the magnetization has made a 180° rotation and $\overline{M}_o$ and $\overline{H}_1$ are aligned.

Still another technique initially has $\overline{H}_0$ set sufficiently far off resonance so that the effective field lies approximately along the z direction even with $\overline{H}_1$ on. The system is then brought into resonance by changing $H_0$ slowly enough so the magnetization can follow the effective field until the system is on resonance and $\overline{M}$ is aligned along $\overline{H}_1$. This is the adiabatic fast passage introduced in Section I.B. except that the passage is stopped when the resonance condition is reached. We sketched the field variation for this scheme in IV.B.2. For a review of experimental methods for $T_{1\rho}$ measurements, see Ailion (1971) and references therein.

If $H_1$ were zero after the alignment, $T_{1\rho}$ would equal $T_2$. Similarly, if $H_1$ were equal to $H_0$ in magnitude, then $T_{1\rho}$ would be $T_1$. So, $T_2 \lesseqqgtr T_{1\rho} \lesseqqgtr T_1$. Furthermore, even when $T_2 \ll T_{1\rho}$ the magnetization does not disperse in the rotating frame in the time $T_2$ until <u>after</u> the locking field is turned off. As seen from the rotating frame the magnetization is locked along $\overline{H}_1$ and does not disperse for the same reason that it does not disperse when it is aligned along $\overline{H}_0$.

Note that before the 90° pulse is applied, the spin temperature equals the lattice temperature. If the process of spin locking is achieved in a time short in comparison to $T_1$ or $T_{1\rho}$, $M_o$ is preserved in magnitude. If M is now to be described by Curie's Law as $M_o = CH_0/T_L = CH_1/T_S$, $T_S$ must be much lower than $T_L$ because $H_1$ is much smaller than $H_0$. As long as the system can be described by a spin temperature, M can change its magnitude by spin-lattice relaxation processes and $M_0$ will relax to $CH_1/T_L$ or alternatively $T_S$ will warm up to $T_L$ with a time constant $T_{1\rho}$.


## REFERENCE

D. C. Ailion, "NMR and ultraslow motions" in Advances in Magnetic Resonance, vol. 5, edited by J. S. Waugh (Academic Press, New York, 1971), pp. 177-227.


## IV.C.2.   THE STRONG AND WEAK COLLISION REGIMES

The laboratory frame spin-lattice relaxation time $T_1$ is related to the molecular rotation correlation time $\tau_c$ by the expression

$$\frac{1}{T_1} = \frac{2}{3} \langle \Delta\omega^2\rangle \left[ \frac{\tau_c}{1 + \omega^2\tau_c^2} + \frac{4\tau_c}{1 + 4\omega^2\tau_c^2} \right]$$

as discussed in III.C.1. and IV.A.2. This expression predicts a curve of $T_1$ vs. $1/T$ which has a minimum near $\omega\tau_c \sim 1$ and the method is most sensitive to molecular motions in the region of the minimum, i.e., where $\tau_c \sim 1/\omega$. Since the usual applied field is in the tens of kG with the corresponding resonant frequency in the tens to

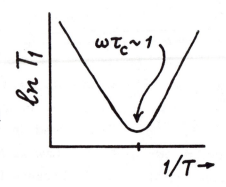

hundreds of MHz, the method is maximally sensitive to motional frequencies roughly in this range. Expressions for the relaxation time of this form are called BPP expressions after their originators, Bloembergen, Purcell, and Pound. Simple variations on this functional form can be used for translational diffusion as well.

In order to study molecular motions effectively in systems in which they occur in the kHz region or below, the obvious thing to do is to lower $H_0$ and thereby the resonance frequency. This method is only useful over a limited range of $H_0$, typically down to a Larmor frequency of a few MHz, because the sensitivity is approximately proportional to $H_0^{3/2}$ so that sensitivity suffers at

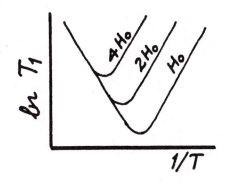

small $H_0$. As noted in Section IV.C.1., the rotating frame relaxation technique was developed to do low field relaxation while preserving the sensitivity of the high field method. However, in calculating the rotating frame relaxation, one cannot simply take over the high field expression by substituting $\omega_1 = \gamma H_1$ for $\omega_0$.

The BPP expression was derived under the assumptions that the dipolar interactions formed a perturbation on the Zeeman levels and that the time dependent part of the dipolar interaction could be treated by time dependent perturbation theory or equivalently the density matrix approach to determine the relaxation expression. This does not rule out a BPP type expression for relaxation in the rotating frame. Specifically, a BPP type approach can be used to derive the following expression for $T_{1\rho}$ due to rotational motion under basically the same assumptions, namely, $H_0 \gg H_1 \gg H_{loc}$, $\omega_0 \tau_c \gg 1$, and $\tau_c \ll T_{2RL}$, where RL stands for rigid lattice and $T_{2RL}$ is the spin-spin relaxation time for the system when the motion being studied is frozen out.   Then

$$\frac{1}{T_{1\rho}} = \frac{3\tau_c \gamma^2 (1-q) H_L^2}{1 + 4\omega_1^2 \tau_c^2} \quad , \quad \omega_1 = \gamma H_1, \quad H_{loc} = S_2/3 \quad .$$

In the expression above, $(1-q)$ is the fraction of the mean square local field which is time dependent and therefore participates in the relaxation process.   When $\omega_0 \tau_c \lesssim 1$, the $T_{1\rho}$ expression is more complicated.   See Look and Lowe (1965) for a derivation.   The restriction $\tau_c \ll T_{2RL}$ where $\tau_c$ is the rotational correlation time is that of the weak collision regime. Specifically, it requires that, on the average, many motions occur before the nuclei on a given molecule relaxes.

The opposite case of strong collision can be derived under certain restrictive assumptions.   When the Zeeman field is large in comparison to the local field, it has the effect of decoupling the Zeeman reservoir from the dipolar (that is, the local field) reservoir.   In the strong collision regime, the applied field ($H_1$ in the rotating frame experiment) is not large with respect to the local field $H_{loc}$ so that the Zeeman and dipolar reservoirs become strongly coupled and achieve a single spin temperature. In the strong collision case each event, in this case a translational jump, relaxes the spins involved and the system returns to a common spin temperature after each such event.   This requires that $\tau_c \gg T_{2RL}$ since $T_{2RL}$ is the approximate time to achieve a spin temperature.   If we assume that these translational jumps are the only relaxational mechanism, then on exact resonance

$$\frac{1}{T_{1\rho}} = \frac{2(1-q)H_L^2}{H_L^2 + H_1^2}\frac{1}{\tau_c}$$

which has no resemblance with the BPP type form of the weak collision case. For rotational motion rather than translational motion, the result is identical except that it is a factor of 2 smaller. In both the strong and the weak collision cases, it is essential to be familiar with the exact calculation of 1-q and the region of applicability of the expressions. For more details, see Ailion (1971) and references therein.

A relaxation time versus temperature plot for cyclohexane, is sketched below. It reveals two minima, one influenced predominantly by molecular translation and the other by molecular rotation. The strong collision regime is applicable below 170° and the weak collision regime above this temperature (Roeder and Douglass, 1970).

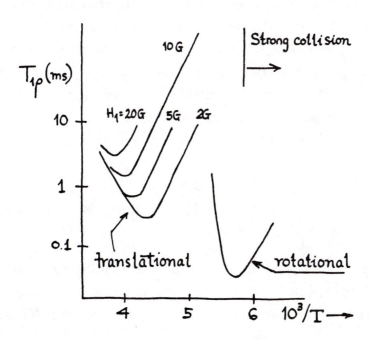

REFERENCE

D. C. Ailion, "NMR and ultraslow motions" in Advances in
Magnetic Resonance, vol. 5, edited by J. S. Waugh (Academic
Press, New York, 1971), pp. 177-227 (1971).

D. C. Look and I. J. Lowe, "Effect of hindered molecular
rotation between unequal potential wells upon nuclear magnetic
resonance spin-lattice relaxation times and second moments,"
J. Chem. Phys. 44, 3437-3441 (1966).

S. B. W. Roeder and D. C. Douglass, "Molecular motions in
several solids studied by nuclear magnetic relaxation in the
rotating frame," J. Chem. Phys. 52, 5525-5530 (1970).

## IV.D.    SPIN TEMPERATURE

Consider a spin-$\frac{1}{2}$ system in a static magnetic field $\overline{H_0}$.
For each spin, we will denote the high and the low energy
states by the labels b and a so that a spin in state b is
antiparallel to the static field and vice versa. The difference
in energy between the two states for one spin is $\Delta E = 2\mu H_0$.
We sketch the spin population associated with the spin states
in the following way where each x represents a spin.

b ————————x————

$\updownarrow \Delta E$     $N_b/N_a = \exp(-\Delta E/kT)$

a ————x—x—x—x—x

If the spin system is in thermal equilibrium with the
surroundings (which we will call the lattice), the ratio of the
populations in the two states will be given by the Boltzmann

$$N_b/N_a = \exp(-\Delta E/kT)$$

relation where T is the temperature of the lattice. Let us say that the sketch above represents a typical case of nuclear spins in a laboratory field at some normal temperature even though $N_b/N_a$ should be very close to one under these conditions. You can check this by remembering that $\Delta E = h\nu$ where $\nu$ is the Larmor frequency which is of the order of $10^8$ Hz and T is around 300K at room temperature.

Suppose that the spin populations are now disturbed so that some additional nuclei in state a are promoted to state b. While the system may no longer be in thermal equilibrium with the lattice, we can still describe the populations by the Boltzmann relation provided that we change the definition of T to keep the relationship correct. Since the ratio $N_b/N_a$ is now greater than it was when the system was in thermal equilibrium with the lattice, the new temperature $T_S$ must be greater than the lattice temperature T. $T_S$ will recover towards T by giving up thermal energy at a rate characterized by $1/T_1$ as discussed in III.A. We define $T_S$ as the <u>spin temperature</u> for the particular spin system and note that a two-level system can always be described by it. The sketch below shows the qualitative change in the populations for the system at a higher $T_S$.

If we have a system consisting of more than two energy levels per spin, there is no guarantee that the spins will be describable by a unique spin temperature. In the sketch below, system A might be characterized by a spin temperature but system B cannot be.

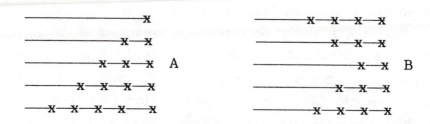

For the remainder of this section, we will only consider two level systems for simplicity.  The generalization to a more complex system, if it has a unique spin temperature, is straightforward.  We now consider several situations in order to understand the connection between the spin populations and $T_S$.  First, suppose that all the spins are in the lower energy state.

Since $N_b/N_a$ is zero, the Boltzmann relation requires that $T_S=0$.  Even though the third law of thermodynamics states that we can never achieve this temperature, we note that when $T_S$ approaches zero the system is approaching its lowest energy configuration.

Next, heat up the spin system so that some of the spins are now in state b, even though most are still in state a.  Now

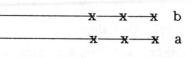

$0 < N_b/N_a < 1$ so that $T_S$ has a finite positive value.

What if we keep putting more energy into the spins?  We will eventually reach the situation shown at right where $N_b/N_a=1$.

In NMR, we say that the spins are saturated and this corresponds to the situation immediately after a $\pi/2$ pulse.  From the Boltzmann relation, $T_S=\infty$ and what's more $T_S=-\infty$ is a valid solution, too. Now this is getting interesting. What happens if we heat up the spins even further?

If we can have an excess of spins in state b, $N_b/N_a > 1$. But, in order for that to be true the exponent in the

—————————x—x—x—x b
—————————————x—x a

Boltzmann equation must be positive or $T_S$ must be negative! So, now it seems that the spins are hotter than they were at $T_S=\pm\infty$ but the temperature is negative.

What is the most energetic arrangement for the spin system? By now it is clear that it is when all the spins are in state b,

x—x—x—x—x—x b
————————————— a

as shown at right. In this case, $T_S=(-0)$ so that the spin temperature has steadily gone from $-\infty$ to $-0$ as we went from saturation to the most energetic configuration. Thus, we can conclude that a negative temperature is hotter than infinity and it can be achieved starting from its positive counterpart by a $\pi$-pulse which inverts the populations.

Another plausibility argument for negative temperatures is to consider mixing equal amounts of two spin systems at $T_S=T$ and $T_S=-T$. The resulting mixture will obviously have a temperature intermediate between the two starting spin systems and you should be able to convince yourself that its temperature will be $\infty$.

Negative temperatures are not known in ordinary life. What distinguishes systems like these from the usual ones to allow the existence of negative temperatures? There was a clue in the last case considered above, namely the one at $T_S=(-0)$, when we implied that the spin system cannot be heated up any further. The existence of negative temperature is contingent on the existence of a maximum energy.

For further discussion, see Landau and Lifshitz (1958).

A spin system in equilibrium with a static magnetic field $\overline{H}_0$ produces a magnetization according to Curie's law $\overline{M}=C\overline{H}_0/T$. If the system can be characterized by a spin temperature $T_S$ which is not necessarily equal to the lattice temperature $T_L$, the Curie's law still holds so that $\overline{M}=C\overline{H}_0/T_S$. All other thermodynamic relations hold, as well, so that spin temperature can be used to calculate the polarizations of interacting spin systems especially if they are relatively isolated from the lattice. Such spin calorimetry can be extremely useful, either in calculations or in giving physical insights where applicable. See, for example, the discussion on cross-polarization in IV.E.3.

We already pointed out that it is much less likely for systems with $I>\frac{1}{2}$ to be characterized by a spin temperature. Even for $I=\frac{1}{2}$ systems, the existence of a uniform spin temperature is not automatic. What is necessary is some mechanism by which a large number of spins can come to thermal (in the spin temperature sense) equilibrium with each other in times short compared with the spin-lattice relaxation time $T_1$. Such a coupling between spins usually exists in solids as dipolar coupling which can be characterized by the spin-spin relaxation time $T_2$. In most solids, $T_2<T_1$ so that such internal equilibration is assured. This, however, does not rule out the possibility of another system in the sample with a different spin temperature. In nonviscous liquids, the condition $T_1\cong T_2$ prevents the existence of a homogeneous spin temperature. For more on spin temperature and how it can be used, see Hebel (1963), Redfield (1969), and Goldman (1970).

The foregoing suggests that as long as the spin system comes to a spin temperature rapidly compared with spin-lattice relaxation rate, the details of the spin-spin interactions will not affect the spin-lattice relaxation process. Thus, the existence of a spin temperature implies that the system is quite straightforward, at least insofar as studying spin-lattice relaxation goes.

## REFERENCES

M. Goldman, Spin Temperature and Nuclear Magnetic Resonance in Solids (Oxford, London, 1970).

L. C. Hebel, Jr., "Spin temperature and Nuclear Relaxation in Solids," in Solid State Physics, vol. 15, edited by F. Seitz and D. Turnbull (Academic Press, New York, 1963).

L. D. Landau and E. M. Lifshitz, Statistical Physics (Addison-Wesley, Reading, Massachusetts, 1958), section 70.

A. G. Redfield, "Nuclear spin thermodynamics in the rotating frame," Science 164, 1015-1023 (1969).

## IV.E.  HIGH RESOLUTION NMR OF SOLIDS

In a normal solid, the NMR line is likely to be quite broad because of dipolar interaction between the nucleus of interest and other magnetic nuclei in its neighborhood. It was recognized long ago that dilution of the neighboring magnetic spins would effectively reduce the dipolar linewidth. Lauterbur (1958) was able to measure the chemical shift rotation pattern for the naturally abundant $^{13}C$ in a single crystal of calcite $(CaCO_3)$, which contained no abundant spins. In the presence of abundant spins, it may be possible to obtain a high resolution (i.e., without dipolar broadening) spectrum of low abundance spins, such as naturally abundant $^{13}C$, by simply decoupling the abundant spins in analogy to solution NMR. In both of these cases, sensitivity would be a problem. Under certain conditions, cross polarization to transfer polarization from other nuclei can help the sensitivity greatly.

Another method for achieving high resolution NMR in some solid samples entails complicated pulse sequences to make the spins think that the interaction is being modulated in such a way that the NMR line is narrowed in analogy to motional narrowing discussed elsewhere (IV.A.2.). A third category of experiments to artificially narrow NMR lines in solids is to physically spin the sample in a specific way to eliminate the broadening. This is known as magic angle spinning (MAS).

With the advent of superconducting magnets, there is now another method for reducing the relative effects of the dipolar broadening and that is to simply use a very strong static magnetic field. Since the chemical shift is proportional to the field while the dipolar shift is independent of the field, it is possible to obtain spectra in which the chemical shift effects are dominant over dipolar broadening for certain nuclei, namely those with relatively large chemical shifts. See, for example, the work by Schweitzer and Spiess (1974) in which they determined the nitrogen-15 chemical shift anisotropy in pyridine.

Some of these diverse methods will be discussed qualitatively in the following few sections wherein several excellent reviews such as those by Haeberlen (1976), Mehring (1976), and Slichter (1978), are referenced. Before going on, we stop to consider one last question: Why do we want to try high resolution in solids?

There are several reasons for performing high resolution NMR in solid samples. A pragmatic reason is that the sample preparation would be easier if the compound of interest need not be dissolved in a solvent to obtain a high resolution NMR spectrum. A corollary is that the S/N could be much better in a solid sample if its solubility is very small in the appropriate solvent. A more fundamental variation of this theme is that the physical and even chemical properties of a molecule will, in general, not be the same between a solid and a dissolved sample. Thus, an uncomfortable extrapolation may be necessary to draw conclusions about molecules in a solid from a high resolution NMR spectrum in solution.

There is an even more fundamental reason for doing NMR in solids and this is related to the obvious difference between a solid and a liquid, that is the lack or the presence of rapid molecular motion. Consider, for example, a benzene molecule which is a planar hexagon. At the corner of the hexagon, where there is a carbon atom, the effects of the chemical bond should have a directional character governed by the shape of the molecule. The covalent bond which holds the molecule together is made up of electrons and the nuclei on the molecule feels the effects of the bonding (and other) electrons in the form of additional small magnetic fields which give rise to the well known chemical shifts. Since the site at the corner of the hexagon is not spherically symmetric, the chemical shift must depend on the orientation of the molecule with the external magnetic field due to the magnet in the laboratory. To put it in mathematical language, we say that the chemical shift is characterized by a tensor rather than a scalar.

In the particular example of benzene, there are three inequivalent principal directions for the tensor at the carbon nucleus. It should be quite clear by inspection that the three directions are defined by 1) the line from the center of the hexagon through the corner, 2) the line through the corner and perpendicular to the plane of the hexagon, and 3) the

line mutually perpendicular to
the first two.   In contrast,
carbon monoxide, for example,
is characterized by only two
principal directions:   one is
along the molecule and the
other is perpendicular to it.

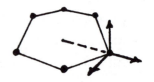

In general, there are six com-
ponents to a chemical shift tensor [see, for example, Chapter
2 in Mehring (1976) referenced in IV.E.3.].   Thus the chemical shift tensor at a site on a molecule can tell you a great
deal about the site's immediate surroundings, possibly quantitative indication of the bonding, in addition to the site symmetry.

In solution NMR, the chemical shift ends up being a
scalar, which means just a number, because of the rapid rotation of the molecule.   For example, as far as an individual
carbon at a corner of a benzene molecule goes, it sees the
external field orient equally probably along the three principal
axes as do all the other carbons so the resulting NMR spectrum consists of just one line, if the effects of protons are
ignored.   Therefore, high resolution solution NMR yields relatively little information compared to what might be available if
we could get the full chemical shift information by some means.


## REFERENCES

P. C. Lauterbur, "Anisotropy of the $C^{13}$ chemical shift in
calcite," Phys. Rev. Letters 1, 343-344 (1958).

D. Schweitzer and H. W. Spiess, "Nitrogen-15 NMR of pyridine in high magnetic fields," J. Magn. Resonance 15, 529-539
(1974).

## IV.E.1.  MAGIC ANGLE SPINNING (MAS)

In liquids, isotropic and rapid tumbling averages out interactions such as the direct dipolar coupling and the chemical shift anisotropy by a phenomenon called motional narrowing. In solids, these interactions are usually not completely averaged out owing either to the absence of molecular tumbling and diffusion, or to their insufficiency for complete motional narrowing.

The dipolar interaction between a pair of nuclei i and j with a separation $R_{ij}$ is of the form

$$(A/R_{ij}^3)(3\cos^2\theta - 1)$$

where $\theta$ is the angle between the internuclear vector $\overline{R}_{ij}$ and the magnetic field $\overline{H}_0$. [In the more sophisticated books, this kind of an interaction is described in terms of a Hamiltonian, but never mind. You can simply think of a term like the one above as a potential energy of the interaction between the two nuclei which depends on the separation $R_{ij}$ and the angle $\theta$.] This kind of an angular dependence is that of a Legendre polynomial of second kind $P_2(\cos\theta)$. It is easy to see that any interaction with this angular dependence vanishes for $\theta = 54.74°$, the magic angle, so that if all internuclear vectors can line up together at this angle, the dipolar interaction vanishes.

There are occasional cases in nature in which the internuclear vectors are already lined up. One example is oriented samples of fatty acids such as oleic acid. These molecules can be made into (oleic acid)/$D_2O$ bilayers and oriented by glass plates. They are already rapidly tumbling about their axes and diffusing parallel to the surface of the plates. The rapid diffusion averages out intermolecular dipolar coupling and the rapid rotation about the longitudinal, i.e., molecular, axes averages out the angular dependence about those axes

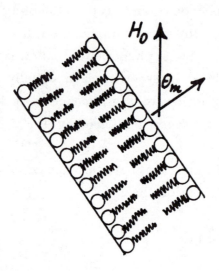

leaving only the θ dependence in the dipolar coupling without any mechanical spinning (Roeder, 1975).  The duration of the FID can be greatly increased by merely orienting the glass plates so that θ is at the magic angle.

However, in most solids, one nuclear magnetic dipole is coupled to a number of neighbors and if the crystal is oriented so that one pair is at the magic angle, the other pairs usually will not be.  How are we going to get rid of the dipolar interaction between all pairs at once?  It is clear that we must resort to some clever manipulation and one is the so-called magic angle spinning (MAS).

Consider a nucleus fixed in a lattice.  Suppose that we rotate the sample about an axis which does not go through the nucleus.  Then the nucleus executes a circular motion about the axis of rotation and if the motion is fast enough (by some criterion which we leave until later), the nucleus can be considered to be at the center of the circle, that is, on the axis.  By the same token, when a pair of nuclei with an internuclear vector not parallel to the rotation axis is rotated rapidly, the average internuclear vector will now coincide with the rotation axis. Thus, the rotation, in effect, makes all internuclear vectors parallel to each other and to the rotation axis. When the rotation axis is chosen to be at the magic angle, all interactions of the form $3\cos^2\theta-1$ will vanish and this is called magic angle spinning.

This technique has been used successfully to reduce the dipolar interactions in a solid but it is only successful when

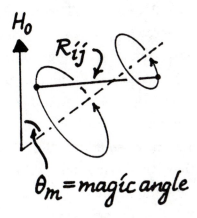

$$\theta_m = magic\ angle$$

the rate of rotation is rapid in comparison to the linewidth (in Hz) that you are attempting to reduce (Lowe, 1959; Andrew, 1971).

In macromolecules the molecule may already be tumbling in solution but not sufficiently rapidly to average out all dipolar interactions. Under these conditions, spinning at a few kHz may not significantly contribute to further averaging.

Under general spinning, the expansion for the FID as a function of time,

$$f(t)=1-S_2 t^2/2!+S_4 t^4/4!-\ldots$$

becomes

$$f(\theta,\tau)=1-(S_2\tau^2/2!)(3\cos^2\theta-1)^2/4$$

$$+(S_4\tau^4/4!)(3\cos^2\theta-1)^4/16-\dots$$

The observed increase in the duration of the FID for $\theta$ close to the magic angle is consistent with this expression, in which the effective time scale of the experiment has been increased by the relation $\tau=t/P_2(\cos\theta)$. Theoretically it is possible to achieve infinite narrowing when $P_2(\cos\theta) \propto 3\cos^2\theta-1 = 0$.

The early MAS line narrowing experiments were all cw NMR. Lowe (1959) was the first to use pulsed NMR with MAS. Kessemeier and Norberg (1967) found that in the absence of MAS the FID observed was Gaussian (consistent with having a Gaussian lineshape in a solid as the Fourier transform of a Gaussian is another Gaussian) but with MAS the FID became an exponential (yielding the Lorentzian lineshape commonly observed in liquids). In studying the relaxation properties of solids undergoing MAS, they found that both $T_1$ and $T_2$ depend on the spinning speed, generally increasing as the speed increased. This is not surprising considering that $T_1$ and $T_2$ are influenced by dipolar interactions.


## REFERENCES

E. R. Andrew, "The narrowing of NMR spectra of solids by high speed specimen rotation and the resolution of chemical shift and spin multiplet structure for solids," Prog. NMR Spectroscopy **8**, 1-39 (1971).

H. Kessemeier and R. E. Norberg, "Pulsed nuclear magnetic resonance in rotating solids," Phys. Rev. **155**, 321-337 (1967).

I. J. Lowe, "Free induction decays of rotating solids," Phys. Rev. Letters $\underline{2}$, 285-287 (1959).

S. B. W. Roeder, unpublished results (1975).

## IV.E.2.  HIGH POWER DECOUPLING

A common situation is exemplified by the usual organic solids in which there are protons in profusion and carbon-13 only in natural abundance of about 1%. For the NMR spectra of $^{13}C$ in such a sample, most of the broadening comes from the heteronuclear dipolar interaction between the $^{13}C$ and the protons and very little from the homonuclear dipolar interaction between the carbons. Thus, if it were possible to eliminate the heteronuclear dipolar interaction, you could have a high resolution $^{13}C$ spectrum even though its S/N may be quite poor because of the small number of $^{13}C$.

The obvious way to eliminate the heteronuclear contribution to the carbon linewidth is to decouple the protons in analogy with the liquid experiments. This, in fact, is possible to do but it is quite difficult because of the large magnitude of the decoupling magnetic field required in the typical solids. A crude estimate of the decoupling field intensity can be made by our noticing that the "flip" rate of the protons must be comparable to the dipolar linewidth of the carbons in order for the decoupling to be effective. Since the protons "flip" at the rate of 4.3 kHz/G and since a typical solid dipolar linewidth is a few kHz, it requires a decoupling field of the order of a gauss to do the job. Considering that a one gauss field can rotate a proton 90 degrees in 60 µs, you can see that this is a fairly robust field, especially for a de-coupling field which has to stay on for a long time compared to a transmitter pulse. Another limitation of high power de-coupling as being described here is that it does not work for homonuclear decoupling. You can use it to decouple abundant protons from the rare $^{13}C$ but not protons from protons, for example.

So, in this section and in the last, we have discussed two line narrowing methods which are fairly difficult to im-plement in typical solids where the lines are quite broad to begin with. It turns out that a combination of both tech-niques, i.e., MAS and high power decoupling, can work quite well in reducing the linewidth because each technique reduces the linewidth enough to make it easier for the other technique to work. The sensitivity problem, due to the fact that the nucleus of interest is rare, still remains and will be addressed in the following section.

## IV.E.3  CROSS POLARIZATION (CP) EXPERIMENTS

We now want to consider observing the high resolution NMR spectrum of a dilute and/or insensitive nucleus (usually $^{13}C$) in a solid with enhanced sensitivity by taking advantage of the dipolar reservoir of an abundant nucleus in the same system. These experiments are an outgrowth of two lines of development. One of these is the attempt to suppress dipolar coupling in a solid sample. The second resulted from the experiments of Hartmann and Hahn (1962) in which the detec-tion of a dilute or insensitive spin was made possible through a double resonance experiment.

The basic experiment, stated in terms of observing naturally abundant carbon-13 in the presence of an abundant proton spin system in the solid is as follows. The protons are
initially prepared in a state with its spin temperature $T_P$
(IV.D.) very low. The small carbon system, initially at the
lattice temperature $T_L$, is made to contact (thermally, in a
spin temperature sense) the large proton reservoir in order to
cool it to a temperature close to the initial proton temperature
$T_P \ll T_L$. Assuming the thermal connection between the protons
and the carbons, when made, is much better than the connections from the protons or the carbons to the lattice, the carbon
temperature decreases much more than the protons' increase
because the "specific heat" of the proton reservoir is much
larger than that of the carbons.

Recall that from the definition of spin temperature a low
temperature implies large polarization of the magnetic moments.
Thus, low carbon spin temperature is what is desirable just
before inducing an FID. Because in the basic cross polarization experiment as described you can gain S/N according to
the ratio $T_P/T_L$, the improvement depends on how well the
protons are cooled initially. Some of the methods commonly
used to do this will be described later in this section.

In order to get a high resolution carbon-13 spectrum, we
also have to get rid of the heteronuclear dipolar interaction
between the carbon and the neighboring protons. (The homonuclear contribution is already small because the carbon-13
nuclei are dilute.) This will be done by decoupling the
protons, that is, by applying rf power at the proton Larmor
frequency so that the protons will be "stirred" up and the
net dipolar field at the carbon-13 from each proton will be
very small.

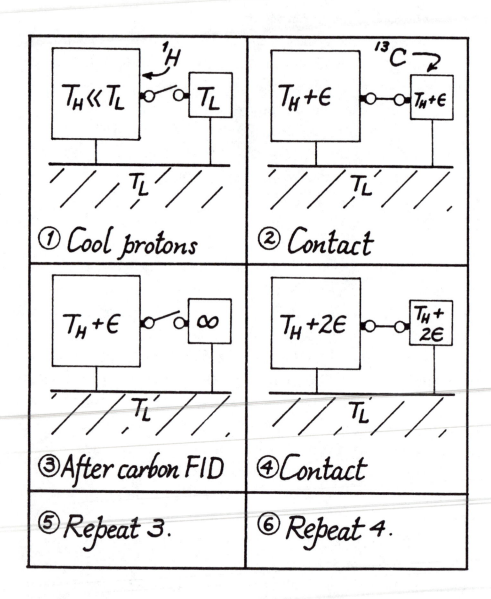

What has been described so far is a single contact cross polarization experiment. Although we have not even said how the carbon FID will be induced, when it does occur it leaves the carbon spin temperature very high. Ordinary spin-lattice relaxation process (via the line joining the carbon reservoir to the lattice) has been assumed to be quite slow so that it would take a long time, say seconds, for the carbon temperature to come down even to $T_L$ by this process.  A much quicker way of cooling the carbons would be to make the thermal contact with the still cold protons again, i.e., repeat the cross-polarization experiment.  In fact, this cycle can be repeated over and over until the proton spin temperature rises sufficiently so that it is not enough cooler than the carbon temperature for further cross-polarization to do any good. This is the multiple contact cross-polarization experiment and, in favorable cases, it is possible to realize a S/N gain of two orders of magnitude by co-adding the carbon FID's after each contact for approximately as many times as the ratio of the abundant to the dilute spin, which is on the order of 100 for naturally abundant carbon in an organic solid.

The cross-polarization process, which utilizes the polarization of the abundant spins to enhance the S/N of the rare spins, can be combined with decoupling and MAS, discussed earlier, to yield high resolution NMR spectra of rare spins in the presence of abundant spins.  We now discuss some of the specific techniques used in these experiments.

### How to cool the protons

There are several ways in which the abundant spins can be prepared in a state of low spin temperature.  Qualitatively, the spin temperature is low when there is a large polarization in a small field.  A simple way to accomplish this is to spin-lock, i.e., to rotate the initial magnetization so that it is along the $H_1$ field in the rotating frame.  The different ways to spin-lock were discussed in IV.B.2. and IV.C.1.

One is to apply a $\pi/2$ pulse to the protons and immediately follow it with an rf field which is phase shifted by $\pi/2$ with respect to the $\pi/2$ pulse.  Thus, in the rotating frame,

the field for the first pulse, say along the x'-axis, will rotate
the magnetization from the z'-axis to the y'-axis and then an
rf field is rotated to be along the y'-axis to spin-lock the
magnetization. This same result can be accomplished by using
adiabatic fast passage (I.B.) until the magnetization has been
rotated by $\pi/2$ and then stopping the sweep.

### How to make thermal contact: The Hartmann-Hahn condition

For nuclear spins, what does thermal contact mean?
From the definition of spin temperature, we know that the
temperature is correlated with the polarization of the spin
system for a fixed field strength. Therefore, thermal contact
implies the ability to propagate through the sample a specific
polarization of the spins. We know from Chapter I that the
energy required to flip a magnetic moment $\bar{\mu}$ in a field $\bar{H}$ is
$2\mu H$ and such a spin rotated against the field needs to get rid
of that potential energy in order to return to the favored low
energy state. Thus, in order for the spins to be in thermal
contact, one spin must be prepared to accept a specific quantum of energy ($2\mu H$ in our example) to make a transition from,
for example, spin up to spin down, while an adjacent spin
makes the opposite transition. The requirements for such a
mutual spin flip are that the quantity $2\mu H$ be the same for the
two nuclei in question and that the two nuclei be close enough
so that the energy can actually be exchanged to cause the
mutual flip.

You will recognize that the condition for thermal contact
as stated above is identical to that for the existence of a spin
temperature already discussed in section IV.D. For identical
spins, all that is required, then, is that the field be uniform
enough so that the energy associated with the flip of a nucleus
be the same for any adjacent nuclei. How do you make thermal

contact between two different nuclear spin systems, for example, carbons and protons? Obviously, thermal contact is impossible for systems of different $\mu$'s if the field H is the same for both systems so something clever is required. Slichter, in Chapter 7 of his book (Appendix A), described the solution to this problem as a magic trick performed by the Wizard (E. L. Hahn) and the Sorcerer's Apprentice S. R. Hartmann (1962). Slichter himself, together with Lurie (1964), also made a major contribution in this area.

From the discussion above, it is clear that we need different field intensities for the two different nuclei so that the Larmor speeds are the same for the different nuclei. Thus, for the carbon-proton system, we need the fields for the carbons and the protons $H_{1c}$ and $H_{1p}$ to satisfy the condition $\gamma_c H_{1c} = \gamma_p H_{1p}$ where $\gamma_c$ and $\gamma_p$ are the gyromagnetic ratios of carbon and proton, respectively. The only way (thought of so far) to accomplish this is to do it in the rotating frame. For two different nuclei, we can define two different rotating frames with freedom to choose the magnitide of $H_1$ in each frame. Thus, the carbon rotating frame would rotate at approximately one fourth of the angular velocity as that for protons. If now the two $H_1$'s are adjusted so that $\gamma_c H_{1c} = \gamma_p H_{1p}$, the two magnetizations will each precess around their own $H_1$ at identical angular velocities so that each nucleus will see an oscillating z component of the magnetic field at its precession frequency due to a dissimilar neighbor. The crucial point is that the oscillating field seen at one nucleus due to another kind of nucleus has just the correct frequency to cause the first nucleus to flip (or more scientifically, undergo a transition) and vice versa. Thus, it is possible to induce mutual spin flips of dissimilar nuclei in this way and the necessary condition $\gamma_c H_{1c} = \gamma_p H_{1p}$ is the Hartmann-Hahn condition.

famous Hartmann-Hahn condition.

## CP experiment: Implementation

We now briefly describe the actual sequence of the rf excitations used for CP experiments as performed initially by Pines, et al. (1972, 1973). In the sketch below, we indicate the amplitudes of the rf pulses and the proton and carbon magnetizations during such a sequence.

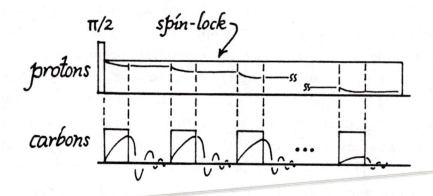

Note that after spin-locking the protons, the proton transmitter is simply left on. During the times that the carbon transmitter is on, the two rf fields satisfy the Hartmann-Hahn condition and the two different kinds of spin reservoirs are in thermal contact as described earlier. When the carbon transmitter is off, the carbon FID's are being recorded and the proton transmitter must be left on so that it can perform as a decoupler. [The amplitude of the spin-locking "pulse" is reduced compared to the initial $\pi/2$ pulse in order to keep the transmitter power level at a reasonable value.]

For the details and more explanations of this very important experiment, the readers are referred to several reviews (Mehring, 1976; Slichter, Appendix A) and references therein as well as the detailed article by Pines, et al. (1973).

We should also be aware that the dipolar interaction is not the only one used to transfer polarization from one spin system to another. Recently (as of 1981) there have been successful experiments using the electron coupled spin-spin interaction (the J-coupling) for performing CP in liquids in order to enhance the signal of small γ nuclei in the presence of large γ nuclei. See, for example, the aricle by Chingas, et al. (1980), Barum and Ernst (1980), and references therein. One particular form of such a signal enhancement scheme is known in INEPT (Morris and Freeman, 1979).

We end this section with a brief description of the apparatus required for CP experiments. Since we need to irradiate two different kinds of spins, a necessary item is a second transmitter. The transmitter system, starting from the signal sources must be very stable not only in frequency but in amplitude in order to maintain the Hartmann-Hahn condition. The easiest way to design the probe is to have one coil for carbon to both transmit and receive, i.e., to make it up like a single coil probe, and have a second coil specifically for the proton irradiation. The pulse generation should not be a problem with any modern pulse generator for NMR but data acquisition may be if the $T_1$'s are short. Basically, you do not want to spend a lot of time transferring and accumulating data between the FID's because the protons may warm up through its $T_1$ interaction with the lattice.

# REFERENCES

D. P. Barum and R. R. Ernst, "Net polarization transfer via a J-ordered state for signal enhancement of low-sensitivity nuclei," J. Magn. Resonance 39, 163-168 (1980).

G. C. Chingas, A. N. Garroway, W. B. Moniz, and R. D.

Bertrand, "Adiabatic J cross-polarization in liquids for signal enhancement in NMR," J. Am. Chem. Soc. 102, 2526-2528 (1980).

S. R. Hartmann and E. L. Hahn, "Nuclear double resonance in the rotating frame," Phys. Rev. 128, 2042-2053 (1962).

F. M. Lurie and C. P. Slichter, "Spin temperature in nuclear double resonance," Phys. Rev. 133, A1108-1122 (1964).

M. Mehring, High Resolution NMR Spectroscopy in Solids, vol. 11 of NMR: Basic Principles and Progress, edited by P. Diehl, E. Fluck, and R. Kosfeld (Springer-Verlag, Berlin, 1976).

G. A. Morris and R. Freeman, "Enhancement of nuclear magnetic resonance signals by polarization transfer," J. Am. Chem. Soc. 101, 760-762 (1979).

A. Pines, M. G. Gibby, and J. S. Waugh, "Proton-enhanced nuclear induction spectroscopy. A method for high resolution NMR of dilute spins in solids," J. Chem. Phys. 56, 1776-1777 (1972).

A. Pines, M. G. Gibby, and J. S. Waugh, "Proton-enhanced NMR of dilute spins in solids," J. Chem. Phys. 59, 569-590 (1973).

## IV.E.4.  MULTIPLE PULSE LINE-NARROWING IN SOLIDS

A hint that manipulating the rf excitation can produce suppression of dipolar interaction in solids in analogy with MAS came in an experiment by Lee and Goldburg (1965) although the possibility had been recognized 10 years earlier by Redfield (1955) as an "amusing" observation. At the same time, it had long been recognized that multiple pulse sequences such as the CPMG discussed in I.C.2., can refocus magneti-

zations to effectively narrow liquid lines. Mansfield and Waugh
with their respective coworkers (Mansfield and Ware, 1966;
Ostroff and Waugh, 1966) applyied analogous sequences to
solids in order to narrow the dipolar broadened lines. These
groups, together with that associated with Vaughan, have
created sophisticated multiple pulse sequences which in suppress homonuclear dipolar broadening in solids. We will give
an elementary description of how these experiments work and
leave the details to the comprehensive reviews (Mansfield,
1971; Ellett, et al., 1971; Vaughan, 1974; Haeberlen, 1976;
Mehring, 1976; Slichter, 1978).

In a nutshell, the function of the pulse sequence is to
rotate the nuclear spins by large amounts in such a way that
over a cycle (consisting of 4 to as many as 52 pulses at present) the average dipolar interaction vanishes. The reason it
is possible to accomplish this is that the sign of the dipolar
field from a magnetic moment depends on the position around
the dipole. Such an averaging of the dipolar interaction is
called spin-flip narrowing by Slichter (Appendix A) and we
now paraphrase his simplified description.

Consider a magnetic mo-
ment $\overline{M}$ at the origin of a cartesian coordinate frame. If $\overline{r}$
is the vector between the ori-
gin and a point at which the
field is to be sampled, the
dipolar field at $\overline{r}$ is

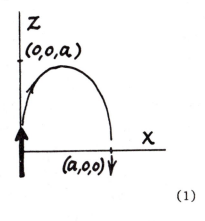

$$H(r) = \frac{3(\overline{M}\cdot\overline{r})\overline{r}}{r^5} - \frac{\overline{M}}{r^3}. \tag{1}$$

In particular, if $\overline{M}$ lies along the z axis, the field at distance
"a" along the x, y, and z axes are

$$\overline{H}(a,0,0) = -\overline{M}/a^3; \quad \overline{H}(0,a,0) = -\overline{M}/a^3;$$

and $\overline{H}(0,0,a) = 2\overline{M}/a^3.$ (2)

Now, if we put another identical moment $\overline{M}'$ at, say, $(0,0,a)$ on the z-axis we can compute the magnetic dipolar interaction between the two moments as a function of the orientations of the two moments. We now consider the dipolar interaction between two parallel spins at $(0,0,0)$ and $(0,0,a)$ when the spins are successively oriented along the three principal axes.

First consider the case where both spins are pointing along the x axis. The interaction energy between $\overline{M}$ and $\overline{M}'$ is $E=-\overline{M}\cdot\overline{H}$ where $\overline{H}$ is the field due to $\overline{M}'$ which is $-\overline{M}'/a^3$ from eq. (2). By symmetry, the interaction is the same when the spins are along the x or the y axis and is equal to $E=MM'/a^3$. In contrast, when the spins are pointing along the z axis, the interaction energy is $E=-2MM'/a^3$. Thus, it is possible to average the dipolar interaction to zero by suitably rotating these spins between the three orientations considered here. One obvious scheme is to orient the spins an equal amount of time along each of the three principal axes which can be accomplished with appropriately phase-shifted $\pi/2$ pulses.

The two most commonly used multiple pulse sequences for line narrowing in homonuclear solids are the four pulse cycle called the WAHUHA or WHH-4 sequence proposed by Waugh, Huber, and Haeberlen and the eight pulse cycle called the MREV-8 sequence proposed by Manfield and independently by Rhim, Elleman, and Vaughan. Only for your curiosity, we state that the WAHUHA sequence is a series of 4 repeating $\pi/2$ pulses with phases and intervals $(270\text{-}2\tau\text{-}90\text{-}\tau\text{-}0\text{-}2\tau\text{-}180\text{-}\tau)_n$ and the signal is sampled during one of the long $(2\tau)$ windows.

These sequences are described in detail by Haeberlen (1976) and by Mehring (1976). An early review of the necessary instrumentation was written by Ellett, et al. (1971).

# REFERENCES

J. D. Ellett, Jr., M. G. Gibby, U. Haeberlen, L. M. Huber, M. Mehring, A. Pines, and J. S. Waugh, "Spectrometers for multiple pulse NMR," Adv. Magn. Resonance 5, 117-176 (1971).

U. Haeberlen, High Resolution NMR in Solids: Selective Averaging (Academic Press, New York, 1976).

M. Lee and W. I. Goldburg, "Nuclear-magnetic-resonance line narrowing by a rotating magnetic field," Phys. Rev. 140A, 1261-1271 (1965).

P. Mansfield, "Pulsed NMR in solids," Prog. NMR Spectroscopy 8, 41-101 (1971).

P. Mansfield and D. Ware, "Nuclear resonance line narrowing in solids by repeated short pulse r.f. irradiation," Phys. Letters 22, 133-135 (1966).

M. Mehring, High Resolution NMR Spectroscopy in Solids, vol. 11 of NMR: Basic Principles and Progress, edited by P. Diehl, E. Fluck, and R. Kosfeld (Springer-Verlag, Berlin, 1976).

E. D. Ostroff and J. S. Waugh, "Multiple spin echoes and spin locking in solids," Phys. Rev. Letters 16, 1097 (1966).

A. G. Redfield, "Nuclear magnetic resonance saturation and rotary saturation in solids," Phys. Rev. 98, 1787-1809 (1955).

R. W. Vaughan, "Application of nuclear magnetic resonance to solids: High resolution techniques," in Annual Review of Materials Science, vol. 4, edited by R. A. Huggins, R. E. Bube, and R. W. Roberts (Annual Reviews, Inc., Palo Also, 1974), pp. 21-42.

CHAPTER V

NMR HARDWARE

## V.A.  BASIC SPECTROMETER CONSIDERATIONS

In the following sections, we consider in fairly general terms, the actual hardware which make up NMR spectrometers. Details of components are deferred to V.C.

## V.A.1.  PULSE NMR SPECTROMETERS

Let us consider an analogy. In electronics courses, it is a common laboratory exercise to provide each student with a closed box with terminals. His job is to determine what is in the box (without opening it) by sending in signals and seeing what comes out. For example, he might hook a frequency

Eiichi Fukushima and Stephen B. W. Roeder, Experimental Pulse NMR: A Nuts and Bolts Approach

generator to the input and examine the output while the frequency is changed from dc to the maximum available. The purpose of this test is to learn what is called the "transfer function" of the device. As an alternative, one could put in a transient signal and see what the response is at the output. In these ways we can deduce something about what is in the box. In NMR, the box corresponds to the sample and the spectrum is the transfer function. The response to a pulse input, i.e., the FID, contains the same information as the response to a cw input. In NMR we obtain a non-zero transfer function only when the sample is exposed to a magnetic field $H_0$ and the angular velocity of the applied rf field $H_1$ satisfies $w_0 = \gamma H_0$. It so happens that the corresponding frequency $v_0 = w_0/2\pi$ is in the rf region. The signal generator corresponds to a radio transmitter, and as the output from the sample is always small, the output stage consists of preamplifiers, amplifiers, and a detector; in other words, a radio receiver. A block diagram of the major components for an NMR apparatus looks very much like one used for a radio system.

We can learn more from this analogy between radio systems and NMR spectrometers. In radio, the desired information in the audio frequency range is combined with the carrier wave which is in the rf range. This process of combining the lower frequency information with the carrier is called modulation. The modulated carrier wave is sent out over the air and received by a receiver tuned to the carrier frequency. After amplification, the carrier wave is removed from the total signal by a process called detection thereby leaving the desired signal (called the audio signal because you can usually hear it -- it should be pointed out that it is possible to hear NMR signals). It is then fed to speakers for final processing (by the brain). In NMR, the carrier frequency is the Larmor frequency. The modulation (the information desired) is the free induction decay (FID) envelope for pulse NMR or the signal amplitude as the resonance is traversed for cw NMR.

The carrier frequency is not at all important, in principle, for radio communication, i.e., the same message can be sent on another carrier just as well (except for propagation properties and such things). In NMR, the carrier frequency is important in a special way. The carrier is modulated by the sample ONLY near the Larmor frequency and it is this modulation which is the desired effect both in pulse and cw NMR. In cw NMR, the amplitude of modulation is measured as a function of the magnetic field (or the carrier frequency) while in pulse NMR, the equivalent quantity is measured in the time domain (the FID). Since the exciting energy must be coupled to the sample and the resulting signal must be extracted from the sample, some additional paraphernalia are needed (the probe and associated networks).

Transmitter

The transmitter consists of a frequency source and amplifiers to boost the rf signals to the desired level. For pulse NMR, the irradiation must be modulated into pulses. The simplest scheme is to have an oscillator which works at the correct voltage level and frequency which can be turned on and off. Such a scheme might be illustrated as in the figure below. Gated oscillator transmitters like this have been used for a long time in acoustic attenuation and velocity measurements as well as in NMR. Their main advantages are their simplicity and attendant low cost. Their main disadvantage is that the phases of different pulses are random with respect to a continuously running oscillator at the same fre-

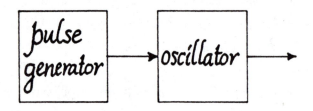

quency because the rf cycle starts the same way each time and the phase of the pulse relative to this oscillator depends on when it starts. (This is why an oscillator like this is called an incoherent transmitter or a gated oscillator.) This is a disadvantage if phase sensitive detection or any other manipulation in the rotating frame is desired.

The more common scheme in pulse NMR is to form the pulses at low power by gating a continuous rf signal on and

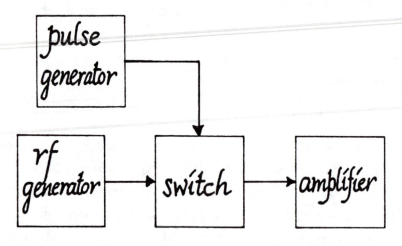

off with a switch and then feeding them through amplifiers to attain the desirable level. This can be sketched as shown above and is called a frequency coherent transmitter.

The power required for typical pulse NMR experiments is between 100 and 1000 watts and design and construction of amplifiers for this kind of rf power is not common, though not difficult, outside of a few NMR people and many more amateur radio operators (or ham gear manufacturers). Therefore, it is tempting to try to use amateur radio apparatus for NMR and this has been done innumerable times with success. The main problem with using ham transmitters and amplifiers for NMR is that the pulses cannot be turned on and off (mostly off) quickly enough because the hams do not have a critical requirement for this parameter. In fact, the opposite is true in the sense that, if a radio signal is turned on and off too quickly, key clicks are generated and they are annoying, at least, and sometimes just make it impossible to listen to the message. So, at least minor modifications are usually needed to convert ham gear to NMR gear but this still remains an economical way of putting together a pulse NMR transmitter (usually below 30 MHz but also convenient near 52, 144, and 228 MHz).

The "switch" shown in the last figure usually consists not of a single switch, as illustrated, but a distributed switch in several stages including some sections acting on the amplifier so that it becomes a gated amplifier. A good switch (meaning large impedance when opened as well as good conductance when closed) is needed because a substantial rf leakage through the switch to the sample between pulses can compete with the relaxation process and can also attenuate the signal by saturating it. This can show up in the form of inaccurate $T_1$ or NOE. The effect depends on the rate of tran-

sition caused by the leakage compared with the natural relaxation rate.  Clearly, the effect of the leakage is more important for small relaxation rates so that the need to have a good
switch is greater for samples with long relaxation times.  It
has also been noticed that accurate NOE values from gated
decoupling experiments are possible only when the transmitter
on-off ratio is sufficiently high.

Typical on-off ratios quoted in the literature for transmitters are in the neighborhood of 80-120 dB (a voltage
attenuation factor of four to six orders of magnitude; see
VI.E.3.) which is large enough that it is not easy to measure.
This on-off ratio is usually achieved with a combination of a
standard switch (built, for example, out of commercial double
balanced mixers) with 70-80 dB attenuation and additional
gating of the amplifier(s).

Receiver

In complete analogy with radio receivers, NMR receivers
consist of amplifier stages to make the rf signals large enough,
followed by a detector which removes the rf carrier so that
only the signal envelope, which is the desired information,
remains.  Usually, the detected signal (often called the audio
signal which is a direct carry-over from radio language) is
further amplified in order that the signal will be the right
size for a digitizer.  A functional diagram of a full receiver
will, therefore, look like the sketch below.

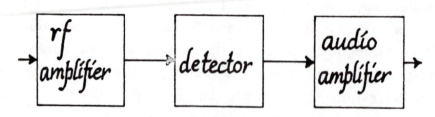

Because NMR is a very insensitive experiment in terms of being able to detect numbers of atoms (due to the small size of nuclear magnetic moments) compared with many other spectroscopic techniques such as electron paramagnetic resonance, the prime job of the receiver is to amplify and detect the signal without introducing distortion or additional electrical noise (the static in the radio analogy). A comprehensive review of receiver design has been written by Hoult (1978).

Radios are designed to produce the clearest signal in part by having amplifier stages amplify only those frequencies which contain the signal from the desired station and suppressing all other frequencies. Aside from the obvious necessity for cutting out all other stations so that the desired message is isolated, this has the effect of removing all the noise with frequencies outside the immediate region of interest. The noise is generated by all kinds of things like the transmitter and the receiver, the NMR coils (which are analogous to the radio antennas), the sample, and external sources such as electrical noise in the laboratory due to flickering lights or even lightning. Often, the noise has frequency components that are not bunched up at certain frequencies (although there are some systematic sources of noise with characteristic frequencies) so that it is impossible to discriminate against it except by making a window in the receiver so that only the desired signal and the corresponding part of the noise spectrum will pass through the window. It is obvious, then, that the window should be set as narrow as the signal but no narrower; more noise comes through a wide window used to admit a broad NMR line (or a lot of narrow lines which are spread out in frequency) than a narrow window for a narrow line. This is the only reason, in principle, that narrow lines are easier to see than broad lines in NMR.

This process of setting a window to limit the range of frequencies admitted to the receiving system is called narrow-banding and can be accomplished in several ways. For fixed frequency operation, it is easy to construct a narrow-band amplifier with a window centered at the desired carrier frequency. This is frequently done, even in commercial spectrometers, and is sufficient if either no other nuclei need to be examined at that field intensity or if no frequency dependent parameters need to be measured. On the other hand, a variable frequency operation can be implemented in several ways. The first is to make (or buy) as many fixed frequency units as needed. This is a simple solution if there is no need for a continuous frequency range and if only a few discrete frequencies are adequate for one's needs. Another is to make the spectrometer tunable, but to keep it a narrow band device. This means that each transmitter and receiver section has to be made tunable, and it is a fairly complicated operation. (The third way is to make it tunable but by a technique known as heterodyning discussed a bit later.)

There is another way to make a straightforward (as opposed to more complicated schemes discussed below) receiver sensitive to different frequencies and that is a so-called broadband receiver. Now a truly broadband receiver suffers from the S/N ratio problem because it admits the noise over a needlessly broad frequency range as mentioned before. Therefore, the usual broadband receiver in NMR is a combination of a broadband amplifier and a filter which cuts out the noise outside of the frequency range of interest. The only qualitative difference between the two kinds of receivers discussed so far is that in the latter there is only one section which limits the frequency range while in the former all stages have more or less the same window characteristic.

The advantage of the broadband scheme is that it is much easier to use. One needs to assemble enough broadband amplifier stages for sufficient gain and then get one band-limiting stage which is usually a filter.

Broadband amplifier stages have much lower gain per stage than the same stage designed to have a narrow pass band because a device like a transistor or a tube can have either lots of gain or bandwidth but not both. Therefore, in the old days (before transistors) broadband amplifiers were uneconomical because many stages were needed for adequate gain and tubes were relatively big, power consuming, and expensive. With the advent of solid state devices like transistors and integrated circuits, however, the need to reduce the number of amplifier stages vanished and broadband amplifiers are now very common.

There is yet another way to construct receivers so that they have a narrow window through which the signal is received. This is the way superheterodyne receivers work, and simply put, involves shifting the frequencies to a more convenient one before further processing takes place. There are several advantages to this scheme. One is that it may be easier to filter (band limit) the signal if it is shifted to a lower frequency so that the window width (to be determined by the filter) is a larger fraction of the total frequency. A well-known example of this is the lock-in detector which is universally used in broadline cw NMR. Another advantage of heterodyning arises from the fact that signals at any carrier frequency can be heterodyned to the same fixed frequency so that variable frequency operation can be achieved with the main amplifier tuned to a fixed frequency called the IF frequency for intermediate frequency. [IF is an only partially appropriate name because it implies a frequency intermediate between the carrier and the detected signal whereas many modern electronics use an IF which is higher than the carrier. An example of such a scheme is given a little later.]

The accompanying figure shows the basic building blocks of a simple heterodyne spectrometer system.  (See V.C.1. for an example.)

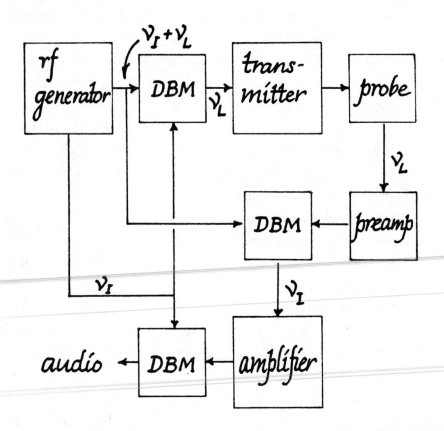

Each of the mixers (called phase sensitive detectors, mixers, or double balanced mixers abbreviated as DBM in the figure) will form the sum and difference of the input frequencies.  The IF amplifier is tuned to $\nu_I$ and all other frequencies will be discriminated against.  The result is that the signal being amplified by the IF amplifier is always $\nu_I$ regardless of the Larmor (or the carrier) frequency $\nu_L$. Therefore, $\nu_L$ may be changed without ceremony at least as far as the receiver is concerned except for some minor adjustments such as of the phase of IF coming through the amplifier with respect to its phase at the source.

The third advantage of the heterodyne system is the possibility of performing phase shifting at the IF rather than at the carrier frequency. This is because the phase of the heterodyned signal is just the difference of the phases of the two original frequencies so that when the IF phase is changed by a certain amount, so is the phase of the resultant. Therefore, we can set up a phase shift network, say, with 0°, 90°, 180°, and 270° phase shifted outputs at the intermediate frequency once and for all and use them at any Larmor frequency provided that we adjust the overall phase at each frequency.

There are variations and embellishments possible to this example of a simple heterodyne system. Perhaps the most significant of these is due to Santini and Grutzner (1976)

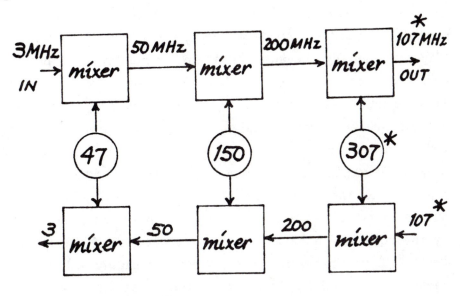

the essence of whose scheme is illustrated in the accompanying sketch. It is a triple conversion system with the first IF chosen to be above the carrier. In addition, the first IF is derived by the use of a mixing frequency which is the sum of the carrier and the IF, rather than the difference as has been more traditional.

Let us consider a concrete example. In their system, tritium has a resonance at 107 MHz. In order to get to an IF of 200 MHz, a local oscillator frequency of 307 MHz is used. The main reason for doing this is that the mixing frequency at 307 MHz is higher than any frequency of interest so that it, or its harmonics, cannot possibly interfere with any part of the desired signal.

The first IF is chosen to be higher than the carrier (in a process called up-conversion) for a similar reason, namely to get it away from other signals and noise which might inter-fere with it. Besides, in an up-conversion scheme we need not worry about resonances which might occur very close to the IF as we do in a down-conversion scheme.

Diode detectors of the sort used in radios [see, for example, The Radio Amateur's Handbook] are still used in NMR occasionally because of their simplicity and they are good detectors so long as their limitations are recognized. Their main disadvantage is that they are not linear (their outputs are not proportional to their inputs) under some con-ditions depending on the mode of operation and S/N because they are amplitude detectors. A detector which is only sensitive to the signal amplitude cannot cope with a signal buried in noise because the output is then mainly determined by the noise which has a non-zero average amplitude (Stejskal, 1963). This means that even signal averaging will not help because the random character of the noise will not tend to cancel, a negative excursion being indistinguishable from a positive one.  These detectors <u>do</u> work well for signals larger than noise.  It is still a good idea to calibrate the detector to make sure what the output means in terms of the input.

Phase sensitive detectors (PSD), on the other hand, are linear over a wide range of signal size.  Its use does require a reference signal but this is always available for the common NMR setups wherein a free running rf source is used together with gated amplifiers.  (The exception is the use of a high level oscillator like the one mentioned at the beginning of the discussion on transmitters.)  Small high quality broad band PSD's are available commercially at reasonable prices so that they should be utilized whenever possible.

There is a special situation in which a diode detector works much better than a PSD.  Suppose we want to measure a $T_1$ for a liquid on an NMR machine which is usually used for solids with broad lines.  Typically, such an apparatus has a field regulated magnet with moderately poor short term field stability so that an FID with a long $T_2$ is not reproducible. One solution is to sample the FID near its beginning before the field has had a chance to change.  If the sample has a poor S/N, however, you can use diode detection which makes the FID relatively insensitive to small field fluctuations because it gives you the FID envelope without the beats.  Then it is possible to integrate over most of the long FID for a significant gain in S/N.

## Matching Networks

Now we consider the network surrounding the NMR probe and the probe itself, at least as far as their electrical properties are concerned.  The actual construction of the probe will be discussed in V.C.4. and V.C.5. while the

matching network is discussed in V.C.7. and its operation in
V.A.4., as well.  The basic function of these networks is to
deliver the transmitter power to the sample and then to pick
up the NMR signal from the sample and couple it to the receiver.  This sounds not too difficult until one realizes that
the NMR signal is in the microvolt range while the transmitter
signal is hundreds of volts or more.  The difficulty stems
directly from the fact that the smaller signal is the wanted
one and the larger one is the unwanted one.  Therefore, an
essential property of these networks, while performing the
basic functions described above, is to decouple the receiver

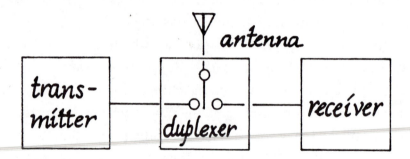

at the receiver.  For this reason, the circuitry for these networks is sometimes called a duplexer, another carry-over from
communications language; a duplexer in an amateur radio
transceiver is a circuit which switches the antenna between
the transmit and the receive modes.

One method of separating the transmitting and the receiving functions is to use two different coils, one for transmitting
and one for receiving.  A fair amount of rf isolation between
the two coils can be achieved by mechanically aligning the two
coils to be orthogonal to each other.  This scheme is the
cross-coil probe.  Its main advantages are that the separate coils may be easily optimized independently, so that the

transmitter coil has fast recovery and the receiver coil
yields the maximum S/N, and
the quasi-Helmholz configuration
of the transmitter coil provides
good $H_1$ homogeneity. Its
disadvantages are 1) that probe
construction is complicated
because of the presence of two
coils with the necessary mechanical alignment; 2) the probe

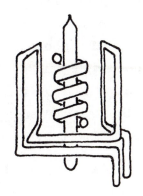

is quite susceptible to mechanical vibrations and to temperature variations since any change in alignment of the two coils
will show up as a change in the signal; and 3) it wastes
transmitter power since the transmitter coil is larger than the
receiver coil for purely geometric reasons (the two coils cannot
occupy the same space and the receiver coil needs to be the
smaller one in order that the filling factor is large for maximum
S/N) and the intensity of the magnetic field in a coil decreases
with increasing volume of the coil for the same amount of input
energy.

So this leads to the other scheme of using the same coil
for both transmitting and receiving, i.e., the single-coil duplexing scheme. There are several different ways in which
these duplexing systems can be designed. We will make a
distinction between those single coil setups using some kind
of a circuit necessitating a dummy load to cancel the effect of
the NMR tank circuit in the absence of an NMR signal and
those single coil circuits which do not require a dummy load.
Bridge circuits of one type or another are in the former
class. One problem with most bridges is the need to construct
them in such a way that their sensitivity to mechanical vibra-

tion (microphonics) and temperature variation is minimized.
[This is analogous to the susceptibility of the cross-coil
probes to microphonics and temperature variation compared
with the single-coil probes as discussed above.]     Another
disadvantage is that the dummy circuit acts as an additional
load to the transmitter or as an additional source of noise to
the receiver or both.

REFERENCES

D. I. Hoult, "The NMR receiver: A description and analysis
of design," Prog. in NMR Spectroscopy 12, 41-77 (1978).

R. E. Santini and J. B. Grutzner, "A broadband system for
the observation of NMR spectra of any resonant nucleus," J.
Magn. Resonance 22, 155-160 (1976).

E. O. Stejskal, "Use of an analog-to-digital converter in
pulsed nuclear resonance," Rev. Sci. Instrum. 34, 971-975
(1963).

V.A.2.   MAGNETS

Magnets are essential to most magnetic resonance experiments.   (The exceptions include NQR and zero field NMR
such as in ferromagnetic samples but an auxiliary field is
quite often desirable even then.)    There are four kinds of
magnets to consider for use in magnetic resonance:   air-core
electromagnet, permanent magnet, iron-core electromagnet,
and superconducting magnet.

    The air-core magnet is used seldom in magnetic resonance

except to supply field modulation, auxiliary perturbing field in NQR, or other low field applications. An application which is expanding is in zeugmatography wherein air-core magnets are used to get a very large sample space (for example for a human subject).

Permanent magnets are popular in fixed field applications where moderate field strengths are required, hence they are often used in the less expensive analytical spectrometers. No utilities are required but the temperature of the room must stay constant for good long-term stability. Short-term stability is excellent.

Electromagnets (meaning iron-core) still comprise the majority of magnets for research NMR although the superconducting and permanent magnets are making inroads at the high and low field ends, respectively. Regulation of electric current for field stabilization is used in virtually all electromagnets for high resolution work requiring good field stability. Because the magnetic field intensity depends on the permeability (which is temperature dependent) of the iron in the magnet for a given electrical current, good thermal stability is required for good field stability as with permanent magnets. This is accomplished by using a well-thermostated room with well-regulated cooling water. It may also help to insulate the magnet by enclosing it within a covered frame.

Because it is difficult to achieve the full field stability necessary for typical high resolution work, additional coils are usually used to "lock" the field by a feedback system using an NMR signal. Such a lock system is usually not very versatile so that typically only a few fixed fields (quite often only one) are available for locking. NMR locks work by tracking a resonance and controlling the current in the power supply to maintain the resonance condition. As such, they are excellent at preventing slow drifts but a rapid transient of the field may cause the system to lose track of the resonance and become unlocked. Therefore, NMR locks are usually used in conjunction with a flux stabilizer (often called a "superstabilizer"). Any change in the magnetic field induces a voltage in the coils wound around the magnet pole pieces which causes a feedback circuit to send a current to correction coils which compensate for the change in field. As the induced voltage is proportional to dH/dt, this circuit acts like a low pass filter to damp out rapid fluctuations in the field. The flux stabilizer does not interact with the magnet's power supply at all. When used in conjunction with an NMR lock, it serves to provide a stable field for both the short and long term.

The alternative regulation scheme (to current regulation) is field regulation. The magnetic field is sensed by a non-NMR probe (usually a Hall-effect probe) and the power supply adjusts the current accordingly. This scheme should work very well, in principle, being independent of the permeability variation with temperature and it should be possible to make a magnet system for high resolution NMR this way, too. Unfortunately, the commercial field-regulated magnets familiar to the authors suffer from excessive field ripple (at 120 Hz) so they are only being used in lower resolution NMR (such as much of the conventional work in solids) and in EPR at this writing. The great advantage of field-regulated magnets over current regulated magnets lies in their easy and reproducible field change (usually by simply dialing the field) and the enormous field sweep range (up to 100% of the field). The long term stability of field regulated magnets can be improved by an NMR lock in a manner described by Wing, et al. (1970).

Electromagnets rely on electric current flowing through

coils to produce the field. For all but superconducting coils, a great deal of heat is generated in this process so that, in addition to supplying the electricity, the laboratory environment must get rid of this heat. Most iron-core electromagnets are water cooled while the power supplies are either water or air cooled. A typical 15 inch (38 cm) diameter polepiece electromagnet with a solid state power supply dissipates 24 kW of heat, mostly into the cooling water, so we are concerned with a nontrivial problem. In addition, the cost of the electricity to generate this much heat (with the magnetic field as a by-product) is not small. At the present prices in the U.S., the electricity cost to continually power such a magnet will equal the purchase price in about three years!

The science and technology of superconductivity will continue to develop over the foreseeable future, but even the superconducting (SC) magnets commercially available today (in 1981) offer an unbeatable combination of stability and field strength at costs which are not out of reason and, as we shall see, at a very much lower operating cost compared to the traditional electromagnets.

A superconducting magnet is either persistent or non-persistent. The former is always (so far) wound with mono-filament wire having virtually no resistance at the junctions so that the current can run around the coil without any driving force (of the power supply) once it is there. The latter is usually wound with a multifilament wire which is capable of carrying larger currents (and thus generating higher magnetic fields) but has a greater electrical resistance due to the difficulties in making good joints. An NMR magnet made with a multifilament wire requires an extremely stable power supply for high resolution work. It is the only way available at present to do NMR at fields greater than about 70 kG ($\equiv$7T).

Persistent mode magnets now typically can have decay rates of $10^{-8}$/hour or better which make them ideal for high resolution NMR (even without locking). In the usual mode of operation, the coil at liquid helium ($\ell$-He) temperature is slowly energized by a power supply. The persistent mode is achieved with the use of an internal SC switch to complete the loop whereupon the power supply can be disconnected (and even hauled away). Part of the electrical leads going into the main coil (and SC shim coils for high uniformity of the field) can usually be removed to reduce the $\ell$-He consumption by reducing the heat flow into the dewar. The $\ell$-He boil-off rate of a modern persistent mode magnet can be so small that it is now a common procedure to keep the magnet perpetually energized. With a $\ell$-He capacity well in excess of $50\ell$ and a boil-off rate of a fraction of a liter per day, it is easy to achieve a holding time of two months or more. One of the authors uses such a superconducting magnet for which, at the present costs for the magnet and the coolant, the time required for the coolant expense to equal the purchase price is 20 to 50 years in contrast to the large electromagnet considered above for which the period was about 3 years.

An important parameter for the operation of a SC magnet is $\ell$-He holding time which is related to the $\ell$-He rate of consumption and the dewar volume. A very low boil-off rate will be to no avail if the holding time is fairly short because a nontrivial amount of $\ell$-He is lost during each transfer. In addition, a large dewar is more efficient than a small one. We would consider transferring less than $50\ell$ of $\ell$-He to be uneconomical and, therefore, suggest a greater than $50\ell$ capacity for the dewar housing the solenoid. In such a system, about $50\ell$ of helium will be needed to fill a nearly empty magnet, say, once every eight weeks. Since $\ell$-He is a fairly expensive coolant, additional steps can be taken to minimize the operational cost. For example, the solenoid dewar may be designed in such a way that the $\ell$-He consumption is decreased at the expense of $\ell$-$N_2$ consumption (through a judicious choice of the radiation shield temperature, for example). Such a design is very sensible since $\ell$-$N_2$ is much cheaper and easier to handle than $\ell$-He.

Standard NMR SC magnets usually achieve moderately good homogeneities (say 1 part in $10^6$ over a large sample volume) with SC correction (shim) coils which do not have to be adjusted very often (and also are persistent). Higher homogeneities can be achieved with ambient temperature correction coils. A useful addition for some SC magnets is an SC sweep coil wound coaxially with the main solenoid which can be used for searching resonances, fast passage, straight cw experiments, etc. Typically, such a coil can sweep 800G in one second with about 10 amp current while the main coil is persistent. Of course, the $\ell$-He consumption will increase during such a mode of operation.

In all of these magnets, a general rule is that the larger the sample space, the smaller the maximum field strength for a given cost. So one should not order magnets with available space larger than necessary. Bear in mind, though, that many features like sideways coils and magic angle spinning can be implemented easier if there is more space.

So, in summary, we have considered several kinds of magnets with emphasis on the iron-core electromagnet and the persistent mode superconducting solenoid. In our view the latter is superior to the former in almost every way except possibly in the accessibility of the working field region and in the ability to change fields at which the field-regulated iron-core magnet is without peer.

REFERENCE

W. H. Wing, E. R. Carlson, and R. J. Blume, "NMR stabilization of a Hall effect regulated electromagnet," Rev. Sci. Instrum. 41, 1303-1304 (1970).

## V.A.3.  TRANSIENT DATA ACQUISITION SYSTEM

In order to signal average a transient signal or to Fourier transform it, it is necessary to digitize the data. [There are some analog methods as well but there are very few, if any, reasons for someone starting out in this field to choose analog methods, now.] The digitizing process can be divided into fast and slow domains. In high resolution NMR the maximum digitizing rate used is tens of kHz. It turns out that this is a convenient limit since there is a practical limit to the number of points into which the digitized FID can be stored. If an FID lasts for 2 seconds and it is digitized at 10 kHz, the total number of points recorded with single precision is 20,000. Any combination of appreciably longer FID or higher digitizing rate would easily increase the record length to an awkwardly large size, say, 32 K or more.

Analog to digital converters (ADC) which perform at tens of kHz are very common and quite inexpensive even with very high resolution, say, greater than 12 binary bits. Indeed, there are many commercial digitizers on the market which will digitize and store an incoming signal at these rates so that it can be transferred to a computer on command. At these rates, though, it may be easy enough to perform the accumulation in the computer itself without having to use a

separate digitizer. For this criterion, the cutoff is roughly
40 kHz acquisition rate, i.e., anything slower can be accumulated in an average minicomputer. The problem arises when
you try to signal average in double precision. Then the
maximum rate which can be handled will be cut down by about
a factor of two.

The problem is much more acute, however, for digitizing
FID's of solid samples (without artificial, or other kinds of,
narrowing) and the reason is two fold. First, fast ADC's are
very much harder to make than slower ones. At the time of
writing, there are new commercial modular units which convert
analog signals into 8-bit digital words at 10 MHz at moderate
(but not cheap) cost and the field is rapidly advancing.

The second difficulty is more serious. What do we do
with the digitized data streaming out of that ADC at 10 MHz
or more? The usual memories, be they core or semiconductor,
can be accessed at only 3 or 4 MHz. Therefore, one must be
clever in the design of the memory system into which the
freshly digitized data may be written. The simplest way to
accomplish this (short of inventing very fast memories and/or
paying for them) is to use n blocks of memory so that each
block stores every n-th point, although it is not necessary to
have actually separate blocks of memory. We have used a
scheme with a 6 MHz 8-bit ADC together with a 1K × 16 random access buffer memory having an access rate of 3 MHz.
Every sequencial pair of incoming 8-bit words is made into one
16 bit word and then written into the memory at half the
original incoming rate. When the digitized data are transferred from the buffer memory into the computer, they are
unravelled into the original 8-bit words before being accumulated. Therefore, the buffer memory thinks that a data block
for an FID consists of 1K (1024) 16-bit words coming in at 3

MHz whereas the actual FID is made up of 2K (2048) 8-bit words at 6 MHz. This scheme can be extended to higher rates in an obvious way.

There are now commercial digitizers which work well in excess of 20 MHz at 8-bits. It is no longer worthwhile even thinking about designing your own digitizer for this reason.

Just what is the highest digitizing rate we might need for solid state NMR? At first sight, it might appear that it should be much lower than the 10 and 20 MHz numbers that we have been talking about. After all, the published solid FT spectra rarely exceed 100 kHz in width so that it should be possible to do all kinds of good work with, say, a 500 kHz or a 1 MHz digitizer. The statement is true enough as it stands but there are a couple of reasons why a higher digitizing rate is useful.

The first is that it is possible to get very fast echoes, for example, for some quadrupolar nuclei or metallic samples exhibiting large inhomogeneous broadening. In fact, many examples are known where the echo width is determined by the transmitter pulse length, i.e., the pulse irradiates only a small part of the broad line, and the frequency or the field strength must be swept to trace out the lineshape. In such a case, a 1 µs transmitter pulse would give rise to an echo with approximately the same width and the echo might be defined by only one or two points with a digitizer running at 1 MHz.

The second reason is that a high digitizing rate permits you to move the Nyquist frequency sufficiently far away from the spectral features. If you make the Nyquist frequency just barely larger than the frequency range of interest and put in a low pass filter to coincide with the Nyquist frequency, two problems develop because the filter is not ideal. First, there is noise folded back into the spectrum from that part of the frequency range which the filter is not quite up to filtering. If you try to avoid this by setting the corner frequency of the filter closer to the spectrum, you may distort the spectrum. With a faster digitizer, you can move the Nyquist frequency and the corner frequency of the filter farther out so that noise will not be folded back into the spectrum nor will there be any distortion from the filter. See the related discussion in VI.C.2. as well as the third sketch therein.

Let us consider this problem in the time domain. Suppose we do an experiment in which the digitizer runs at 10 kHz per channel which means that the second point comes 100 µs after the first. If the filter is sitting right on the Nyquist frequency (10 kHz), there will be a substantial distortion of the second point because the filter being set at 10 kHz means that its characteristic time constant is 100 µs. Depending on the filter, the second point may be off by as much as 33% leading to a distorted baseline in the frequency domain. Even worse, if the filter time constant is, indeed, equal to the delay for the second point, the FID distortion will usually be much worse than we have been discussing because the initial rise in the FID has frequency components much higher than the Nyquist frequency. Therefore, you should set the digitizing rate (and the corresponding filter cutoff) high enough to record the early part of the FID faithfully.

For a similar reason, it is very important that the bandwidth of the digitizer input circuit be much greater than the digitizing rate. You can always narrow the bandwidth by putting an additional analog filter in the input and digitally filtering the recorded data but, if the digitizer has insufficient bandwidth to begin with, there is nothing you can do to rectify the situation.

The overriding considerations in trying to choose the digitizing rate and the filter cutoff are the following. First, you do not want to add any noise to the spectrum of interest by folding them in from beyond the Nyquist frequency. Second, you do not want to introduce any distortion of the FID especially near the beginning. Such distortions can arise from either the analog filter in the input circuit or anything else which might limit the high frequency end of the signal such as an imperfect amplifier at the front end of the digitizer.

We feel strongly enough on this that we are going to belabor the point some more. Usually the frequency response of the input amplifiers is given by the 3 dB points of its characteristic response curve. If a 10 MHz digitizer has a front end with a bandwidth of 5 MHz (which is the highest frequency allowed by the Nyquist criterion; see Section II.A.3.), that is not good enough to see even a 5 MHz signal accurately since its amplitude will be off by 3 dB, a non-trivial amount. Therefore, the bandwidth must be well in excess of the desired spectral width, a condition not always fulfilled by commercial digitizers, if the signal is not to be distorted.

Another important specification, especially for solid state NMR, is the overload recovery of the receiving system including the front end of the ADC or digitizer. This means looking for very small baseline hiccoughs on poor S/N FID's which have to be accumulated a large number of times. You want the baseline artifact (remember, we are now talking about the baseline in the time domain) to be small compared to the weakest FID you might be interested in. Since the average manufacturer does not specify this point in a satisfactory

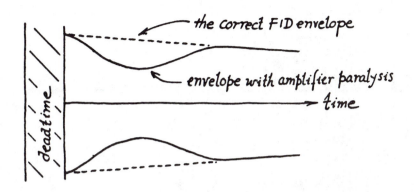

way, the best thing for you to do is to try it out. Signal average the baseline, by pulling out the sample or going off resonance to see how flat it is immediately after the transmitter pulse. Then check also for amplifier paralysis by accumulating long, weak, slightly off-resonance FID's. The amplitude of this beat note should be unaffected by the transmitter pulse. We cannot overemphasize the need to check out the digitizer with very weak signals. After all, you want the digitizer to be useful for poor signals and the poor signals are the ones which get clobbered by "minor" overload recovery problems whereas any digitizer is likely to do just fine on big signals.

An external trigger capability for the digitizer is useful for at least two reasons. One is to be able to sample the magnetization at the correct points in a multiple pulse experiment. (An example would be to sample just the tops of the echoes in a CPMG experiment as described in I.C.2.) The other reason is to allow for synchronization of two digitizers, for example, in quadrature detection.

Sometimes, an "enable" pulse which is required before the trigger pulse can tell the digitizer to record the information proves to be very useful. (This control is sometimes called "arm.") For example, suppose we wish to measure a

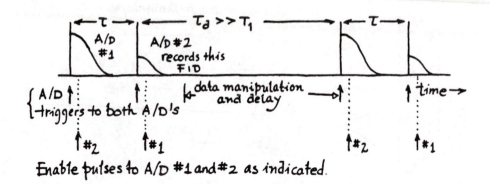

Enable pulses to A/D #1 and #2 as indicated.

$T_1$ in a solid sample with a $\pi/2$-$\tau$-$\pi/2$ sequence and try to save time by recording both the first and the last FID. If the pulse pairs were spaced far apart compared with $T_1$, the first FID would measure the fully recovered magnetization and the second FID the partially recovered magnetization as discussed in III.D.1. It is often much easier to generate two separate enable pulses in the computer than to generate two separate trigger pulses in the pulse programmer if the two digitizers can separately "talk" to the computer as they can in CAMAC systems.

CAMAC systems are worth considering as the basic interface connecting the computer with the experiment. (See V.C.2. and V.C.3.) There are several ADC's and digitizers available as CAMAC modules which are as good as the non-CAMAC units which makes the whole thing very attractive because of the ease of adding units and of the additions being of modest cost.

V.A.4.   RECOVERY TIMES IN PULSE NMR

The problem:  Broad lines in the frequency domain decay quickly in the time domain.  In pulse NMR experiments there is always a deadtime associated with each pulse, during which no signal is detectable, due to some or all of the following causes.

(1)  The receiver will be paralyzed by the transmitter pulse which gets through to it.  In the majority of spectrometers, the transmitter pulse feedthrough is limited by crossed silicon diodes which means that the receiver could see about 0.5 volt during the pulse.  Since the receiver is designed to amplify a signal in the microvolt range, you can imagine the problems caused by such an overload pulse.

(2)  The rf excitation due to the transmitter pulse will take about 20 tank circuit time constants to decay to amplitudes comparable to NMR signals regardless of the time constant.

(3)  There will always be deadtimes of at least one-half of the transmitter pulse length because the signal cannot be seen during the pulse and the time origin is in the middle of the transmitter pulse.

There are two parts to solving the receiver paralysis problem.  One is to reduce the pulse feedthrough and the other is to make the receiver recover quickly.  The feedthrough may be reduced by a well designed duplexer together with a set of crossed diodes shunting the receiver input.  The commonly used crossed diodes limit the input to the amplifier to only about 0.5 volt and no less.  A modification to reduce this threshold has been suggested by Stokes (1978).

The receiver paralysis can be reduced by a better receiver design.  The amplifier bandwidth must be wide enough to respond to the quick turn-off of the transmitter pulse. The amplifier also must not have large capacitors (usually to bypass unwanted signals on the power supply lines) which will be slow to discharge after being charged by the pulse feedthrough. One subtle manifestation of such an effect is a time dependent gain as opposed to the expected obliteration of the signal. A well known commercial preamplifier suffers from this effect for as long as a millisecond. It is easy to check by putting a steady signal on the input (for example, by a grid-dip meter near the probe) and then pulsing to see if the envelope of the cw signal "buckles" as sketched near the end of last section.

In the usual scheme, there are two regimes to the tank circuit ringdown. The excitation in the tank circuit will last according to the unloaded Q of the circuit after the series crossed diodes (see V.C.1., V.C.2., and V.C.7.) stop conducting when the excitation decays to around 0.5 volt. How long it takes for these diodes to stop conducting depends on the output circuitry of the transmitter. If its impedance is high without the transmitter pulse, the circuit is likely to be overcritically damped and this portion of the deay may take about 7 time-constants. (This will happen with amplifiers in which the final tube is completely cutoff between pulses.) In other transmitter schemes, this may take much longer. The crossed diodes protecting the preamplifier do not contribute to speeding up the decay because they present a high impedance to the tank circuit if they are not degrading the transmitter pulse.

For transmitters which do not have rapid ringdown, Spokas (1965) proposed a passive device which used two TV damper diodes in such a way that it presented a low impedance to the transmitter output for small outputs and high impedance for large outputs. It is most suitable for a cross coil probe because it does the most good in cutting off the long tail of the tansmitter pulse which does not affect single coil setups utilizing series crossed diodes.

A novel suggestion by Hoult (1979) to make the transmitter pulse decay rapidly is to reverse the phase of the end of the transmitter pulse in order to cancel the ringing. This is attractive because no additional circuitry is needed. Most modern spectrometers already have the capability of phase shifts and complex pulse spacing necessary to implement this method.

We now consider a duplexer like that described on pp. 395-397. The main elements are the two sets of crossed diodes and the λ/4 line. After the cross diodes stop conducting, the decay rate will be determined by the unloaded Q of the tank circuit in the absence of any active damping. In a high Q circuit, especially at low frequencies, this could entail an intolerable wait, e.g., in excess of $T_2$. Lowering the tank circuit Q, while allowing the observation of the decaying signal sooner, degrades S/N. This trade-off, shown in the accompanying sketch, is treated by Pollak and Slater (1966) and Hoult (1979).

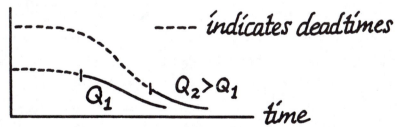

---- indicates deadtimes

$Q_2 > Q_1$

$Q_1$

time

The tank circuit Q can be adjusted with damping resistors or the coil may be wound with less than ideal geometry, e.g., non-ideal turns spacing or the use of a ribbon instead of wire. In the same paper, Hoult points out

a negative feedback amplifier whose noise figure is independent of the feedback. He advocates the use of a receiver
preamplifier with a capacitive negative feedback which leads
to a low input impedance to help in the damping.

An active Q-spoiler circuit which acts as a shunt to
drain off the oscillating electric current at the input of the
preamplifier is, in principle, the best solution of the tank
circuit recovery problem. With such a circuit, a gate pulse
will actively control the initiation and duration of the damping. The most common problem encountered with Q-spoilers is
that the act of de-Qing can set off ringing in the tank circuit
which could be as bad as the ringing which we were trying to
damp out in the first place. The solution involves de-Qing
gently which means avoiding a voltage step associated with the
undamping. The usual culprit is the gate current in the
de-Qing device which turns into a voltage pulse by Ohm's law.

There have been many attempts to design and build
"gentle" Q-spoilers. In an unusual scheme, A. A. V. Gibson
of Texas A&M University uses a Teledyne Crystalonics FF-412
photo-FET (a field effect transistor activated not by a gate
voltage but by the amount of light impinging upon the device)
across the preamp input. It is controlled by a Fairchild
FLV-104 LED (light emitting diode) whose intensity is controlled with a 1 kΩ variable resistor. Because the conductivity of
the FET depends on the amount of light, this acts as a
damping control. The optimum decay should occur at critical
damping when the damping time constant is equal to $1/w_0$
where $w_0$ is the Larmor speed (Millman and Taub, 1965).

In a single coil NMR experiment, the damping has to
attenuate the ringing by as much as nine orders of magnitude
which will take about 20 time-constants at critical damping.
Since the critical damping time-constant is $1/w_0 = 1/2\pi v_0$ where

$\nu_0$ is the Larmor frequency, we have a rule of thumb that the best damping we could ever hope for would take about $3/\nu_0$.

Unfortunately, critical damping is hard to achieve with a $\lambda/4$ duplexer if the damping device is located near the pre-amplifier because the duplexer will present the tank circuit with a critical damping resistance only at one frequency, namely the carrier. The problem seems to originate in the 10% bandwidth of the $\lambda/4$ line. Nevertheless, adjusting the impedance seen by the tank circuit during de-Qing does optimize the tank circuit recovery to within a factor of four or five of the critical damping limit. With such a technique, Kaufmann, et al. (1979) achieved a recovery time of 3 µs at 4 MHz. D. Brown at Los Alamos has similarly achieved a 35 µs dadtime at 500 kHz. It is possible that a Q-spoiler which works directly across the tank circuit is necessary for critical damping. For a detailed discussion of a similar setup without $\lambda/4$ lines, see Section IV of Clark and McNeil's paper referenced in V.C.7.

We have already said that the recovery time is most critical at low frequencies. At higher frequencies, problems other than the rate of damping of the tank circuit (such as amplifier saturation) would take over.

In summary, what we have described is a system which recovers from the transmitter pulse in two distinct steps. The first part takes place through some low impedance path usually, but not always, associated with the output circuitry of the transmitter. In order to improve the S/N, the trans-mitter circuit is passively disconnected by a set of series crossed diodes after which the second phase of recovery proceeds according to an active damping circuit which, if present, leads to better S/N because of the improved recovery even though it is a source of additional noise. Neither sets of crossed diodes directly contribute to speeding up the recovery process; one set protects the receiver while the other decouples the transmitter for improved S/N.

Lastly, the contribution to the deadtime from the second half of the transmitter pulse is minor compared with the other contributions. If the transmitter pulse is reasonably short, the problem will not be noticeable unless the system recovery is outstanding. See VI.D.4. for a related discussion.

## REFERENCES

D. I. Hoult, "Fast recovery, high sensitivity NMR probe and preamplifier for low frequencies," Rev. Sci. Instrum. 50, 193-200 (1979).

E. N. Kaufmann, J. R. Brookeman, P. C. Canepa, T. A. Scott, D. H. Rasmussen, and J. H. Perepezko, "Observation of pure quadrupole resonance in zinc metal," Sol. St. Commun. 29, 375-378 (1979).

J. Millman and H. Taub, Pulse, Digital, and Switching Waveforms (McGraw-Hill, New York, 1965), pp. 56-61, 795-797.

V. L. Pollak and R. R. Slater, "Input circuits for pulsed NMR," Rev. Sci. Instrum. 37, 268-272 (1966).

J. J. Spokas, "Means of reducing ringing times in pulsed nuclear magnetic resonance," Rev. Sci. Instrum. 36, 1436-1439 (1965).

H. T. Stokes, "Tuned limiter for receiver amplifier in a fast-recovery pulsed NMR spectrometer," Rev. Sci. Instrum. 49, 1011-1012 (1978).

## V.A.5.  TIME-SHARE MODULATION

The oldest modulation scheme in NMR is field modulation in cw NMR.  A periodic signal is applied to coils wound coaxially with the $H_0$ field, either on the magnet or on the probe.  The field is usually sinusoidal which leads to a sinusoidal amplitude modulation (AM) of the signal while the $H_0$ field is slowly swept through a resonance.  In such a scheme, modulation serves two purposes.  If the NMR signal comes into the sample coil with some known modulation signal superimposed, the output can be phase (lock-in) detected with respect to that modulation signal resulting in enhanced S/N.  Simply put, we selectively observe only that part of the signal which is at the modulation frequency and discriminate against the rest.  In addition, when the carrier component has been removed from the NMR signal by the detector, the resulting signal, in the absence of modulation, is sufficiently close to dc in character as to make amplification difficult.  The presence of an audio modulation simplifies the electronics by permitting ac coupling between amplifier stages.  In cw NMR, the dc magnetic field, the transmitter frequency, or the time during which the rf is on can be modulated.

Another development in modulation as applied to NMR is time-share modulation (Anderson, 1962; Grunwald, et al., 1962; Baker, et al., 1965; Boden, et al., 1968).  This is a variation of amplitude modulation (AM) in which the transmitter produces an rf pulse while the receiver is pulsed off.  Then, when the rf pulse is off, the receiver is turned on to detect the NMR free precession signal.  Although time-shared modulation is a hybrid between cw and pulse methods, we discuss it because the apparatus for pulse NMR can be used for it with trivial modifications.

All modulation schemes produce sidebands that are removed from the main signal by the frequency $\Delta v$ of the
modulating signal.  Of course the frequency separation between
the centerband and the sidebands must be at least as large as
the spectral region you wish to observe.

In time-share modulation, illustrated below, rf switches
are employed to alternately turn on the transmitter and the
receiver.  (A slight delay between turning the transmitter off
and the receiver on may be necessary to eliminate any pulse
feedthrough into the receiver.)  In this method of modulation,

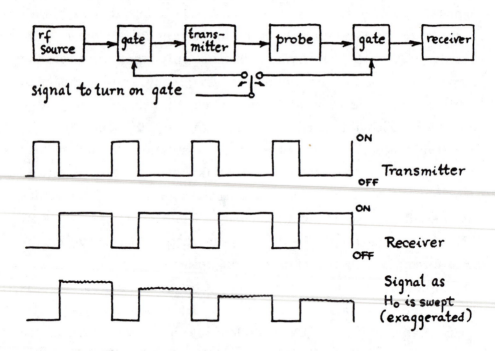

the pulse initiates an FID and the receiver detects the start
of the FID.  Narrow lines will result in long FID's which
require a relatively slow pulsing frequency.  (Note however,
the lowest frequency of pulsing permitted is that sufficient to
move the first sideband out of the spectral region of interest.)
In order to observe wide lines, the pulsing frequency must

be increased, so that it is at least the width of the line being sought.

When field modulation is applied to cw NMR, the spectrum is subject to distortion if the amplitude of the modulating field becomes an appreciable fraction of the line width. Small modulation amplitude is required for accuracy, but large modulation is required for sensitivity. Also, large amplitude modulation often dictates a slower frequency of modulation which can lead to modulation distortion as the sidebands invade the signal and this becomes an additional problem. These conflicting requirements lead to compromises which result in less than the desired sensitivity or in spectral distortion. Time-share modulation avoids some of these problems. Also, a time-share modulation produces the signal lineshape instead of its derivative although that is not necessarily an advantage. As no modulation coils are required, the probe can be simpler.

Pulse FT NMR does not suffer from the same problems as the standard cw NMR but it suffers from another: the early part of the FID is lost because of the finite recovery time of the system after the application of a high power pulse. (See Section VI.D.4. for other ways of recovering this lost piece of information.) Time-share modulation alleviates some of this problem by using much smaller amplitude excitation than the standard pulse NMR experiment. So, the method is really a hybrid between cw and pulse NMR with the attendant advantages of quite high sensitivity and accuracy but without the multiplex advantage of FT NMR. In time-share modulation, the operating parameters can vary widely. We refer you to the references for details.

Time-sharing can be used in a broader context than just providing the necessary modulation. For example, it can be used for homonuclear decoupling with Fourier transform, in which the decoupling is performed by time-sharing the decoupling oscillator with the receiver. The decoupler is turned on while the receiver is off and then the receiver monitors a small portion of the FID before the decoupler comes on again. See the discussion on homonuclear decoupling at the end of II.C.

A modern NMR spectrometer has at least three channels: analytical (observe), lock, and decoupling (or double resonance). With field modulation, all of these must use the same modulation parameters. Time-share decoupling permits independent modulation of each channel. An additional advantage of time-share modulation is the very good isolation between transmitter and receiver. In the usual cw setup, any leakage from the transmitter into the receiver may cause baseline problems. The time-share scheme (normally added to the usual cross-coil or bridge type probes) provides sufficient isolation to eliminate this problem. Some modern spectrometers are built with time-sharing in their design and it may be a valuable feature to look for in special situations.

REFERENCES

W. A. Anderson, "Application of modulation techniques to high resolution nuclear magnetic resonance spectrometers," Rev. Sci. Instrum. 33, 1160-1166 (1962).

E. B. Baker, L. W. Burd, and G. N. Root, "Time-sharing modulation at 200kc applied to broad and narrow line NMR for base-line stability," Rev. Sci. Instrum. 36, 1495-1498 (1965).

N. Boden, J. Capart, W. Derbyshire, H. S. Gutowsky, and J. R. Hansen, "Frequency sweep, field-frequency stabilized double resonance spectrometer," Rev. Sci. Instrum. 39, 805-816 (1968).

E. Grunwald, C. F. Jumper, and S. Meiboom, "Kinetics of proton transfer in methanol and the mechanism of the abnormal conductance of the hydrogen ion," J. Am. Chem. Soc. 84, 4664-4671 (1962).

## V.B.  PURCHASING A SPECTROMETER

We now consider some practical aspects of buying a commercial spectrometer or components. Much of this is common sense as you would invoke when purchasing a house or an automobile.

## V.B.1  BASIC CONSIDERATIONS

The following are considerations in purchasing a basic pulse machine. The additional considerations for Fourier transform instruments are discussed in the next section. This section may be educational even if you are not in the market.

### Transmitter

Two important specifications are the maximum rf field intensity $H_1$ and the duty cycle. The 90° pulse duration $\tau$ (specified, for example, for protons) is related to $H_1$ by $\gamma H_1 \tau = \pi/2$. $H_1$ is a measure of the power delivered to the sample coil but also depends on the Q and the volume of the transmitter coil. For a given transmitter output power, one might be able to do much better (or worse) with different probe parameters. For example, a cross-coil probe needs much higher power than an equivalent single-coil probe to achieve the same 90° pulse time. The power delivered to a dummy load is usually considered to be a measure of the transmitter capability, although what really counts is the power to a particular load which maximizes the current through the load. A minimum of a few hundred watts is desirable in any pulse machine used for solids. A 1 μs $\pi/2$ pulse for protons in a 10 mm sample tube with a recovery time of 10 μs at 15 MHz is a performance not difficult to attain even without an active Q-spoiler. For lower power experiments, the operator can reduce the power to the desired level in virtually all transmitters. A minimum of 1 kW should be specified for samples of several cubic centimeters, especially with nuclei other than protons and fluorine.

Duty cycle is the percentage of the time the transmitter is on. With a transmitter having an inadequate maximum duty cycle, the rf power can decay during a long pulse or during a pulse train resulting in various difficulties. If the rf pulse amplitude droops during spin-locking in a rotating frame experiment (see IV.C.1.), the results may be misleading or uninterpretable. Also, this can lead to a continuous phase change during the rf pulse making it impossible to keep the magnetization locked. A high duty cycle is also needed for decoupling, for example, in double resonance experiments. A reasonable specification for duty cycle might be the ability to maintain a constant $H_1$ field of 5 gauss for 1 second for an instrument to be used for $T_{1\rho}$ experiments. For conventional $T_1$ and $T_2$ experiments, the duty cycle requirement is far less demanding; one part in $10^3$ at full power might be quite acceptable.

A second consideration is rf isolation -- the ability of the transmitter not to leak rf power to the receiver with the pulses nominally off. Isolation is an important parameter with long $T_1$ samples because leakage irradiation can saturate the spin populations between pulses. Usually more than 80 dB of rf suppression is desirable. [A dB is $20\log_{10}(V_{on}/V_{off})$ where $V_{on}$ and $V_{off}$ means voltages with the transmitter pulse on and off respectively. Thus, 80 dB equals $10^4$ in voltage. See VI.E.3. for more.] As a good test, one should be able to hook the output of the transmitter to the input of the receiver and not detect any rf with the pulses off. The receiver output should be sampled before the detector to see the rf; otherwise, in the absence of modulation, no signal will be seen, anyway. (Don't turn the pulse generator on during this test!)

A frequency coherent transmitter is one that operates from a continuously running oscillator. Therefore, the pulses have well-defined phase relationships to one another permitting phase sensitive detection and experiments such as the Carr-Purcell Meiboom-Gill, the $T_{1D}$, and the $T_{1\rho}$. An incoherent spectrometer has a gated oscillator which is turned on with each pulse so that the rf phase for each pulse is random with respect to a continuously running oscillator (see section V.A.1.). This is still useful for some experiments like $T_1$ measurements and for obtaining spectra. The reason for the existence of incoherent transmitters in magnetic resonance is that they are easier to build especially at high frequency. We know, for example, of a commercially available 700 MHz gated oscillator which can be used for NQR.

Pulse sequence and phase capabilities

The pulse sequence and phase capabilities should, of

course, depend on the experiment you want to do. However, there are many marvelous recent developments, e.g., 2-D spectra, selective excitation, high resolution in solids, etc., which are possible only if the spectrometer is versatile. Unless you want to shut yourself off from these new experiments, we suggest that you try for the maximum versatility in your spectrometer. Incidentally, a digital pulse sequence generator is far better than an analog one because the timing is easily resettable and it can easily be interfaced to a computer. (See the example in V.C.2.)

The pulse sequence can be generated by a separate hardware generator or by the computer if one is used in the experiment as it is for an FT experiment. For the ordinary high resolution NMR, the latter course may be satisfactory but for any experiments requiring quicker timing an independent digital pulse sequence generator is far better because the pulse timing and duration is not limited by the cycle time of the computer which is of the order of a few microseconds at the present time. Furthermore, a hardware pulse sequence generator can be very flexible in its output capability. It is possible to trigger several devices nearly simultaneously, for example, a transmitter, a receiver damping circuit, an oscilloscope, and a digitizer. Therefore, if you think you might be pushing the computer's capability to generate suitable pulse sequences, we suggest that you get a hardwired pulse generator such as the timing simulator described in V.C.2.

Some modern (new in 1981) commercial NMR spectrometers are extremely versatile with computer controlled pulse sequence generators which are easily programmable even for complex sequences. Thus, even for a homemade machine, you should seriously consider purchasing the operating system software with a computer and a pulse sequence generator.

As two phases in quadrature will allow Carr-Purcell Meiboom-Gill, $T_{1D}$, and $T_{1\rho}$ or other spin-locking experiments, we recommend this as a minimum. However, providing different rf phases for the transmitter pulses is not difficult and this capability should not increase the cost of the spectrometer very much so you might as well go for four orthogonal phases. As discussed elsewhere, the phase shifts can be performed in the IF stage rather than at the carrier frequency so that one phase shifting network can suffice for all carrier frequencies.

### Frequencies

In purchasing a new spectrometer, you should specify the desired frequency(ies) which will be dictated by the magnet available and the nuclei and experiments desired. If at all possible, a spectrometer which can operate over a large frequency range (a factor of 25 or more) should be chosen. This will enable you to study a variety of nuclei under conditions of optimum magnetic field. It also permits the study of relaxation times and lineshapes as a function of frequency. (This is an important method in identifying relaxation mechanisms and in separating different contributions to the lineshape. See III.C.2. and VI.D.1.)

If, however, a spectrometer built for one frequency is to be chosen, how should it be done? For looking at only one nucleus, we should choose the highest frequency consistent with the highest magnetic field, in order to maximize sensitivity and resolution. In addition, choosing the highest frequency will result in the shortest dead time because recovery time is inversely proportional to operating frequency for the same Q circuit. However, more power is needed at higher frequencies to drive the same Q circuit for the same volume.

These remarks concerning the desirability of the highest field possible are true for a range of magnetic fields up to what used to be thought of as high field. Beyond some value, however, the expected sensitivity and resolution improvements will not be realized.

Such improvements were contingent on the chemical shift being proportional to the field $H_0$ and all other interactions being independent of the field. At low fields, this is, indeed, true to a good approximation and the expected improvement is realized.

Suppose, however, that there is a small contribution to the linewidth besides the chemical shift which varies linearly with the field. Then at some high field, it will become as large as the largest of the field independent linebroadening mechanism and any increase in the field beyond that point will not improve the resolution. Suppose, further, that there is a small interaction which is proportional to $H_0^2$. Then, beyond some high field, the resolution will even become worse.

Such an interaction might be the linebroadening due to chemical shift anisotropy modulation which was described in III.C.1. Another possibility is that for high enough fields, the usual dipole-dipole, chemical shift anisotropy, etc. interactions, may no longer be in the extreme narrowing limit so that $T_2$ will be less than $T_1$. The NOE might become smaller, as well (Shaw, 1978).

The optimum fields for the various nuclei in various situations are not known at the present time. For many systems, the highest field being used now represents an improvement from the lower fields although for some other systems the optimum field is much lower.

These remarks also apply to high resolution experiments in solids. VanderHart et al., (1981) have shown that the expected resolution improvement is not present in $^{13}C$ MAS spectra of certain organic solids even in going from 1.5 to 5 T.

On the other hand, for looking at a variety of nuclei with one frequency, a choice must be made. For a 50 kG variable field magnet, a spectrometer frequency of 50 MHz will enable us to observe, $^{1}H$, $^{3}He$, $^{11}B$, $^{13}C$, $^{19}F$, $^{23}Na$, $^{27}Al$, $^{31}P$, $^{51}V$, $^{55}Mn$ and some other higher Z nuclei. The proton sensitivity at 50 MHz is greatly reduced from that at 200 MHz but is still greater than it is for the other nuclei at this frequency. Going to higher frequency will improve the resolution, S/N, and recovery, for the fewer nuclei which will be accessible.

An additional consideration is the cost of additional frequencies (if available) and the ease of changeover. This will be limited mostly by the probes, as discussed later. The most desirable alternative is to use the frequency corresponding to the highest field for each nucleus. A heterodyne system combined with a frequency synthesizer is the best way to go, both for cost and simplicity. Keep in mind, too (as pointed out elsewhere), that it is not prohibitively difficult to engineer one's own heterodyne system or probes.

Receiver

The receiver should include a phase sensitive detector. The receiver should recover from saturation in a few microseconds at least for low resolution experiments. For high resolution work, the output should be dc coupled. The output level should be adjustable by as much as the full ADC input range. For maintenance and tune-up, it is a great convenience to have access to the amplified rf signal before the detector. This has the added advantage of being able to use your own detector and also being independent of any

built-in filtering circuits, if any.  Lastly, quadrature detection
(see II.A.3.) is a must for a modern FT machine.

## Probe

The first consideration is whether to choose a cross-coil
or a single-coil probe; the former has higher $H_1$ homogeneity
but will deliver a much smaller $H_1$ field for the same transmitter power compared to a single-coil probe.  Single-coil
probes are less susceptible to microphonics and temperature
dependent instability than cross-coil probes and this is especially true in variable temperature operation where the receiver coil may not be at room temperature.  For most applications, the single-coil has everything going for it over the
cross-coils.

The next consideration is the sample size to be accommo-
dated by the probe.  A probe built for a large sample will
have a smaller $H_1$ field for the same power because $H_1 \propto V^{-\frac{1}{2}}$
where V is the coil volume.  Since S/N is proportional to the
number of nuclei and the square root of the filling factor,
such a probe will give good S/N with that particular large
sample, but if the sample is small, the fill factor will be poor
and will result in an even lower S/N than the same small
sample in a probe designed for that particular size tube.
Therefore, if you are going to run small samples for lack of
material most of the time, do not get a probe designed for
larger samples.

The probe should not give rise to spurious signals, for
example from mechanical oscillations of an inadequately secured
rf coil or electromagnetic generation of acoustic waves.  (For
these and other sources of spurious ringing, see VI.B.5.).
The probe should also not give rise to NMR signals which will
interfere with the desired signal.  For example, beware of

probes with parts made of Teflon when you study fluorine NMR and similarly for protons and epoxy, especially if the desired signals have linewidths similar to Teflon and epoxy. You can check for these signals by signal averaging without a sample after optimizing the spectrometer settings with a sample.

The ringdown time of the probe should be specified at the working frequency although we have already pointed out elsewhere (V.A.4.) that this is a function not only of the probe but also of the entire system containing the transmitter, the coupling circuitry, the probe, and the receiver. For broadline work in solids, a ringdown time of 10 μs or better from the end of the pulse is very desirable at any frequency. For high resolution work in liquids, a few hundred microsecond recovery time can be tolerated because of much longer $T_2$'s. For example, if the digitizing rate is 1 kHz, the second data point is 1 ms later than the first, and only the first point is lost if the deadtime is less than 1 ms.

The 90° pulse duration and probe sensitivity involve the probe and the transmitter or the receiver, and these parameters involve tradeoffs. A high Q probe results in high sensitivity and large $H_1$ field at the expense of long ringdown time which can obscure much of the high sensitivity signal (see section V.A.4.). The best overall S/N occurs for some intermediate Q which has to be chosen according to the sample $T_2$.

Often probes are built to permit insertion of a variable temperature dewar inside the receiver coil. In this configuration, changing samples is simple and there is not much need to retune the tank circuit because it remains at ambient temperature. The main disadvantage is the small filling factor. Other variable temperature probes may be designed to let the sample fit snugly in the coil to maximize the filling factor (and thus the S/N) and with the coil always at the sample temperature.

For a superconducting magnet, the most common high resolution NMR probes at the present time use saddle coils (see section V.C.5.) which allow very convenient access for samples into the probe at the expense of poorer S/N. Therefore, if you are contemplating experiments requiring much signal averaging in a superconducting solenoid, opt for a probe with a "sideways" solenoid rather than a saddle coil. At this writing, sideways coils are not commercially available so you have to engineer them yourself. See section V.C.5. for references. Of course, one of the two major attributes of doing NMR in a superconducting magnet is the greater resolution due to the expanded chemical shift caused by the higher field, and if that is your only reason for going to such a magnet the saddle coil probe will work just fine.

Lastly, we discuss the frequency range to be covered by an individual probe. Probes that can be tuned over a wide frequency range, of course, are cheaper per frequency. Their disadvantage is that the S/N can be optimized only at one frequency. Usually, a given probe works best at the highest frequency it is designed for. Thus, if your interest is limited to only a few nuclei, use probes designed for those particular frequencies.

## Data system

There are minicomputers on the market which were developed for pulse/FT NMR use and contain a complete interface and have a library of programs appropriate for NMR data manipulation. An alternative is to purchase a CAMAC crate as an interface to a standard minicomputer and then

develop a program library for it. This will be discussed in
V.C.2. and V.C.3. A third alternative is to "build" your
own interface which, though straightforward, requires substantial time and effort.

Currently, minicomputers commonly used in NMR have
word lengths of 12-, 16-, 20-, and 24-bits. The larger word
size permits more FID's to be added without overflowing
memory, but more important, it permits more powerful in-
structions in the computer's language. Computers can be
used in double precision for signal averaging, which means
that two words are strung together to create a word twice as
long to accommodate larger numbers. Sixteen-bit computers
are by far the most common and 12-bit computers should be
avoided because of the limitations of such a small word size.
Detailed discussions of using minicomputers is contained in a
book by Cooper (Appendix A).

Most of the commercial data systems for NMR are developed for high resolution Fourier transform spectroscopy and,
therefore, have an analog to digital converter (ADC) whose
shortest dwell time may be 20 μs per point or longer. While
this is fast enough for most high resolution applications, it is
often inadequate for FT work in solids which typically requires
a time resolution down to 1 μs per point. (Occasionally, even
faster digitizing rates are necessary, e.g., for capturing very
fast echoes in metals or in NQR.) Such digitizing rates in
excess of those which can be handled by the computer are
handled with a digitizer, also called a fast transient recorder.
The information of a single scan is digitized and stored in the
memory of the digitizer at the high rate and then transferred
to the computer at a slower rate for the accumulation. Finally,
quadrature detection will require two digitizers or one digitizer
multiplexed to handle two independent channels of data.

An additional convenience of purchasing a minicomputer system developed specifically for NMR is the large number of programs available which are specifically written for NMR applications. The library developed by the manufacturer is often supplemented by a users' group library of programs. While such programs are generally available for most of the popular minicomputers, they will usually not have been written expressly for NMR use. Although one of the authors has spent many weeks and months writing operating system programs for doing NMR experiments in the past, we believe that the time has come (in 1981) when the commercial operating system can be so good that serious consideration should be given to purchasing commercial software written for NMR.

## Magnet

The major considerations concerning magnets have already been discussed in V.A.2. To measure or observe NMR lineshapes directly either by cw or FT techniques, the magnet homogeneity over the volume of the receiver coil and stability during the observation time must be better than the narrowest line to be observed. This means that the observed decay constant $T_2^*$ must be determined by the $T_2$ of the sample and not the magnet. In solids and viscous liquids with lines broader than 100 Hz this is not a difficult condition to meet but in nonviscous liquids this usually requires a high resolution magnet equipped with shim coils to create homogeneity of 3 parts in $10^9$ or better, i.e., a 0.3 Hz magnet-limited linewidth at 100 MHz which corresponds to a $T_2^*$ of $1/(0.3\pi)$ seconds.

In many solids, the FID persists for less than 100 μs making very little demand on the magnet's homogeneity or stability. In many cases, even for narrow lines, a variety

of measurements can be made without field homogeneity sufficient to observe the details of the lineshape. In order to
make $T_1$ and other relaxation measurements, it is not necessary
to have a homogeneity better than the linewidth except for
resolving overlapping lines. $T_2$'s in liquids and solids can be
measured with echo trains which avoid the limitations of an
inhomogeneous magnet. Also, by Fourier transforming the
echo envelope, spectra can sometimes be obtained with spectral
features no longer obscured by poor resolution. (See the
discussion of J spectra in II.D.4.) Finally, there are hole
burning experiments which can detect the intrinsic linewidth
within the inhomogeneous line (see IV.A.1.).

Magnet stability is another consideration. Signal averaging requires that the magnet instability not significantly
broaden the observed resonance during data acquisition.
Persistent SC magnets are now available with drift rates of
$10^{-8}$/hr or better. Long term averaging of narrow lines
(e.g., 1 Hz or less) in an iron-core electromagnet requires the
use of a flux stabilizer, often called a superstabilizer and an
NMR lock. For long term averaging of broad lines in solids,
an electromagnet equipped with a Hall effect probe or a current
regulator in a temperature stable room will be sufficient.

When purchasing a magnet, bear in mind the largest
sample size or apparatus it might have to accommodate. For the
superconducting solenoid, sample access is more of a problem
than for an iron-core electromagnet and the possibility of
orienting samples perpendicular to the axis of the bore and/or
spinning samples at orientations other than axial should be
seriously considered in determining bore size and homogeneity
requirements. The access to the field should be as convenient
as possible. Make sure, for example, that the legs of the SC
magnet are long enough for you to be able to insert any con-

ceivable probe from the bottom.

Finally, it is desirable to have a magnet whose field is adjustable from nearly zero to its maximum field with good homogeneity to study field dependent quantities or have a wide range of observable nuclei with one frequency. To a significant extent, the lock system limits the adjustability in high resolution electromagnets. Although the SC magnets are harder to adjust in the persistent mode operation, their field homogeneities are much less sensitive to the value of the field than iron-core electromagnets, so that it is practical to consider using them in variable field experiments provided there are power supplies with them. (We mention this because the magnet power supply is often sold as an option these days since many persistent magnets are used only at one field. Our advice is to buy the supply if you think there is any chance that you will want to change the field or that you might run out of liquid helium.)

## Miscellaneous

Additional features to look for depend upon your needs and budget. Items to be considered are: (a) an NMR lock, driven from the same basic frequency source as the transmitter (a lock is mandatory for high resolution work in an iron-core electromagnet) and easy to operate reliably; (b) a stable frequency source such as a frequency synthesizer or a phase-locked-loop stabilized oscillator, even for low resolution work; (c) modularity in construction to permit modifications; (d) a spectrometer built on a heterodyne principle; (e) a decoupler for decoupling or other double resonance experiments; (f) automated data acquisition and spectrometer control; and (g) ease of operation.

The ability to do spectrometer service "in house" may be

important to avoid high cost, long shutdowns, and the worry over the company's reliability. Most spectrometers are sufficiently straightforward that much of the servicing is not beyond the capabilities of even a modest electronics shop. It is important to insist on receiving the <u>correct</u> circuit diagrams and, if possible, with the desired performance indicated at each stage as a part of the package. As an invaluable service aid, some time should be devoted to the compilation of typical parameters (voltage, current, waveform, etc., at various points as well as signal parameters like S/N and deadtime) <u>while the apparatus is functioning</u>.

## Summary

In purchasing a pulse NMR spectrometer and magnet, all of these general considerations as well as those discussed in V.B.4. about dealing with manufacturers are applicable. It is wise to buy as flexible an instrument as one can afford. The field of pulse NMR spectroscopy is changing rapidly and it would be tragic to be limited by some needless instrumental characteristics. In addition, it seems to be a fact of life that we never can think of all of the uses of device capabilities until we actually use them. A serious alternative to purchasing a complete spectrometer is to construct one out of commercial components and the reader is referred to section V.C. for further discussions.

## REFERENCES

D. Shaw, "Experimental methods" in <u>NMR and the Periodic Table</u>, R. K. Harris and B. E. Mann, editors (Academic Press, New York, 1978), pp. 21-48.

D. L. VanderHart, W. L. Earl, and A. N. Garroway, "Resolution in $^{13}$C NMR of organic solids using high power proton decoupling and magic angle sample spinning," submitted to J. Magn. Resonance (1980).

## V.B.2.   DESIRABLE FEATURES IN HIGH RESOLUTION SPECTROMETERS

In this section, we continue the discussion of the previous section as they apply to high resolution spectrometers. In the next section, we extend the discussion even further as they apply to cross polarization magic angle spinning (CP/MAS) spectrometers.

### Probe

There is always the question of sample size, as already discussed in the last section. There is the usual trade-off involved here, too, but in ways different from those already mentioned. In any magnet, the region of uniform magnetic field is finite in size. Therefore, there will be a point where increasing the sample size further will degrade the apparent homogeneity of the magnetic field and will give you broader lines. This is true in an electromagnet as far as the sample diameter is concerned and you should insist that sample spectra be run with different size sample tubes to really determine for what sample size the inhomogeneity starts to nullify the S/N gain.

In a superconducting magnet, this argument holds as well. For sideways coils, however, the problem is more severe because the dimension perpendicular to the magnetic field is likely to be significantly greater for the same sample diameter. Thus, if any large sample tube experiment or any sideways configuration is considered, choose a large enough bore (or gap, in the case of an electromagnet) in the magnet so that you will not be limited that way.

For carbon-13 NMR requiring proton decoupling, a very useful feature is the ability to look at both the proton and the carbon spectra without changing probes. This means that either the coil is double tuned to carbon and proton or that there are two coils in the same probe for the two nuclei. Then the proton spectrum can be run before the mode is changed to carbon with the appropriate proton lines to be decoupled identified. One easy scheme to accomplish this is to use the proton decoupling coil as an NMR coil to look at the protons and then use it as the decoupling coil during the carbon-13 observation. The proton spectrum will not have the best S/N but it can be quite adequate for the purpose at hand.

Lastly, the business of changing probes would be much less painful if the magnet shimming required upon changing the probes were small. Some commercial spectrometers have an automatic shimming capability controlled by the computer. In the least, you should insist on the shim settings to be reproducible for each different probe with a superconducting magnet so that the correct shim parameters need to be measured only once (and written down!). Unfortunately, this is beyond the capability of most iron electromagnets.

### RF system and decoupling

For true multinuclear operation, it is very important to specify an adequately short $\pi/2$ pulse length for some small gyromagnetic ratio nuclei, especially one that falls near the bottom end of the frequency range for any probe. The multinuclear capability is of little use if it works poorly for the low γ nuclei. Note that if a quadrupolar nucleus is chosen, the meaning of the $\pi/2$ pulse length is modified with respect to γ according to the discussion in II.D.1. Thus, nitrogen-15 with spin 1/2 might be a good standard to use.

The decoupling modes should include gated and pulsed decoupling capabilities. The former is necessary to obtain decoupled spectra without NOE as well as a fully coupled spectra with NOE by gating the decoupler on only during the FID and turning it off only during the FID, respectively. The latter also allows homonuclear decoupling by pulsing off the potentially troublesome decoupling signal only during the sampling period of the ADC, i.e., for an FID digitized at 8K points the decoupler is gated off 8K times.

One of the more difficult specifications to evaluate is the decoupler efficiency, that is, how effective the decoupler is for a given amount of power into the decoupler coil. (See section II.C. for further discussion.) This is an important parameter especially at high frequencies (at few hundred MHz) because much of that rf power could go into the sample as heat. The problem is especially acute in ionic solutions. If the heating effect is significant, it will be difficult to minimize the temperature gradient in the sample regardless of any efforts to cool the sample so that the advantages gained by going to the higher field will be nullified. Besides, when the decoupler coil (and the transmit/receive coil) gets hot, it can detune itself. So the trick is to design decoupler schemes so that the fewest number of watts is required for the most decoupler field intensity. The best way to evaluate this potential problem is to actually run some samples which might exhibit this problem.

## Magnet

Finally, we touch on the magnet, once again. At the risk of repetition, we point out that the higher fields and larger bores may be very desirable but at some price to be paid. Aside from the literal meaning of the last phrase (and the price increment for higher field or larger bore is significant), both higher fields and larger bores usually entail greater field drifts (i.e., non-persistence) as well as greater probability of the magnet quenching. We guess that this is a manifestation of the old adage "you can't get something for nothing." Be sure to read the caveat on going to the highest possible field and frequency under Frequencies in V.B.1.

## V.B.3.   CP/MAS SPECTROMETERS

At this writing (1981), cross polarization/magic angle spinning (CP/MAS) experiments have yielded a multitude of useful information mostly from, but not limited to, carbon-13 spectra of organic solids. The commercial companies are still developing foolproof spectrometers which utilize this method and the basic problem seems to rest with the MAS portion of the apparatus. Simply put, it is hard to spin samples fast enough in a routine way to be free of sideband problems for carbon-13 spectra at resonance frequencies much higher than 25 MHz. As pointed out in the section on high speed spinning (VI.D.2.), it is now possible to spin samples at rather high speeds, but it takes ideal samples to do this. The routine spinning of carelessly packed powder is still limited to something of the order of 2 to 3 kHz which limits the carbon-13 Larmor frequency to the aforementioned 25 MHz or even lower. Thus, for the time being, CP/MAS of carbon-13 requires a low field magnet, either an electromagnet or a widebore low field supercon; the latter is a much better choice unless you want to utilize an existing electromagnet.

It should be pointed out, though, that the acceptable spinning rate depends on the interaction you are trying to average out. For nitrogen-15, for example, the chemical shift anisotropy is quite a bit less than it is for carbon-13 so that a field higher by a factor of approximately two is tolerable and is desirable from the standpoint of compensating for the smaller $\gamma$. In fact, even for carbon-13 it is possible to obtain very good CP/MAS spectra at quite high fields provided that the chemical shift anisotropy is not large for that particular sample or that the particular sample can be spun well. Thus, the inadvisability of high fields depends on the sample and applies mainly to routine operation of arbitrary carbon-13 containing samples and an ideal CP/MAS multinuclear spectrometer would be a variable field instrument which can best cope with the problem of optimizing the chemical shift anisotropies to match the available spinning speed. See, however, the discussion on high field effects in general under <u>Frequencies</u> in V.B.1.

## V.B.4   DEALING WITH INSTRUMENT COMPANIES

Ask for a users' list. Contact those users for their experiences with the machine, the delivery, and the service.

Question the availability and quality of service in your area; service might be good in some areas and poor in others.

Request to be taken to a demonstration center; request that the instrument features which are particularly important to you be demonstrated. You should bring samples to run. At least one of them should have very poor S/N. Another should have as broad a spectral range as you might want to study. If homonuclear decoupling is important to you (or decoupling protons from fluorine), try out this feature, without fail.

Draw up a detailed specification list. Do not leave any stones unturned. Since the proof of the pudding, so to speak, is in the actual NMR data, state as many specifications in terms of what the spectrometer will be able to do with actual samples.

Determine whether you have to pay an import duty on a machine of foreign manufacture or whether you can get an exemption because it is being purchased by a public university or a governmental agency or because it offers a capability not available in domestic instruments. (This point is especially applicable to U.S.A. users.)

The computer is an integral part of the instrument. Is there a good software and is it growing? What will the software updates cost? Who will service the computer? The interface? Does a single warranty cover the entire package including the computer?

Look carefully at the instrument's utility requirements. For example, can you meet the temperature and volume requirements for raw cooling water or will a water chiller be necessary? What about the requirements for compressed air? Do you have special requirements because of your special environment, for example, high altitude? Do you have a high enough ceiling for transferring liquid helium into the SC magnet?

If you cannot afford all of the instrument and its accessories this year, check to see if the company will loan or lease the additional units as an incentive for purchase. Such loan agreements should be spelled out in their bid.

Insist that the warranty on the entire spectrometer system begin only after the entire instrument is operating and checked out, including all accessories. This provides an incentive for the company to deliver all the accessories in a reasonable time. Furthermore, some accessories or capabilities may work well only in the absence of other accessories. Statements such as warranteed for 1 year after system installation but not for more than 15 months after initial delivery of equipment should not be accepted.

Insist that the contract provide a date by which all accessories are delivered. It may be possible to insert a penalty clause in the contract for delays in delivery either for reduction in payment or for additional equipment.

Insist that all operating manuals and service manuals be complete, be for the specific model instrument you have, be in English if you are English speaking, and have them delivered with the instrument or before. You will need them.

A point to be negotiated under terms and conditions will be the question of who pays for the shipping. The difference between FOB factory and FOB destination could be very significant for something as big as an NMR spectrometer.

All too frequently, manufacturers promise new capabilities before they are ready to deliver them and they may be highly optimistic in their sales talk. One should be wary of this and expect such newly developed accessories to have many bugs in them.

The terms and conditions should be worked out so that only a part of the payment is required upon delivery of the system and the remainder upon acceptance when everything is working. In addition, avoid phrases which may be construed to mean that some payment is required for every individual item recieved rather than for an agreed upon package in case the order is not received all at once.

Be reasonable about agreeing to delivery dates suggested by the company. Otherwise, you may receive a system which is poorly made and the problems can linger through a finite warranty period and beyond.

Instrument representatives are not miracle workers and can be caught in between the customer and the factory. Work with them closely and they should treat you fairly because they want to make the sale.

## V.C.  SPECTROMETERS AND COMPONENTS

In this section, we discuss quite extensively the at right hardware components which are used in spectrometers. In many ways this is one of the most unique and, hopefully, useful parts of this book. We start with some examples of spectrometers we have assembled, not especially because of their good designs but as examples of what you might be able to do. There are many variations to NMR spectrometers and you have to choose the combination of components and, indeed, design elements which suit your needs the most.

## V.C.1.  AN EXAMPLE OF A SIMPLE LIQUIDS MACHINE

While the instrument whose block diagram is shown at right was specifically intended to do $^{13}$C Fourier transform NMR at 15.1 MHz in liquid samples, it could be used for a variety of nuclei in liquid and solid samples. It uses a Varian DP-60 magnet, normally operated at 14.1 kG.

The instrument is built on the heterodyne principle with 10 MHz as the intermediate frequency. Both the lock frequency and the spectrometer frequency are derived from the synthesizer, which ensures that any source frequency drift would appear equally in both the lock and the observe channels thus resulting in maintaining the resonance condition.

A major benefit of using the heterodyne principle in a spectrometer is the ease of changing frequencies at moderate cost. In this system, the spectrometer frequency can be changed by about a factor of two by simply dialing a new frequency (Larmor frequency plus 10 MHz), retuning the series inductor, and changing the λ/4 cables, without having to change the probe or the impedance matching transformer.

The probe uses a single-coil for 20 mm spinning samples at room temperature. There are no tuning elements in the probe -- rather the tuning is by a variable inductor at the end of a λ/4 cable as described in the discussion of quarter wave cables in V.C.8. The transformer at the end of the second quarter wave cable matches the impedance between the tank circuit and the broadband preamplifier. The output

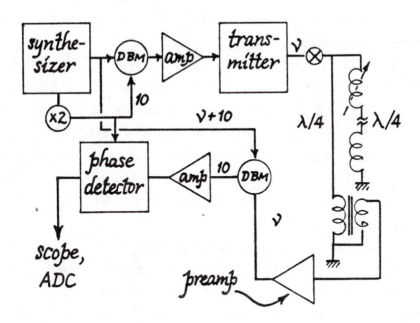

of the preamplifier is beat down (heterodyned) to 10 MHz which is the center frequency of the main amplifier.

The homebuilt computer interface contains a 12-bit ADC (DATEL) multiplexed to handle analog control inputs (such as phasing knob signals) as well as the NMR signal and 10- and 8-bit digital to analog converters for outputting information to a recorder and an oscilloscope, respectively.  The rf for the transmitter is obtained by mixing 10 MHz and ν+10 MHz from the synthesizer to create ν MHz and the level is subsequently boosted using a broadband amplifier to a level adequate for the transmitter. The transmitter is a gated modulator consisting of gated rf stages and power amplifier stages.  The computer provides all the gating pulses and delays although the transmitter pulse sequence and the rf pulse durations can be controlled on the front panel of the gated modulator.  The latter is a convenient feature for setting up the spectrometer and troubleshooting since it allows the apparatus to operate without the computer.

We list the specific components used for operation at 15.1 MHz, without any endorsements, implied or otherwise.

The frequency synthesizer is a General Radio (now GenRad) model GR 1164-A with the range 0.01-70 MHz.  It is used to provide the frequencies referred to as ν+10 MHz and 5 MHz. The 5 MHz is then doubled by a doubler (Relcom D8E) to provide the 10 MHz intermediate frequency.  The double balanced mixers are Vari-L model DB-100B.

The transformer is homemade with 4 and 9 turns on Amidon core as indicated while the cables are 50 ohm coaxial except for the $93\Omega$ $\lambda/4$ cable to the receiver.  The crossed diodes are 1N3731 and in the transmitter cable there are 6.8 volt rf Zener diodes (Unitrode) with silicon diodes in series to reduce capacitances. (See V.C.6.)

The transmitter is a MATEC gated modulator, model 5100, with rf gated amplifier plug-in model 515 (2-30 MHz;

1.5 kW peak; duty cycle 0.5%). Pulse lengths available are 1-100 μs at high power and 10-1000 μs at lower power. We use a broadband amplifier in the transmitter rf source line just to get the rf level correct for the transmitter while keeping it optimal for the phase detector reference, a very important consideration. It is an IFI wideband amplifier model 395A, an old amplifier which happened to be around.

The probe is homemade with a free standing 20 mm i.d. coil with 4 turns of #14 bare copper wire and a Wilmad 20 mm spinner. The NMR lock is a Bruker NMR stabilizer B-SN15 (pulse lock).

The receiver system consists of a preamplifier (Radiation Devices model BBA-1P, bandwidth 3-500 MHz at 15 dB gain and 3 dB noise), a 10 MHz amplifier (F+H Instruments mixer-amplifier, model 39 A/M), and a phase detector (F+H Instruments phase detector 1-30 MHz).

The computer is a Data General Nova 1210 with 24K memory and a homemade interface.

## V.C.2. AN EXAMPLE OF A HIGH POWER MACHINE FOR SOLIDS

This is an example of a general purpose high power spectrometer for solids. As sketched, it was set up for proton $T_1$ measurements at 63 MHz but it has been used at various frequencies between 2 and 85 MHz by changing the transmitter and the receiver preamps as well as by using different probes. We do not heterodyne as in the previous example because the receiver is broadbanded. This does mean that the phase shift must be adjusted for each frequency.

The MATEC transmitter is a tuned class-C amplifier that is designed as a pulse transmitter. It can put out pulses of peak power around 1-1.5 kW but cannot sustain such a power level over a long time. Even lower power pulses cannot be maintained without drooping for more than some milliseconds. Thus, it is not a good transmitter for $T_{1\rho}$, ZTR (VI.D.4.), or high power decoupling. This kind of transmitter, however, is a very good value for the money for high power at low duty cycle applications such as simple FT lineshape experiments or $T_1$ measurements. The transmitter has two amplifier plug-in units to cover the range between 2 and 160 MHz. One has nominally a 50 Ω output while the other has an impedance adjustment in the output circuit. The homebuilt probe has an impedance matching network consisting of a variable capacitor in series with the parallel resonance tank so that it is possible to match the impedance of the probe to either amplifier modules. Details of the probe are given in section V.C.4.

The MATEC receiver has a desirable provision of making the signal accessible between the amplifier stages. We insert an rf gate after the first such stage (with a nominal gain of up to 26 dB) to disable the receiver during the transmitter pulse. We find that putting such a gate farther upstream in this system does not make it work any better (i.e., the later stages are the ones which saturate first) and it degrades the S/N. The gate itself is a double balanced mixer gate using two Hewlett-Packard 10534B DBM's. With 560 Ω current limiting resistors, a 5v (TTL) signal will turn on the gate with a few dB attenuation while the attenuation is about 90 dB without any gating signal. We do not think the choice of the components or the configuration is too important here and we have not tried other possibilities.

The phase detected signal from the receiver goes to two

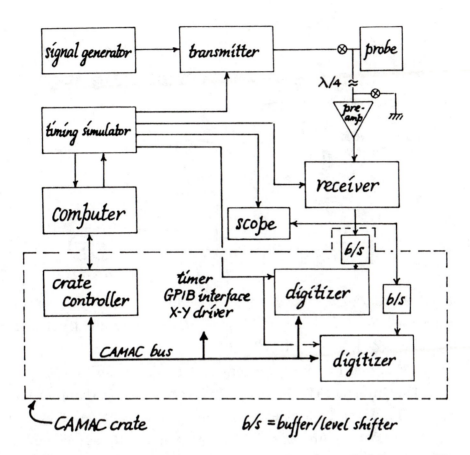

level shifter/buffers before going to the fast digitizers. The
buffering is necessary because the receiver output as well as
the digitizer inputs are designed to be 50 Ω which means that
when two digitizers are connected to the receiver output
without buffers the signal is attenuated by a factor of two.
The level shifting is necessary because the receiver output
has a dc offset in the dc coupled mode (the ac coupled mode
can be used for very short FID's only). Although the LeCroy
digitizers have provisions for level shifting, they are not
easy to use (screwdriver adjustments) and we also wanted
visual indication of the correct offset to match the dc level to
that suitable for the digitizer. Thus, our homebuilt dual

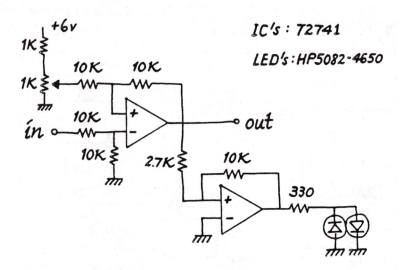

level shifter/buffer is a CAMAC module with a pair of LED
indicators in each channel to show if the incoming level is too
high or too low.   (CAMAC will be described in V.C.3.)   A
circuit diagram for one such channel is shown above.

The Interface Technology RS-648 timing simulator is used
here as a versatile pulse generator.   It is capable of outputting pulses through eight different ports (16 as an option)
for transmitter, receiver damping, digitizer trigger, scope
trigger, etc.   The device forms the pulses by running through
a series of programmable steps like a small computer.   Each
computer "word" contains information on how long that state
should last and which output should be high during that
state.   The longest time programmable is one second with a
resolution of 100 ns.   The length of the program can be as
long as 1024 words although for most NMR applications only
a few words are needed because of a loop cycling feature
which allows each sequence to be repeated up to 1000 times.
It also has the capability of adding a preamble to the loop
cycling such as for a CPMG sequence in which the first pulse

is different from the remainder. Since the timing simulator is computer programmable (an option) it is also possible to change the sequence even during a run.

The timing simulator is programmed for a particular pulse sequence at the beginning of the experiment. The Nova 1200 computer, through a homebuilt interface, sends a pulse to the RS-648 to start the pulse sequence at the appropriate time. The RS-648 takes care of the pulse sequence, triggers the digitizers to record the FID's, and returns the control of the experiment to the computer by a "done" pulse. The computer gets the data from the digitizer and then decides on the next course of action, which usually is to repeat the pulse sequence after a predetermined delay signal from the timing module to be discussed later.

In this spectrometer, the timing and the digitizing functions already mentioned, as well as several other functions are performed by modules located in a CAMAC crate. A discussion of CAMAC will follow in the next section (V.C.3.). Simply described, it is a commercially standardized hardware system interfaced to the minicomputer with a few hundred modules of all kinds available commercially.

The Bi Ra 1303 crate controller acts as the translator between the minicomputer and all the remaining CAMAC modules. In this example, it interfaces the Nova 1200 with the two digitizers, the timer, the X-Y driver, and the GPIB interface.

The digitizers are LeCroy 2256 waveform digitizers which can digitize analog signals with a resolution of 8-bits at speeds up to 20 MHz, i.e., a time between digitization of 50 ns. Such a high digitizing speed is not necessary for the usual FT application because the deadtime for the usual spectrometer will put an upper limit to the digitizing rate necessary. However, we have used the high digitizing rate to our advantage in looking at very broad lines for which spin echoes exist (usually metals) as very sharp spikes in the time domain. A secondary benefit of such a fast digitizer is that its front end amplifier will probably be very good so that no amplifier-induced distortion is expected for normal digitizing rates.

We have not yet mentioned why there are two digitizers although several examples are given in this book with two digitizers. They include quadrature detection (II.A.3.) and removal of coherent noise (VI.C.1.). You may have already noticed that the two digitizers in the figure are apparently connected identically which, of course, can't be true if they are not to work identically. The trick here is that the CAMAC dataway is not a single wire but contains many wires through which module-specific signals are sent by the crate controller. Thus, in a typical sequence, each digitizer responds to the trigger signals from the timing simulator only after a signal is received from the computer through the crate controller activating that digitizer. An example of a particular sequence with the timing information is given in the next section (V.C.3.).

The timer module (Bi Ra 2101) is really a scaler/timer to count nuclear events in a specific timing interval but we use only the timing capability of this unit. It outputs a signal to the computer after a delay originally specified by the computer As we have it set up, the timing range available is 1 ms to 160 s although the entire range can be shifted by 8 orders of magnitude, a capability obviously in excess of what we need. This unit controls the delays between pulse sequences, for example, for automated $T_1$ runs.

We have two output modules in the CAMAC crate. One is a Joerger x-y driver (module XY) which drives a low frequency x-y oscilloscope (Tektronix 503) for displaying computer outputs. The other output module is a GPIB interface (Kinetic Systems 3388) through which the signals from the computer are sent to a digital plotter (H-P 7225A) with alphanumeric capability so it can draw a spectrum and then label it.

There are many other spectrometers in the literature. The first classic paper was by Clark (1964). For subsequent works see, for example, the comprehensive article by Ellett, et al. (1971) as well as the article by Karlicek and Lowe (1978).

## REFERENCES

W. G. Clark, "Pulsed nuclear resonance apparatus," Rev. Sci. Instrum. 35, 316-333 (1964).

J. D. Ellet, Jr., M. G. Gibby, U. Haeberlen, L. M. Huber, M. Mehring, A. Pines, and J. S. Waugh, "Spectrometers for multiple-pulse NMR" in Advances in Magnetic Resonance, vol. 5, edited by J. S. Waugh (Academic Press, New York, 1971), pp. 117-176.

R. F. Karlicek, Jr., and I. J. Lowe, "A pulsed broadband NMR spectrometer," J. Magn. Resonance 32, 199-225 (1978).

## V.C.3.   USE OF CAMAC FOR COMPUTER INTERFACE

Mini- and micro-computers have now become an indispensable part of many NMR spectrometers. A major reason for this, of course, is the ever decreasing cost per performance of the entire computer family. In fact, the hardware cost of the computer itself is now so low that a major cost, either in terms of actual outlay or of time and effort, is that of the computer programs (software) and with the interface between the computer and the experimental apparatus. If a complete data acquisition and processing system is purchased, the software and interface costs are already included in the total cost. If only the computer is purchased the acquisition cost is minimal, but one faces the prospect of devoting much time and effort to developing the software and the interface.

There are several reasons why a researcher may choose to develop his own interfaces and software. An obvious one is to reduce the initial outlay of money and develop the necessary software and interface as an educational experience. Another is to develop a system more suited to the particular experimental needs than to buy a system which lacks the necessary versatility or which has highly developed features which are not needed.

This section points out the existence, apparently not well known in NMR circles, of a middle ground between developing a computer interface from scratch and buying a data system already interfaced to the spectrometer. We have assembled such an interface using CAMAC modules which are extensively used in particle physics experiments and we give a brief description here as an example of what is possible (Fukushima and Swenson, 1980).

The hardware cost associated with CAMAC also represents a middle ground between a homebuilt interface and a commercially built interface. The software must still be written for the particular system as in the case of a homebuilt interface but, due to the standardization of the commands required for the modules, the task is a little easier with the CAMAC interface. Furthermore, since CAMAC modules are commercially available, any programs written for a CAMAC system can be adapted quite easily for another CAMAC system.

CAMAC is an international standard for modular computer instrumentation. It specifies the electrical, mechanical, and functional characteristics of modular instruments to be plugged into a standard multi-receptacle "crate" which, in turn, can be interfaced to a minicomputer. The unique feature of CAMAC systems, as opposed to other modular systems, is that the modules not only get power from the crate but the "dataway" between the modules and the crate "controller" can carry 24-bits of data in each direction in addition to various function, address, and flag signals. The controller resides in the crate as a module and acts as an interface between the modules and a specific computer. Some of the lines in the dataway are location-specific so that a module in a given slot can be addressed by the controller. The modules (except for the controller) cannot communicate with other modules except through the controller.

There are many manufacturers of CAMAC equipment and the number of currently available modules exceed several hundred. Some of the modules are designed specifically for particle physics but many are general purpose units such as transient digitizers, digital to analog converters, oscilloscope drivers, and stepping motor drivers. See Appendix D for a listing of some of the CAMAC manufacturers.

The spectrometer described in the last section (V.C.2.) uses a CAMAC system with mostly commercial components. It consists of a crate with 25 slots, two of which are taken up with a crate controller to interface with a Data General NOVA 1200 computer. The essential modules in the crate, in addition to the controller, are two transient digitizers (each stores 1024 8-bit words recorded at rates up to 20 MHz), one timing module, a GPIB interface, and one X-Y driver. Together with other modules which are primarily used for utility and diagnostic purposes, these modules occupy 15 of the remaining 23 slots leaving some space for expansion within the crate.

The block diagram in the last section shows the spectrometer with the CAMAC modules identified. Several different experiments requiring different pulse sequences can be performed easily with such a system. A moderately complicated example is a spin-lattice relaxation time measurement in the time domain on a polycrystalline intermetallic sample containing I=3/2 nuclei. Since a non-cubic I=3/2 system has unequally spaced levels, special techniques must be used for relaxation time measurements (see III.C.3.) and we adopt the procedure of Avogadro and Rigamonti (1973) to initialize the populations before the magnetization recovery.

The two digitizers are used separately to record the desired FID at the beginning of a comb of 1000 pulses and a background signal to be subtracted which is recorded after the last pulse in the comb. The use of the two digitizers avoids having to interrupt the comb in order to process the first FID. The relaxation time data are taken by plotting the accumulated difference of the two signals as a function of the spacing between the combs.

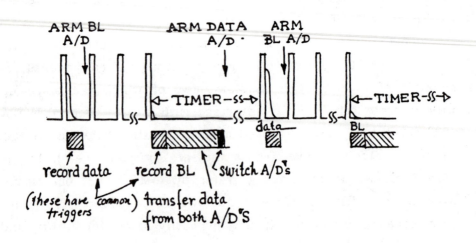

Now we will describe the sequence in more detail with the help of a timing diagram shown at left. A basic comb with the attendant trigger pulses (for example, to fire the transient digitizers after they are individually enabled by the computer) is generated by the timing simulator. The computer starts the pulse sequence by triggering the timing simulator (through a homemade interface since the simulator is not a CAMAC module). When the second digitizer is finished, the computer senses this through the crate controller and data from both digitizers are transmitted to the computer and subtracted before being accumulated in the memory.

The timing module was started at the end of the pulse sequence and the computer now waits for a signal from the timing module before initiating another comb. An advantage of the timing module being a CAMAC module is that since all the control functions are then interfaced through the dataway, the computer can send the desired timing information to the module. See the flow chart for some of the seemingly complex operations we have been describing.

Our procedure for a relaxation time run is to specify the number of points to be taken on the recovery curve, the time interval between those points, the number of FID's to be accumulated per point, and then turn the data acquisition over to the computer. Further embellishments include the fact that we calculate the recovery curve from integrals of selected parts of the on-resonance FID's and that we alternate the functions of the two digitizers with each pulse sequence to cope with the differences between the digitizers. After a specified number of FID's are accumulated at a given value of delay between the combs, a new value of delay is transmitted to the timer and the next point is measured. At the end of such a run, there will be a table of the magnetization recovery as a function of time.

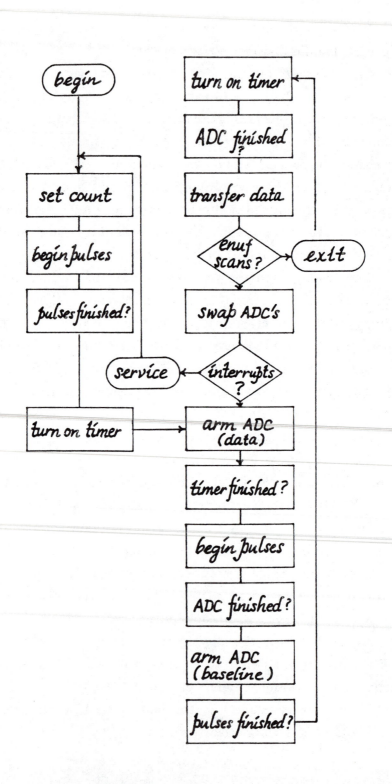

Further processing can proceed from there.

We have found the CAMAC system to be very convenient. After the initial investment for the crate and the controller, modules can be added usually at quite modest cost. In addition, troubleshooting is simplified by the availability of modules specifically designed for such uses. For example, we have a display module which shows the status of all the CAMAC dataway lines at specific times in the CAMAC cycle. The convenience of having a CAMAC system is enhanced considerably if there are other research groups nearby with CAMAC instrumentation so that modules may be shared or exchanged. Most of all, it is nice to have the feeling that we may never again have to design and build an elaborate computer interface, at least not to the extent before CAMAC.

REFERENCES

A. Avogadro and A. Rigamonti, "Nuclear spin-lattice recovery laws and an experimental condition for an exponential decay," in Magnetic Resonance and Related Phenomena (Proceedings of the XVIIth Congress Ampere, Turku, August, 1972), edited by V. Hovi (North-Holland, Amsterdam, 1973), pp. 255-259.

E. Fukushima and K. D. Swenson, "Use of CAMAC for computer interface in NMR," J. Magn. Resonance 39, 325-328 (1980).

V.C.4.   PROBES

One might wish to build a probe to operate at a new frequency, to use a different sample size or geometry, to do new experiments requiring different coils, or to save money. In fact, for the newcomer to NMR, this is likely to be the first instrument modifications he can undertake profitably. It is a good place to start because mistakes are relatively isolated from the rest of the instrument. To the novice probe building may seem to be a black art, but it is not as difficult as it first seems.

General considerations

First, if high resolution spectra are required, it is important not only to have high homogeneity in the $H_0$ field but also to have a sample geometry in which the shape of the sample permits it to be uniformly magnetized. Susceptibility variations owing to sample geometry are absent in only three geometries: (a) an infinite cylinder, (b) a sphere, or (c) a toroid. For most applications, the first two are more practical. This is why high resolution machines use long cylindrical samples which tend to look like infinite cylinders to the receiver coil. When sample size is limited, a micro-sample cell with a spherical cavity must be used to achieve the desired high resolution. For the same reasons, the probe must be constructed so that there are no large susceptibility variations in the probe material near the sample.

For variable temperature work, the coil can be placed immediately around the sample inside the dewar or outside the dewar which then contains the sample. The advantage of the former configuration is a better filling factor resulting in better S/N and shorter pulses per tip angle per power available. The advantage of the latter is that the coil does not need to be retuned as the sample temperature changes since the coil remains at ambient temperature. Usually, the sample access is also easier with this scheme.

The next decision is whether to use single-coil or cross-coil geometry. We already said in V.B.1. that the single-coil is preferable to the cross-coil geometry. The overriding advantages of a single-coil probe are that it is simple to construct and adjust, it is less susceptible to microphonics and temperature induced drifts, and it is economical in transmitter power, i.e., you get much larger $H_1$ amplitude for a given power than in a cross-coil probe. There are two reasons for the last point. One is that the larger size transmitter coil reduces $H_1$ in a cross-coil configuration. The other is that it is geometrically most convenient to split the transmitter coil to fit it around the receiver coil and a split coil is not as good a coil as a solenoid as we discuss shortly. Our subsequent discussion will basically apply to single-coil probes.

We note that it is possible to have a two coil probe in the form of two concentric solenoids as separate transmitter and receiver coils. This still provides the independent tuning advantage, even better $H_1$ homogeneity, and a reduced need for high pulse power. However, the receiver coil would experience a very high rf pulse intensity from the transmitter so you need to use a duplexer as in a single-coil probe so you might as well go to the simpler single-coil system.

## Construction and tuning

The probe enclosure should be very rigid to avoid microphonics, i.e., mechanical oscillations which affect the signal by, for example, modulating the capacitance between the enclosure and the coil. The main part of the enclosure may be milled out of a block of material, say aluminum, or stout parts (1/4 inch stock for the body and 1/8 or 3/32 inch covers) may be screwed together with many screws. It is

possible to purchase extruded sections of aluminum in the
shape of a rectangular cylinder which can be sliced to the
desired thickness to make probe bodies and then screw on
cover plates to the top and the bottom.  Although aluminum
is the usual choice it has certain drawbacks so that the probe
box may be made of other materials.  Brass often has ferromagnetic impurities and should be avoided for high resolution
work.  Before choosing a material, especially if you plan to
operate at low frequencies and large rf power, the discussion
on spurious ringing (VI.B.5) should be read.  An unconven-
tional choice for a probe "box" is a series of concentric glass
tubes appropriately fitted with ground glass joints employing
the cross-coil configuration (Jones, et al., 1965).  Rotating
the tubes relative to one another permits alignment of the
coils.

The capacitors chosen for the tank circuit should not be
ferromagnetic.  (A magnet can tell you immediately.)  Some
ceramic capacitors are piezoelectric and can give rise to transients induced by the rf pulse.  (See VI.B.5. for more on
piezoelectric ringing.)  Piston or vacuum types work well
provided they are not magnetic. Note that it is not necessary
to have any tuning elements in the probe itself, for example,
by having an inductor at the end of a quarter wave cable
(see V.C.6.).  The tuning elements must withstand a high
breakdown voltage.

None of the other components in the probe should be
ferromagnetic, either.  It often happens that only one part of
a component is magnetic and it is possible to replace it with a
non-magnetic part.  Also check the cable connectors for magnetism.  The absence of ferromagnetism in the cable and connectors becomes less important if you are not doing high resolution work and/or if the cable and connectors do not come

near the sample.

Choosing the physical parameters of the coil must be a compromise of several conflicting requirements. For the highest Q (which we do not necessarily want because it will result in the largest S/N but the worst recovery time) we need the largest inductance and the smallest losses. This can be accomplished only as a compromise since large inductance for a given coil size and shape requires the most number of turns, but that means small wire diameter for a given coil volume and high resistive losses. The rule of thumb is to wind a solenoidal coil with its length approximately equal to its diameter with the wire diameter equal to the spacing between the turns. (This is easy to do in practice by using two identical wires to wind a tightly wound coil on a cylindrical form like a dowel and unwrapping one of the wires.) Winding the coil on a threaded rod of the desired diameter and pitch is another common technique. For the highest Q, the coil should be constructed as neatly as possible. A necessary operation to this end is to stretch the wire by clamping one end in a vise and pulling on the other end with a pair of pliers before winding it into a coil. Then, it is easy to deviate from these guidelines to lower the Q, if necessary.

The rule of thumb stated above about making a coil with its length approximately equal to its diameter arises because the Q increases with coil length but quite slowly after the length exceeds the diameter. This rule and the other one about keeping the wire diameter equal to the spacing become mutually exclusive at high frequency where very few turns of coil will be needed. An improvement in Q is possible if some diameter is sacrificed for additional length. All of the foregoing means that solenoids are superior to Helmholz or saddle coils except at very high frequencies where the solenoidal coil

has to be short.  At such frequencies, other configurations
may become favorable such as a slotted line resonator
(Schneider and Dullenkopf, 1977).

For high frequencies requiring small inductance coils,
large diameter wires can be used to form free standing coils.
At the lower frequencies a form, even if it is the sample
itself, must be used to support the coil because the wire size
must be reduced to increase the number of turns in a given
coil volume.  Unstripped magnet wire (wire with a varnish
insulation) should not be used for proton NMR of solids in
order to avoid unwanted signals from the coating.

The strength of the rotating $H_1$ field (in gauss) is
(Clark, 1964)

$$H_1 \cong 3 \ (PQ/V\nu_o)^{1/2} \cong 3.7 \ (PT_R/V)^{1/2}$$

where P is the transmitter power in watts, Q is the quality
factor, $\nu_o$ is the frequency in MHz, V is the volume of the
coil in $cm^3$, and $T_R$ is the ringdown time of the coil in μs.
For high $H_1$ fields, it is essential to keep the coil volume
small, use high power, and have a high Q which usually
implies a long $T_R$.  The Q is given by

$$Q = 2\pi\nu_o L/R$$

where L and R are the coil inductance and resistance.  The Q
can be made large by minimizing the resistance and maximizing
the inductance for a given resonant frequency.  This maximizes $H_1$ during the pulse and increases the sensitivity of the
coil to the NMR signal.  On the other hand, a large Q means
poor pulse ringdown time which is counterproductive.  Generally, one compromises on the choice of parameters to optimize

the performance for the specific type of experiment.

The ringdown and thereby the system recovery time becomes worse at lower frequencies. It is easy to appreciate this fact if one views the damping of the excitation as occurring in a certain number of cycles. Long ringdown times may cause significant problems in observing low frequency nuclei, e.g., deuterium in solids. We refer you to our earlier section on recovery times (V.A.4.) for further discussions.

The following formulas (Landee, et al., 1957; also <u>Radio Amateurs' Handbook</u>) help to estimate how many turns are required for a given solenoidal coil geometry. If "a" is the diameter and "b" the length of the coil in inches, the inductance L in microhenries is given by

$$L = (n^2 a^2)/(9a + 10b)$$

whereas the same formula for metric dimensions with diameter a' and length b' in centimeters is

$$L = (n^2 a'^2)/(3.5a' + 4b').$$

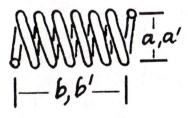

The exact number of turns will have to be determined empirically. The coil inductance can be measured with an inductance bridge or a Q-meter.

Once the coil is wound, it has to be mounted. A freestanding coil can be mounted on two standoffs. However, microphonics or other spurious ringing might exist. During the pulse, a large amount of current flows in the coil. In the presence of a magnetic field, large forces are exerted on the coil causing it to oscillate mechanically as well as setting up acoustic standing waves in the coil. This can cause additional ringing following a pulse, greatly increasing the recovery time. These are frequency dependent and are less of a problem at high frequencies. The existence of these problems can be determined by comparing the ringdown with the magnet on and with it off.

For these problems, one has several alternative solutions. The coil can be potted in a material that does not contain a nucleus of interest. Sulfur has been used to avoid all possible resonance signals from the potting. For non-proton and non-carbon work, Q dope or RTV cement work well and are easier to use than epoxy. It is also possible to wind the coil on a form such as a glass tube or a tube of boron nitride (preferably thin wall to maximize the fill factor unless the coil is wound on the inside) and attach it with epoxy resin, RTV, etc., or with tightly wound Teflon tape. As epoxy, Q-dope, and RTV give proton signals with short $T_2$, it will be satisfactory for observation of protons in liquids but not in solids; similarly Teflon tape will work except when observing fluorine or carbon in the solid state. For observing boron in solids, any inserts or dewars should be made of quartz, not ordinary glass or Pyrex (or boron nitride!). Finally, the wire can be changed, for example, to a stranded wire, to change the mechanical properties of the coil to get away from these unwanted resonances. The reader/probe-builder is once again referred to the discussion on spurious ringing (VI.B.5.) for further information.

Because the sample coil is such an essential part of the

NMR apparaus, it is very important that great care is taken in its design and construction. One of the trivial ways in which the performance of a coil can be significantly degraded is to have defective solder joints. Possibly because of the high currents involved in the transmitting mode, poor joints in a probe tend to deteriorate rapidly. A symptom of this problem is decreased (or no) signal coupled with a significant increase in the spurious noise such as the 60 Hz hum.

Once the coil has been incorporated into a tank circuit, a dip meter can be used to measure the resonance frequency. For a parallel resonant circuit, the dip meter can be used in the usual way, i.e., look for a "dip" indication as a function of frequency. For a series tank circuit, it is best to use the dip meter as an rf source acting like an artificial NMR signal and maximize the receiver output when the receiving system is fully hooked up. This will work also for the parallel tank. (See section V.C.9. on impedance matching.)

The probe can now be tested for $H_1$ homogeneity, $H_1$ strength, and ringdown. One method of measuring the $H_1$ homogeneity is by comparing the size of the signal following a 90° pulse with that which follows a 270° or a 450° pulse. (See VI.B.4. for further details.)

It should be noted that very high $H_1$ homogeneity is necessary only in certain cases. A high resolution spectrum results even if different parts of the sample experience different $H_1$'s. In a cross coil probe, it is easier to achieve uniform $H_1$ while having the sample approximate an infinite cylinder with respect to the receiver coil. High $H_1$ homogeneity is important for multiple pulse experiments, however.

Odds and ends

It is useful to make marks on the side of the probe box

indicating the center and the extent of the rf coil. This permits better alignment of the center of the coil with the center of the pole faces (which, however, may not be the spot having the highest homogeneity) and also indicates the appropriate depth to which the sample should be inserted. It is easy to determine whether the rf coil is centered on the homogeneity shim coils by adjusting the x and y gradients. If the sample is on resonance and centered, the variation of these controls should affect the resonance width but not the position.

A thermocouple in the probe can cause a variety of problems. For example, the rf pulse may disturb the temperature controller hooked to the thermocouple. A more common problem is that the thermocouple can act as an antenna for spurious signals. The thermocouple should be located as close as possible to the sample without introducing these problems. If it is necessary to locate the thermocouple well removed from the sample, a calibration is needed to correct for the temperature difference between it and the sample.

With variable inductors or capacitors of sufficient range, it is possible to make a multinuclear probe which can be converted from one frequency to another merely by changing coils wound on inserts. For single coil probes such inserts can be changed conveniently by modifying commercial rf connectors, for example, BNC, and attaching them to the ends of the inserts.

For a good filling factor, the coil can be wound directly on the sample and affixed with an appropriate dope, tape, or just friction. This is useful when one wishes to make many measurements on a particular (non-spinning) sample and sensitivity is of utmost importance.

One unusual means of generating a very homogeneous and

intense rf field is to wind a toroidal coil around a toroidal sample.

An ingeneous probe which permits a very fast recovery is designed as a lumped parameter delay line (Lowe and Englesberg, 1974; Lowe and Whitson, 1976). Although this design suffers from an unusual sample shape requirement, the general performance is outstanding.

An engineering principle to bear in mind when evaluating various coil configurations for their effectiveness in NMR is the "reciprocity theorem". In practical terms it states that the maximum contribution to the signal from an element of a sample is proportional to the field at that spot if the coil were used as a transmitter coil (Hoult, 1978).

## An example: a 63 MHz probe for solids

We now give an example of a homebuilt probe used in the spectrometer described in section V.C.2. It is a single coil probe with an impedance matching network built in. It is built in an enclosure made from a sawed off section of an aluminum extrusion supplied by Pomona. Its exterior dimensions are 68×105 mm and thickness 40 mm (including two cover plates each of 3 mm thickness). Each cover plate is screwed on to the main body with 8 screws. A sketch of it follows.

The free standing coil consists of 5-1/2 turns of #14 copper wire. Its inside diameter is 19 mm and length 12 mm. Whenever possible, we like to make free standing coils because of their simplicity and the absence of spurious resonances from the coil forms, although it is not possible to make the coils free standing at low frequency because of the need to use finer wire. We would say that #14 wire is about the smallest that can be recommended for a free standing coil of these dimensions.

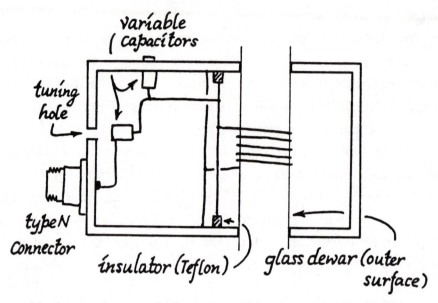

We have also used the same #14 wire to make the connections between the coil, the two capacitors, and the type N
connector.  Mechanical supports were used at both ends of
the wires to which the coil is attached and no part of the
wires is farther than about 20 mm from the closest support.

The high impedance of the parallel resonance circuit is
transformed to 50 $\Omega$ nominally by the second variable capacitor.
Both variable capacitors are Johanson #5341 which are piston
air capacitors with Be-Cu stators and Ag rotors.  They work
very nicely except for the problem of high voltage breakdown.
The dc breakdown under test conditions is listed as 1200
volts at sea level but that number has to be severely derated
for radio frequency and altitude.  (The latter problem is
serious at Los Alamos where the atmospheric pressure is 590
mm Hg and the altitude correction to the breakdown is about
10%.)

A caveat in this type of probe is that the impedance
matching capacitor is floating at high voltage.  Thus, no

attempt to tune it should be made with the power on unless some provision is made to electrically isolate it from the mechanical tuning mechanism. We do not bother with such a provision and make all the impedance adjustments with a low level cw signal.

REFERENCES

W. G. Clark, "Pulsed nuclear resonance apparatus," Rev. Sci. Instrum. 35, 316-333 (1964).

D. I. Hoult, "The NMR receiver: A description and analysis of design," Prog. NMR Spectry. 12, 41-77 (1978).

G. P. Jones, D. C. Douglass, and D. W. McCall, "Apparatus for the detection of ultraslow atomic motion by magnetic resonance techniques," Rev. Sci. Instrum. 36, 1460-1465 (1965).

R. W. Landee, D. C. Davis, and A. P. Albrecht, The Electronic Designers' Handbook (McGraw Hill Book Co., New York, 1957).

I. J. Lowe and M. Englesberg, "A fast recovery pulsed nuclear magnetic resonance sample probe using a delay line," Rev. Sci. Instrum. 45, 631-639 (1974).

I. J. Lowe and D. W. Whitson, "Homogeneous rf field delay line probe for pulsed nuclear magnetic resonance," Rev. Sci. Instrum. 48, 268-274 (1977).

The Radio Amateurs' Handbook (American Radio Relay League, Inc., Newington, Conn., 1973).

H. J. Schneider and P. Dullenkopf, "Slotted tube resonator: A new NMR probe head at high observing frequencies," Rev. Sci. Instrum. 48, 68-73 (1977).

## V.C.5.   PROBES FOR SUPERCONDUCTING SOLENOIDS

The major qualitative difference between an iron-core electromagnet and a superconducting magnet is their geometry. The usual electromagnet for NMR is constructed so that the field goes across the gap and is horizontal as sketched on p. 9.   A solenoidal sample coil perpendicular to the field will, therefore, have its axis in the plane of the gap and in particular the axis can be vertical which is very nice because long sample tubes can be lowered into the coil.   The SC solenoid, on the other hand, has its field along the bore which is nearly always vertical so that a receiver coil must have its axis sideways, that is, in the small dimension of the available space.

The two common ways to do this are either to use a split coil (such as a Helmholz or a saddle coil) or to use a solenoid oriented perpendicular to the bore.   The former scheme allows easy access along the bore for the sample even for a spinning configuration.   A representation of a saddle coil is shown at right.   The problem with such a split coil is 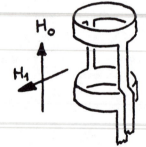 that it is not as good a coil as a good solenoid (Hoult and Richards, 1976).   Recall that (see the previous section) the maximum Q for a coil is obtained with a spacing between turns approximately equal to the wire diameter.   Therefore, any split coil is bound to have a much lower Q than a good solenoid especially when it has relatively few turns as it would at high frequencies.   The saddle coil shown above is the worst case of this since we can think of it as a two turn solenoid with a big gap. This is one of the reasons why many of the new NMR spectrometers with superconducting magnets do not yield the theoretical gain in S/N expected from the $H^{3/2}$ dependence, i.e., the gain in using the higher field is offset by the loss in using a poorer coil. For other effects offsetting the expected gain at high fields, see the discussion under Frequencies in V.B.1.

A sideways solenoidal coil in the bore of the magnet does indeed regain the geometrically lost S/N but now the configuration is less convenient to use. Even for non-spinning experiments, the concept of the usual sample tube must be modified to that of a tube short enough to fit sideways in the available space. Thus, we are talking about tubes about 3 cm long, at most, in any but the widest bore magnets especially if a dewar is involved and, at the same time, the requirement for field homogeneity across the bore becomes more stringent. Furthermore, the tube must be completely filled with the sample for the best S/N with the sideways coil, which may be an inconvenience. For spinning samples, special sideways spinners have been designed and used successfully (Oldfield and Meadows, 1978).

As is often the case, the user must choose between a convenient but less sensitive setup with the saddle-type coil versus a less convenient but much more sensitive set-up with a sideways solenoidal coil. In our experience the inconvenience of using the sideways coil is not major and eventually many of the NMR spectrometers with superconducting magnets may utilize the sideways coil.

REFERENCES

D. I. Hoult and R. E. Richards, "The signal-to-noise ratio of the nuclear magnetic resonance experiment," J. Magn. Resonance 24, 71-85 (1976).

E. Oldfield and M. Meadows, "Sideways-spinning 20-mm-tube probe for widebore superconducting magnet spectrometer systems," J. Magn. Resonance 31, 327-335 (1978).

## V.C.6. CROSSED DIODES AND
## OTHER NON-LINEAR ELEMENTS

In pulse NMR the use of non-linear elements, usually in the form of crossed diodes, is widespread.   Consider a pair of silicon diodes with the diodes antiparallel.

Each diode  has the property that no current can flow against the arrow (the reverse direction) but current can flow in the direction of the arrow (the forward direction) provided that the potential driving the current exceeds a fairly well defined threshold which is about ½v for most silicon diodes.   This means that the crossed diode circuit as pictured above and symbolically   represented   as  shown has the property that it will pass current in either direction if the voltage is greater than ½v or less than -½v; it is a non-linear element because it looks like a good conductor to large incoming signals but like a poor conductor to small signals of either polarity.   Therefore, the crossed diode acts like a switch which is on for large signals and off for small signals.

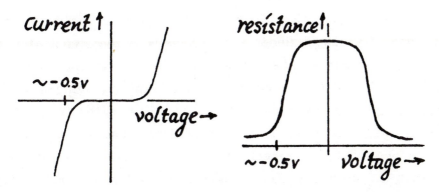

Such a crossed diode unit is obviously a good thing to put across the input of the receiver, as we have pointed out elsewhere. The receiver is all set to receive a very small signal from the nuclear moments in the coil but first it must dodge what's left of the big pulse coming through the matching network. (Even with a good duplexer, the remains of the transmitter pulse is likely to be tens of volts while the desired signal is $10^4$ to $10^5$ smaller.)

A pair of crossed diodes in parallel with the input to the receiver acts as a shunt for signals exceeding the ½v threshold while acting as if it were not there at all (ideally) for the small NMR signal.

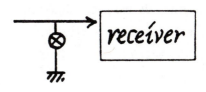

Logically, there could be another kind of non-linear circuit with the opposite characteristic; i.e., a good insulator for large voltages and a good conductor for small voltages. Such a unit could be put in parallel with the output of the transmitter to short out only the low level tail of the transmitter pulse without attenuating the main part of the pulse. Unfortunately, its implementation is limited to cross coil setups and therefore is not used as often. For a tube version of such a device, see Spokas (1965).

Besides protecting the preamplifier input, the most common use of crossed diodes is in series with the transmitter in order to disconnect the transmitter ouput circuitry from the tank circuit and the preamplifier during the receiving mode in order to reduce the noise. If the circuit with the transmitter output still connected is nearly critically damped and if an active Q-spoiler (or conceivably even a passive one) is used so that the circuit is damped adequately after the diodes stop conducting, the level at which the diodes stop conducting is immaterial so long as it is between that of the transmitter pulse and of the NMR signal levels. Since those two levels are separated by some 8 orders of magnitude, the 0.5 volt level at which normal silicon diodes switch is perfectly acceptable. If, for some reason, the circuitry with the diodes still conducting is nowhere near critically damped (possibly because of some peculiarity in the design of the transmitter output circuitry) but the tank circuit can be damped after the diodes cut out, it may be desirable to have series crossed diodes with thresholds for switching considerably higher than 0.5 volt. (This section has presumed that you are familiar with the discussions of section V.A.4.)

Such a device can be made from Zener diodes and

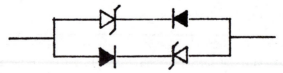

ordinarly diodes as shown. A Zener diode (rhymes with cleaner) acts like a regular diode which also conducts in the reverse direction when the voltage (in the reverse direction) exceeds a specified threshold which can be much higher than the ½v forward threshold voltage for silicon diodes. Such a

combination of a regular silicon
dine and a Zener diode as
shown acts like the regular
diode with the forward conduction threshold given by the
Zener voltage instead of the ½v for the usual silicon diodes.
The crossed diodes made up of these units, then, have thresholds given by the Zener diodes so that appropriate Zener
diodes may be chosen to set the threshold at the desired level.
The extra capacitances of the additional Zener diodes are not
overly bothersome because they are reduced by the series
capacitances of the normal diodes.

It is also possible to make crossed diodes with smaller
than normal thresholds (Stokes, 1978). Such a device will be
very useful to shunt the receiver input because the threshold
voltage of the usual crossed diodes is so big compared to the
NMR signal that the receiver can still become overloaded even
though it is protected.

We have emphasized these passive crossed diodes because
they are easy to implement and are quite effective. There
are more complicated non-linear elements which are active (in
the electrical sense), the most noteworthy ones being PIN
diodes (Kisman and Armstrong, 1974; Hoult and Richards,
1976). Although we have not tried them, they should work
very well so that people who have special requirements ought
to know about them. In particular, ordinary crossed diodes
do not work at high frequencies, say above 100 MHz, because
of their capacitances which cause their impedances to become
too small at these frequencies. PIN diodes, on the other
hand, do not work well at the lower frequencies, say below
100 MHz, so they complement the silicon diodes very nicely.
There are now commercial PIN diode switches which can be
used quite conveniently, although it is still an active com-

ponent requiring an external gating signal.

One more comment about crossed diodes: the series diodes between the transmitter and the probe usually can have much more capacitance than the parallel ones shunting the receiver preamp. Therefore, the former can be fairly husky to handle the higher currents associated with the transmitter pulse, while the latter can be small high speed units with just enough current carrying capacity to clip off the residual transmitter pulse.

## REFERENCES

K. E. Kisman and R. L. Armstrong, "Coupling scheme and probe damper or pulsed nuclear magnetic resonance single coil probe," Rev. Sci. Instrum. 45, 1159-1163 (1974).

D. I. Hoult and R. E. Richards, "An ultra high frequency receiver protection scheme," J. Magn. Resonance 22, 561-563 (1976).

J. J. Spokas, "Means of reducing ringing times in pulsed nuclear magnetic resonance," Rev. Sci. Instrum. 36, 1436-1439 (1965).

H. T. Stokes, "Tuned limiter for receiver amplifier in a fast-recovery pulsed NMR spectrometer," Rev. Sci. Instrum. 49, 1011-2 (1978).

## V.C.7.   SINGLE COIL DUPLEXERS

In section V.A.1 we described the general function of duplexers in single coil NMR. We now give a survey of the more common duplexer schemes. We start, however, with two examples of not so common schemes using commercial components because of their convenience and simplicity.

These schemes involve the use of a commercial magic-tee or a directional coupler for both cw (Klein and Phelps, 1967) and pulse NMR. A magic-tee, described further in V.C.10., can be thought of as a four port (or terminal) device with the property that an rf input at port 1 splits between ports 2 and 3 with nothing going to port 4 if the impedances of ports 2 and 3 are matched. Therefore, it works as an NMR duplexer with the rf input at port 1, a tank circuit containing the sample at port 2, a dummy load to match the tank circuit impedance which is purely resistive at port 3, and the output at port 4. Port 4, then, sees only the differences between ports 2 and 3 and while the two ports would normally look the same, the NMR sample at resonance unbalances the bridge. Schematically, the magic-tee setup looks as shown.

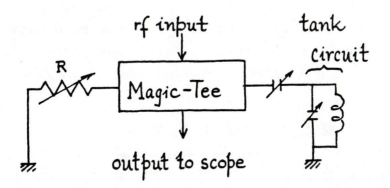

Since the usual magic-tee is a broadband device (like 1 - 100 MHz), only the tank circuit parameters need to be changed upon changing the Larmor frequency. The usual tank circuit for this setup is a series LC tank with a parallel capacitor to adjust the total impedance to the appropriate one for the dummy resistor.   In the example, the required relation is

$$(1 + w^2 LC_1)/[w(C_1 + w^2 LC_1 C_2)] = R$$

in addition to the resonance condition

$$w^2 LC_1 C_2/(C_1 + C_2) = 1 .$$

The magic-tee is inherently noisy because the receiver sees noise not only from the tank circuit but also from the dummy load and because half of the signal from the tank circuit is sent back towards the transmitter.   The easy way out of this limitation is to use a directional coupler instead of a magic-tee.   A directional coupler, also described further in V.C.10., is a three port device which amounts to an asymmetric power splitter and it works well as a duplexer.   The isolation between the receiver and the transmitter is achieved when the tank circuit impedance is adjusted to be a real and specified value, just as with the magic-tee.   The attenuated path is placed between the transmitter and the tank circuit which means that more input power is required than with a magic-tee

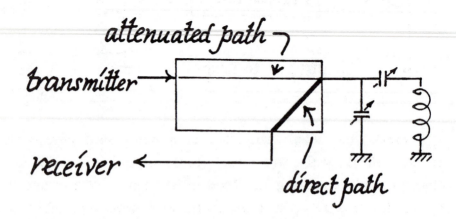

but the receiver has a direct path from the tank circuit which results in better S/N. In effect, the directional coupler duplexer concentrates the "losses" in the transmitting path whereas the magic-tee distributes them symmetrically between the transmitting and receiving paths. The losses in the transmitting path can be overcome simply by more power whereas losses in the receiving path irrevocably degrade S/N.

The most common single coil duplexers are designed with cables of appropriate lengths such that the receiver is at the wrong electrical distance from the transmitter to see the transmitter signal while the probe is at the right distances from both the transmitter and the receiver (Lowe and Tarr, 1968).

An example of such a setup using suitable length cables is given in the next figure (Ellett et al. 1971). The crossed silicon diodes are invaluable in pulse NMR, as already described in V.C.6. They are like switches which will close for large signals but not for small signals. That being the case, we can redraw the figure according to whether the transmitter is on or off, i.e., whether the switches are open or closed, as shown. Now let us move on to the quarter wave or $\lambda/4$ cable. A quarter wave cable acts as a transformer with the property that $Z_i Z_o = |Z|^2$ where $Z_i$ and $Z_o$ are the input and output impedances and $Z$ is the characteristic impedance of the cable which is usually between 50 and 300 ohms. This impedance condition is what makes this work.

During the transmitter pulse, both diode switches are closed. Without the quarter wave line, much of the transmitter power would have been incident on the second diode switch instead of on the tank circuit. With the quarter wave line, however, the low impedance of the diode switch (in the on condition) gets transformed to a high value by the quarter wave line so that, for all practical purposes, there is nothing hooked up to the circuit past the tank circuit.  We can also ignore the resistor which was in parallel with the tank circuit since, at resonance, the tank circuit looks like a very small impedance and the resistor looks like an open circuit in comparison.  When the transmitter pulse is off, both diode switches are open.  As far as the transmitter is concerned, it

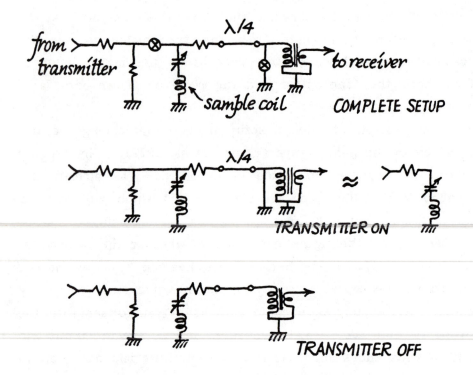

only sees a bunch of resistors as the load and the residual oscillations are damped out in these resistors.

As far as the receiver is concerned, the operation of the circuit is also straightforward.  The low impedance of the tank circuit plus the resistance of the series resistor is transformed by the quarter wave transformer.  The transformed value is matched to the impedance of the receiver by a conventional transformer. This last match is important because the noise generated by an amplifier (such as that in the front end of the receiver system) can be minimized by a choice of the impedance as seen by the amplifier.

As you might guess, there are several problems with using quarter wave lines, even though they work very well under many circumstances and, as already implied, are widely used. One is that they are not broadband devices, that is, a particular quarter wave line is just that for about a 10% variation of the frequency. A second and related problem is that the characteristic impedance Z which governs the transformation property (or the "turns ratio") through the relation $Z_iZ_o$ = $|Z|^2$ is also an inherent property of the cable. Therefore, you have to have a set of quarter wave lines for each frequency as well as for each characteristic impedance which may be acceptable for single frequency operation but not for multinuclear operation. The third problem is that quarter wave lines are very cumbersome and inefficient at low frequencies. With a 50 Ω coax, a quarter wave line is about 3 meters (10 feet) at 15 MHz, which is not bad. At 3 MHz, however, it is 15 meters and that is bad.

The way to get around this problem is to construct a circuit out of lumped components which takes the place of a quarter wave cable. There are many such circuits in the literature and, in fact, many of the single coil duplexers which are not bridges and which do not use quarter wave cables probably utilize such a circuit, intentionally or otherwise. See, for example, the articles by Clark and McNeil (1973) and McLachlan (1980) as well as the widely quoted older works by Gray, et al. (1966) and McKay and Woessner (1966). As far as we know, however, there has been only

one design of a circuit which is exactly equivalent to a quarter
wave line and this is discussed in the following section
(V.C.8).

We give one more example of a duplexer utilizing lumped
components and crossed diodes (Piott and Husa, 1973). The
high impedance of the circuit past the probe during the transmitter pulse is achieved by having L and $C_1$ look like a parallel resonant circuit. The NMR signal couples to the receiver
through a series tank circuit made up of L and $C_2$.

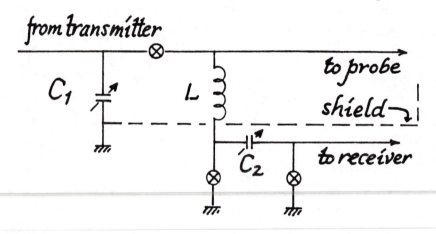

A duplexer based on the bridge-tee (Tuttle, 1940) was
proposed by Jeffrey and Armstrong (1967) and we have used
two versions of it in variable frequency applications up to
about 30 MHz. These compact circuits have worked well down
to about 2 MHz so this scheme represents another alternative
to the quarter wave line schemes at the lower frequencies.
Its input and output impedances are equal and are easily
adjustable to a desired value, for example, 50 Ω.

All of these schemes use the ac characteristics of the rf
and therefore are dependent on the frequency. The transmitter pulse contains many different frequencies besides the
carrier frequency (the only pure monochromatic waveform is

an infinite duration sinusoidal wave) and the duplexer does not block out these other frequencies. (See the sketch on p. 401.) Since the main job of the duplexer is to protect the receiver from the transmitter pulse, there is no problem as long as most of the transmitter power is contained in that frequency component at the carrier frequency. This is equivalent to the statement that the pulse will be long compared with a cycle of rf, an easily satisfied condition most of the time. The components of the pulse at frequencies other than the carrier can be suppressed most easily by a shorted quarter wave line shunt described in the following section.

Finally, we mention a rather unique broad band duplexer by Engle (1980) which is not a passive duplexer like all the others described so far. This duplexer detects the transmitter pulse and uses it to actively switch between the transmit and receive functions. It was tested over the range 5-150 MHz and was characterized by a very good noise figure of 2 dB.

## REFERENCES

W. G. Clark and J. A. McNeil, "Single coil series resonant circuit for pulsed nuclear resonance," Rev. Sci. Instrum. 44, pp. 844-851 (1973).

J. D. Ellett, Jr., M. G. Gibby, U. Haeberlen, L. M. Huber, M. Mehring, A. Pines, and J. S. Waugh, "Spectrometers for multiple-pulse NMR" in Advances in Magnetic Resonance, vol. 5, edited by J. S. Waugh (Academic Press, New York, 1971), pp. 117-176.

J. L. Engle, "Low-noise broadband transmit/receive circuit for NMR," J. Magn. Resonance 37, pp. 547-549 (1980).

K. W. Gray, W. N. Hardy, and J. D. Noble, "Optimized pulsed NMR single coil circuit design," Rev. Sci. Instrum. 37, pp. 587-588 (1966).

K.  R.  Jeffrey  and  R.  L.  Armstrong,  "Simple  bridge  for
pulsed  nuclear  magnetic  resonance,"  Rev.  Sci.  Instrum.  38,
pp.  634-636  (1967).

M.  P.  Klein  and  D.  E.  Phelps,  "Radio  frequency  hybrid  tees
for  nuclear  magnetic  resonance,"  Rev.  Sci.  Instrum.  38,  pp.
1545-1546  (1967).

I.  J.  Lowe  and  C.  E.  Tarr,  "A  fast  recovery  probe  and
receiver  for  pulsed  nuclear  magnetic  resonance  spectroscopy,"
J.  Phys.  E.  :  Sci.  Instrum.,  1,  pp.  320-322  (1968).

R.  A.  McKay  and  D.  E.  Woessner,  "A  simple  single-coil  probe
for  pulsed  nuclear  magnetic  resonance,"  J.  Sci.  Instrum.  43,
pp.  838-840  (1966).

L.  A.  McLachlan,  "Lumped  circuit  duplexer  for  a  pulsed  NMR
spectrometer,"  J.  Magn.  Resonance  39,  pp.  11-15  (1980).

J.  E.  Piott  and  D.  L.  Husa,  unpublished,  1973.

W.  N.  Tuttle,  "Bridged-T  and  parallel-T  null  circuits  for
measurements  at  radio  frequencies,"  Proc.  IRE  28,  pp.  23-29
(1940).

## V.C.8.   QUARTER WAVE LINES
### AND A QUARTER WAVE NETWORK

We  have  already  stated  the  single  most  important  characteristic  as  well  as  some  of  the  drawbacks  of  the  quarter  wave
($\lambda/4$)  cable  in  the  last  section.   We  amplify  some  of  the  concepts  about  quarter  wave  lines  in  this  section  and  then  describe  an  electrically  equivalent  circuit  which  offer  several
advantages  over  the  quarter  wave  line  itself  at  the  lower
frequencies.

The  all  important  characteristic  of  a  $\lambda/4$  line  is  that  it
acts  as  an  impedance  transformer  with  the  transformation  given

by $Z_i Z_o = |Z|^2$, a real quantity, where $Z_i$ and $Z_o$ are the input and the output impedances and where $Z$ is the characteristic impedance of the cable being used. Thus, if we use a cable with a characteristic impedance of 50 $\Omega$, the most common value, it will transform a 50 $\Omega$ impedance into 50 $\Omega$, i.e., it will be a 1:1 transformer. This, in itself, is nothing special since a half or a full wave (or any multiple, thereof) cable has the same property.

Suppose, though, that we now hook up a 10 $\Omega$ resistor to the input. The output impedance is now 250 $\Omega$ and is real. What if the input is shorted (R=0)? The output impedance will be 1/0=$\infty$ which means that hooking up a shorted quarter wave line is like hooking up absolutely nothing! If you don't believe this, try it. Use the scheme for measuring the $\lambda/4$ length described on p. 403.

All bridges and quarter wave lines work only near the design frequency because they are not broadband devices. (The $\lambda/4$ cable has a bandwidth which is about 10% of the de- 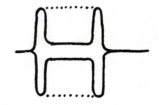 sign frequency.) Therefore, when an rf pulse impinges on a duplexer, what gets through to the receiver might look like what's shown in the sketch. The center of the pulse is attenuated by the duplexer as intended but the ends of the pulse come through because many different frequency components are necessary to form the beginning and the end of the pulses and they are not at the design frequency for the device.

What if we now connect a shorted $\lambda/4$ cable to a part of the circuit which has such a feedthrough signal? Since it acts as an open circuit only for signals at the design frequency, it will attenuate all other frequencies and, therefore, will attenuate the signals coming through at the beginning and the end of the pulse as sketched above. When an FID comes by on the same line, the shorted quarter wave cable will not attenuate it. So, we have a selective filter here. It passes information only at the correct frequency and it is often used, for example, in front of the receiver preamp in addition to the usual crossed diodes to protect the receiver.

We could think of more sophisticated variations such as a quarter wave line terminated in a series cross diodes which now acts like a shorted quarter wave line only for large signals and an open one for small signals. This, then, is a passive nonlinear circuit which selectively attenuates small signals while ignoring the large signals. A further variation is to put a 50 Ω resistor in parallel with the series cross diodes in the quarter wave line above. Now the circuit will look like an open circuit for big signals and a 50 Ω load for small signals.

Now we turn to the question of how to obtain, make, and measure a particular $\lambda/4$ line. We can make a rough estimate of the length by remembering that the speed of light (and rf radiation) in vacuum is $3\times10^8$ m/s so the wave length of rf signal at frequency $\nu$ is $3\times10^8/\nu$ (meters). In a flexible coaxial cable, the dielectric is not vacuum so the velocity of light is considerably slower. With a typical cable, say RG-223/U, the wavelength is more like $1.8\times10^8/\nu$ (meters) so a quarter wave line length in meters is given by $45/\nu$ where $\nu$ is in MHz. Thus, for a Larmor frequency of 90 MHz, the quarter wave line is about 50 cm with this cable.

Because of the transformation properties already discussed, quarter wave lines are very easy to measure. Recall

that an open ended quarter
wave line looks like a short.
So, all you need is a frequency source (with a non-zero
output impedance) like a signal generator, a T-adapter,
and an oscilloscope capable of

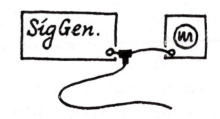

displaying the rf. Connect the generator output to the scope
through the T, connect the cable to be measured to the T,
and change the generator frequency and look for the minimum
signal. This minimum should be quite sharp (as opposed to
the maximum with a shorted quarter wave line in the same
setup) and the generator frequency at the minimum is the
frequency at which the cable is a quarter wave line. Use the
rough estimate as we discussed above to make sure that you
have a $\lambda/4$ line rather than, say, a $3\lambda/4$ cable. People who
are not familiar with $\lambda/4$ lines are quite amazed (and amused)
that hooking up a plain cable with nothing else on it will
attenuate the rf so spectacularly.

A. A. V. Gibson (Texas A & M University) has designed
and used networks made of capacitors and inductors which act
like quarter wave lines insofar as impedance matching is concerned. Such networks have at least two advantages over
$\lambda/4$ lines. One is that the characteristic impedance (as well
as the operating frequency) is adjustable so that one such
network can be adjusted to do the work of many cables. The
other is that we do away with long cables at the lower frequencies which leads to better operation and gets rid of un-
manageable mass of cables.

Consider a $\pi$-network of the form shown below where $Z$
and $Z_1$ are complex impedances. If we attach a load $Z_o$ and
drive the circuit with voltage $E_1$, we can define various loop

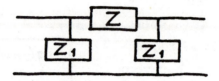

currents and eventually calculate various impedances of the network by using Kirchhoff's rules. Now the circuit looks like this and the equations relating the three currents with $E_1$, $Z_1$, $Z$, and $Z_o$ are

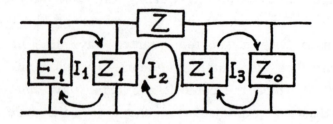

$$E_1 = I_1 Z_1 - I_2 Z_1, \tag{1}$$

$$0 = -I_1 Z_1 + I_2 (2Z_1 + Z) - I_3 Z_1, \tag{2}$$

$$0 = -I_2 Z_1 + I_3 (Z_1 + Z_o), \tag{3}$$

where each equation was obtained by going around each current loop and adding up the voltages using Ohm's law. Now we want to get a relationship between the output impedance $Z_o$ and input impedance $Z_1 = E_1/I_1$ so we want to eliminate $I_2$ and $I_3$ from the three equations. Combining (2) and (3), we can get $I_2$ as a function of $I$ and the Z's and then we can stick the resulting function into (1) to get

$$Z_i = \frac{E_1}{I_1} = \frac{Z_1^2 Z_o + ZZ_1(Z_1+Z_o)}{Z_1^2 + 2Z_1 Z_o + Z(Z_1+Z_o)} \quad . \tag{4}$$

Consider a specific $\pi$-network with $Z_1 = 1/i\omega C$ and $Z = i\omega L$ as shown.

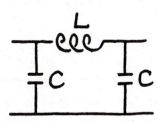

Substitution into (4) leads to

$$Z_i = \frac{Z_o\left[(1-\omega^2 CL)^2 - \omega^2 CL(2-\omega^2 CL)\right] + i(1-\omega^2 CL)\left[\omega L + \omega C Z_o^2 (2-\omega^2 CL)\right]}{(1-\omega^2 CL)^2 - \omega^2 C^2 Z_o^2 (2-\omega^2 CL)^2} \quad . \tag{5}$$

At resonance the imaginary part of $Z_i$ must be zero and this can happen when $\omega^2 LC = 1$. This leads to a simple relation between $Z_i$ and $Z_o$ at resonance

$$Z_i Z_o = 1/\omega^2 C^2 \tag{6}$$

which is identical to a $\lambda/4$ line with a characteristic impedance $1/\omega C$.

Now interchange L and C as shown on p. 406 so that $Z = 1/i\omega C$ and $Z_1 = i\omega L$. We can start with (4) as before and arrive at the same resonance condition $\omega^2 LC = 1$ which leads to

$$Z_i Z_o = \omega^2 L^2 \tag{7}$$

which is identical in form to (6) but with the characteristic impedance $\omega L$.

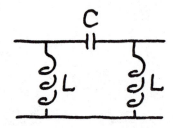

In order to get an idea of what kind of values we need for L and C for various frequencies and characteristic impedances Z, we present the following table derived from (6) and (7). (We state without proof that the same L and C combination can be used interchangeably between the two different $\pi$-networks at the same Z and $\omega$.)

| $f = \omega/2\pi$ (MHz) | $Z = 25\Omega$ | | $Z = 50\Omega$ | | $Z = 100\Omega$ | |
|---|---|---|---|---|---|---|
| | L($\mu$H) | C(pf) | L($\mu$H) | C(pf) | L($\mu$H) | C(pf) |
| 1.6 | 2.5 | 4000 | 5.0 | 2000 | 10 | 1000 |
| 16 | 0.25 | 400 | 0.5 | 200 | 1.0 | 100 |
| 160 | 0.025 | 40 | 0.05 | 20 | 0.1 | 10 |

We see that such a $\lambda/4$ network is easier to implement at the lower frequencies where we already noted that such networks would be more useful. At the higher frequencies we get out of range of the traditional (multiturn) variable inductors, especially for small characteristic impedance Z.

You can make a $\lambda/4$ network from scratch with commercially available variable inductors and capacitors from sources

such as those listed in Appendix D. Because the two components in the legs of the $\pi$ shaped circuit must track with each
other for the above analysis to hold, use identical units even
though the voltage breakdown requirements are vastly different for the two. Because the commercially available variable
inductors tend to be large, the physical size of such a network will not be small. Use a 3½ or 5¼" high cabinet and
arrange the connectors so that you can minimize the external
cable lengths when this unit is inserted between the probe
and the preamp.

Smaller variable inductors can be made by modifying
commercial multi-turn variable resistors. Instead of using
resistive wires, copper or silver wires can be used in the
resistor shell. Although we have never tried the modification
ourselves, we have used inductors made this way commercially
to special order.

In order to simplify the physical construction, we have
made a $\lambda/4$ network as described in this section by modifying
a commercial antenna tuner. Use the type containing a high
voltage variable inductor and modify the capacitors so that
you end up with two equal sections on the same shaft. (Some
manufacturers will sell you the capacitor plates and spacers
so that you can make up any size capacitor you wish, limited
only by the shaft length and breakdown voltage.)

## V.C.9   TANK CIRCUITS, IMPEDANCE MATCHING,
### AND ALL THAT

Most nuclear resonance experiments use a sample coil and

the coil is usually made to be a part of a resonant (tank) circuit. [Exceptions to the former include the use of a cavity (Guenther, et al. 1971; Schneider and Dullenkopf, 1977) and of a capacitor (Gersch and Lösche, 1957) while a transmission line coil (Lowe and Whitson, 1977) is an exception to the latter.] The tank circuit is usually necessary to match the impedance of the coil to the electronic devices (Pollak and Slater, 1966). In this section we shall review some basics of tank circuits and impedance transformations.

A common coil scheme in NMR is a parallel resonant circuit with a capacitor as illustrated.

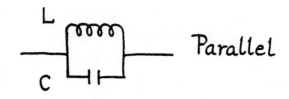

The coil with inductance L has complex reactance $i\omega L$ while the tuning capacitor with capacitance C has complex reactance $-i/\omega C$, when they are subjected to rf at angular speed $\omega$ (which is $2\pi$ times the frequency). The reactance of L and C in parallel, then, is

$$\left(\frac{1}{i\omega L} + i\omega C\right)^{-1} = \frac{i\omega L}{1 - \omega^2 LC}$$

which becomes infinite when $\omega^2 LC = 1$ which is the resonance condition.

Actually this never happens because we do not have ideal components. If the coil has a small resistance r associated with it, the impedance (which is the effect of a reactance and a real resistance) will be

$$\left(\frac{1}{i\omega L + r} + i\omega C\right)^{-1} = \frac{r - i\omega L(1 - \omega^2 CL - r^2 C/L)^1}{r^2 + \omega^2 L^2} .$$

Then the resonance condition (defined to be the condition required to make the impedance real) is

$$1 - \omega^2 CL - r^2 C/L = 0$$

and the impedance at resonance is $(r^2 + \omega^2 L^2)/r$. For ordinary coils $\omega L \gg r$ [for example at $\nu = 15$ MHz and $L = 1\mu H$, $\omega L = 2\pi\nu L \cong 100\Omega$], so the impedance is approximately $(\omega L)^2/r$ which is still very large, though not infinite. It is convenient to rewrite this impedance in terms of $Q=\omega L/r$ which is called the quality factor and is a measure of the efficiency of the tank circuit. The impedance, then, is $Q\omega L$ and the effect of making the coil become part of the tank circuit is to multiply the magnitude of its impedance by $Q\gg 1$. Thus, a parallel resonance circuit is a high impedance tank circuit.

The impedance at resonance can be changed by changing the L-C combination while satisfying the resonance condition. Therefore, a parallel resonance tank circuit can match the impedance of the coil to that of the rest of the circuit with just one additional component, namely a tuning capacitor. This is simpler and more efficient than schemes having more components. [But with a given coil the tank can have a specific impedance only at one frequency. Later in this section, we go into impedance matching requiring additional components.] On the other hand, often the coil has to be located at the bottom of a dewar or in a pressurized chamber (or both) where it is impossible, or at least inconvenient, to locate the variable capacitor especially if the capacitor is to remain variable. It is possible to locate the variable capacitor outside the confining area, but then the analysis of the problem becomes complicated because the capacitance of the cable contributes to the total tank capacitance and, in addition, the cable can reduce the quality factor of the tank circuit. At high enough frequencies (often above 30 or 40 MHz) this cable capacitance becomes the limiting factor in the resonance condition.

A coil in a series tank circuit has the advantage that it is easy to rig in a remote location (as in a dewar tail). Unlike the parallel resonance circuit discussed above, the series tank circuit presents a low impedance to the outside world, i.e., much current with little voltage. [This low impedance could also be an advantage over the high impedance of the parallel tank in dissipating the rf energy after the transmitter shuts off. There are other ways to improve the receiver recovery (see V.A.4.), however, so this in itself is not sufficient reason to choose a series tank circuit.]

C    L        Series

The low impedance of the series tank follows from the fact that the inductive and the capacitive reactances $i\omega L$ and $-i/\omega C$ have opposite signs which means that a current through L and C gives rise to opposing voltage drops by Ohm's law. The voltages are equal and opposite at resonance in an ideal tank circuit resulting in zero impedance. For a real resonance circuit the losses can be represented by a series resistance r which must be the impedance of the tank circuit at resonance. This intuitive conclusion can be confirmed by the rules for summing impedances which are in series so the total impedance is

$i\omega L+(1/i\omega C)+r$.

Since the resonance condition (again to make the impedance real as opposed to imaginary) is $i\omega L+(1/i\omega C)=0$ or $\omega^2 LC=1$, a familiar expression, the impedance at resonance is obviously equal to r.

Now at frequencies below a few hundred MHz, the typical capacitor used in NMR is virtually loss free and r is equal to the resistance of the wire in the coil and possibly of a cable between L and C. Since this is a small number, the series tank circuit requires additional circuitry to raise the impedance to match, for example, to a preamplifier with a 50 ohm input. [The need to present a certain impedance arises because the S/N of the preamplifier depends on this impedance. Thus, this is an important matching condition.]

At this point, we move from a discussion of tank circuits to making the connection to the coil. We have already talked about the effect of the cable capacitance in a parallel resonant circuit. A cable characterized by 50 $\Omega$ impedance has capacitance of approximately 1pF/cm (30 pF/foot) so that it will not affect the resonance condition only if it is short enough so that the capacitance is a small fraction of the tuning capacitance. Another related consideration is the electrical length of the cable; it must be short compared with one-quarter of the wavelength. The electrical length of a cable is not equal to that calculated for electrical propagation in a vacuum. A typical 50 $\Omega$ cable one meter long is a quarter wave line at 45 MHz, whereas a quarter wave for 45 MHz in vacuum is about 1.7 meters. See the previous section for more on quarter wave lines and how to measure them.

A very convenient trick (pointed out to the authors by A. A. V. Gibson) is to connect the sample coil to the rest of the tank circuit with a λ/4 line. Since the coil on the end of a λ/4 cable looks like a capacitor (see sections V.C.1., V.C.6., V.C.7., and V.C.8.), the combination of the coil and the cable can be made into a series resonant circuit by a series inductor.

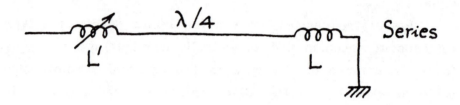

This is especially convenient at medium to high frequencies where the distance from the coil to the first accessible point is equal to or slightly shorter than the length of a λ/4 line. For example, a 42 MHz experiment in a tail of a dewar in which the sample coil is on the end of one meter 50 Ω rigid coaxial cable can be performed by adding an additional cable (in this case about 7 cm long) to complete the λ/4 length and then attaching a variable inductor in series.

Of course, a different solution is called for at a Larmor

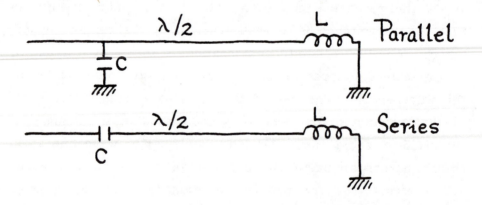

frequency (~85 MHz) where the previous λ/4 line is now λ/2. Since a wave, by definition, behaves the same at a distance λ/2 away (except for a reversal in sign), being λ/2 away is like not having the cable so we can now tune this combination with a series capacitor (or a parallel capacitor to make it a parallel resonant circuit). At an even higher frequency where the cable is 3λ/4, the tuning can again be accomplished with an inductor as with the λ/4 cable.

Now we return to the discussion of tank circuits and, in particular, to that of changing the impedance of resonant circuits to some desired value. First consider the parallel tank circuit for which the impedance at resonance is (approximately) $w^2L^2/r=wLQ$. The combination of L and C can be chosen at a given frequency to get a desired impedance but sometimes there are constraints which will not allow a free choice of L and thus of $wLQ$. A common (and the easiest to implement) way to transform the impedance of a parallel tank ciruit in such cases is to add a series capacitor.

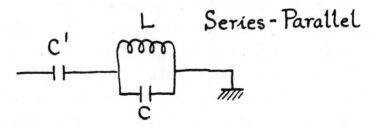

With each additional component, the algebra in calculating the impedance and the resonance condition gets more involved but the principle remains the same. The resonance condition is very close to $w^2L(C+C')=1$ which is our familiar one with C replaced by $C+C'$. The impedance at resonance for this circuit is still $w^2L^2/r$, as in the parallel tank, but the new L will be smaller than the old by a factor $1+(C'/C)$. Therefore, the new impedance will be smaller than the old by $[1+(C'/C)]^2$.

A variant of the above is to make the connection to the coil at an intermediate point in which case the additional

component is a part of the coil.
This is a useful trick to remember also when one wants to
resonate the coil at more than
one frequency.

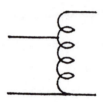

The series tank circuit is most often used in NMR with
the addition of a capacitor in parallel. If C'>>C the resonance
condition remains close to $\omega^2 LC=1$ and the impedance at resonance is $Q\omega LC^2/(C+C')^2$.

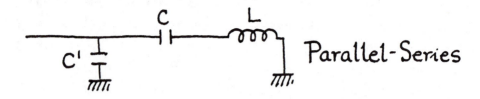

For example, at a Larmor frequency of 100 MHz, if $r=0.1\ \Omega$,
then C'~0.005 μF for the coil to be matched to 50 Ω impedance.
(The resonance condition should be met with the largest possible L and the smallest C, in general, to achieve the highest
Q and also to maintain the condition C'>>C.)

There is another way to deal with the low impedance of a
series tank circuit, and that is to use a transformer. The
usual way is to use a quarter wave (λ/4) cable to get the
signal away from the immediate vicinity of the sample and, at
the same time, transform the low impedance $Z_i$ to some high
value $Z_0=Z^2/Z_i$ where Z is the characteristic cable impedance,
say 50 Ω. Then a transformer can be used to change the
impedance to the desired value (Ellett, et al., 1971). Such a
transformer can easily be made by winding magnet wires on a
ferrite toroid, most easily found in the amateur radio market.
It is preferable to perform these two operations in one shot
with a black box which performs as a λ/4 cable with an adjus-

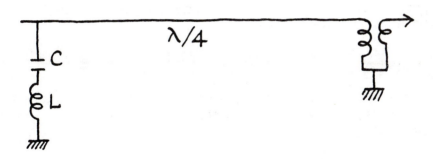

table impedance (see Section V.C.8.).

There are two different strategies for impedance matching. One is to just match everything as necessary at convenient impedance values. The other is to engineer every component so they all have a specific value, say 50 ohms. The latter scheme is more involved but offers great convenience in systems where various components need to be exchanged. It works extremely well provided that all the pieces are matched to 50 ohms. This may take some doing, however, especially with some equipment which is advertized as having 50 ohms impedance but does not.

## REFERENCES

J. D. Ellett, Jr., M. G. Gibby, U. Haeberlen, L. M. Huber, M. Mehring, A. Pines, and J. S. Waugh, "Spectrometers for multiple-pulse NMR" in Advances in Magnetic Resonance, vol. 5, edited by J. S. Waugh (Academic Press, New York, 1971), pp. 117-176.

U. Gersch and A. Lösche, "Untersuchungen der Paramagnetischen Kernresonanzabsorption Im Kondensatorfeld," Ann. Physik 20, 167-172 (1957).

B. D. Guenther, C. R. Christensen, R. A. Jensen, and A. C. Daniel, "Capacitively and inductively foreshortened cavities for magnetic resonance spectroscopy," Rev. Sci. Instrum. 42, 431-434 (1971).

I. J. Lowe and D. W. Whitson, "Homogeneous rf field delay line probe for pulsed nuclear magnetic resonance," Rev. Sci. Instrum. 48, 268-274 (1977).

V. L. Pollak and R. R. Slater, "Input circuits for pulsed NMR," Rev. Sci. Instrum. 37, 268-272 (1966).

H. J. Schneider and P. Dullenkopf, "Slotted tube resonator: A new NMR probe head at high observing frequencies," Rev. Sci. Instrum. 48, 68-73 (1977).

## V.C.10.   USEFUL PACKAGED COMPONENTS

### Double balanced mixer (DBM)

A double balanced mixer is a three port device. When two frequencies are presented at two of the ports, the sum and difference frequencies appear at the third port. A modern DBM is a broadband device, for example 2-200 MHz, and comes in extremely compact packages with or without connectors. DBM's are very useful in a variety of applications.

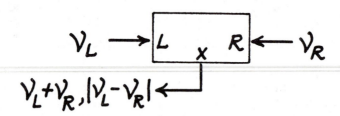

As a _mixer_: A DBM can be used to obtain the sum and difference frequencies in a variety of applications. These are illustrated by the uses that appear in the examples in sections V.C.1 and V.C.2. Any two ports may be used as input and the third as an output. The optimum choice depends on the frequency difference and levels. If the rf to be mixed occurs at two different levels, the optimum down-converter is obtained by applying the low level rf at the R port and the high level signal at the L port with the difference frequency appearing at the X port. For an up-converter, the low level is applied at the X port and the high level at the L port with the sum taken from the R port.

As a phase sensitive detector: When signals at the same frequency are applied to the L and R ports, a dc signal appears at the X port whose voltage varies as the cosine of the phase angle difference of the two sources.

As a frequency doubler: When rf with the same frequency and phase is applied to two ports, a signal at twice the frequency will appear at the third port.

As an rf gate: The mixer may be used to amplitude modulate the rf. If the modulation signal is large enough, the mixer becomes a switch. The most common application of this to pulse NMR is to gate the rf on and off by applying a pulse to the gate. When a bias current flows in the X port, the device passes rf. In the absence of a bias current, the isolation is around 40 dB or better depending on the frequency, so two DBM's in tandem will make a good broadband gate for NMR. In addition to the use as basic rf gates in the transmitter, they may also find applications such as blanking the receiver during the rf pulse (see V.C.2.).

As a phase flipper: There are several applications requiring electronically inverted phases. An obvious example is the phase alternated Carr-Purcell sequence. In the wideband decoupler, commonly used in high resolution FT spectrometers, a bandwidth covering the range of proton resonances is created by flipping the phase of the carrier at pseudo-random times. Frequency independent phase flipping is achieved by reversing the polarity of the current applied to the X port.

## Phase shifter

Phase shifters range from simple delay lines to mechanically or electronically controlled delay lines to those based on phase locked loops or other active circuits. Quadrature and 180° hybrids as well as DBM's can be used as fixed value broadband phase shifters. This last point is important since most continuously adjustable phase shifters are frequency dependent. Phase shifters in NMR spectrometers are required to set the reference phase in a phase sensitive detector and also to set the relative phases of rf pulses as required for various pulse sequences.

## Power divider

The power divider is a broadband multi-port device which permits the rf inserted in one port to be equally divided between the outputs. The outputs have the same phase and the same amplitude and are highly isolated from each other. Conversely, if two rf signals are put into the output ports of the power divider, they will be added vectorially. Hence the

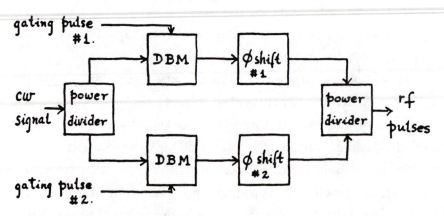

power divider is also a power combiner.

Power dividers are used for splitting and recombining signals to create a train of pulses with different phases. The diagram at left shows a scheme using two power dividers, two DBM's, and two phase shifters, to create such a pulse train. Whenever a DBM is turned on, an rf pulse of corresponding phase will be outputted from the second power divider being used as a combiner.

## Directional coupler

The directional coupler is an asymmetric power divider. It can be used, for example, to divide an rf signal so that a small amount goes to the receiver mixer and the main part goes to the transmitter. A special use for a directional coupler as a duplexer was discussed on pp. 394-395.

## Magic-tee

The magic-tee is also called a hybrid junction or a 180° hybrid. It is a broadband four-port device which is capable of splitting an rf signal into two outputs of equal amplitude, with high isolation between them. If appropriate ports are chosen, the outputs can be either in phase with each other or 180° out of phase. The magic-tee is illustrated at right. An rf input into B results in outputs C and D being identical. An input into A is divided into two ouput signals 180° out of phase at C and D. Thus, the 180° hybrid can serve as a broadband 180° phase shifter.

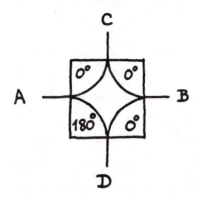

Ports A and B are isolated when rf power is going equally to outputs C and D.  The more interesting applications result when the output ports are unequally loaded.  Load mismatches result in reflected power being directed toward the isolated port.  In this way, the circuit acts like a bridge. Use of this device as a NMR bridge is discussed in section V.C.7.  It is also very useful as a device to set a tank circuit impedance to some real value and this application is discussed in the next section.

In passing, we note that this device can be made into a power divider/combiner merely by terminating the A port and having the rf input at B.  Alternatively, an out-of-phase power divider/combiner can be made by putting the terminating impedance on B and the input at A.  In fact, power dividers are made in exactly this way with the terminating impedance internal to the package.

Quadrature hybrid

The quad hybrid or 90° hybrid is similar to the magic tee.  In this case, an rf input into any port results in two outputs, equal in magnitude but with a 90° phase shift difference be tween them.  If the ports are terminated properly, the fourth port remains isolated from the input port.  Clearly, such a device could be used as a power divider or combiner. 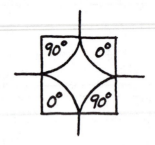 Similarly such a device can be used as a broadbanded 90° phase shifter.  As with the magic-tee, it can also be used as a bridge to monitor the degree of mismatch between the impedances of devices attached to the two ports.

## V.C.11.  HOW TO TUNE A RESONANT CIRCUIT
## AND MEASURE ITS IMPEDANCE

The easiest way to find out the characteristics of a passive L-C-R network in the rf range is to use a vector impedance meter which can tell you the magnitude and the phase of the impedance simultaneously.  Just adjust the components to null the imaginary component and to set the real component to the desired value.  The only disadvantage of this device is that it is expensive.  It turns out that magic-tees can be used to accomplish some of the same goals and they are orders of magnitude cheaper.

As already described, a magic-tee is a four port device in which two of the ports are isolated, i.e., have infinite impedance between them if the impedances at the other two ports are matched.  Thus, the device can be used to match a resonant circuit to 50 $\Omega$ or, for that matter, to measure the impedance at resonance of the tank circuit.  The tank circuit under question will have impedance R at resonance when the output of the magic-tee, easily monitored on an oscilloscope, is an <u>absolute</u> minimum.  (See the figure under magic-tee on p. 393.)  Unfortunately, this is easier said than done because there are relative minima when the capacitors are adjusted to give an impedance which is, for example, equal to R in magnitude but with an imaginary component.  When you play around with a scheme like this for the first time, it will be useful to let R be variable so that it can be adjusted for a null to measure the impedance.   Having an adjustable frequency source can also help (even though at first it might seem like an additional source of confusion).   With practice you will be able to match a probe to a desired impedance in a matter of minutes.

## V.C.12.   HELPFUL HARDWARE IDEAS

We present some hardware features which we have found to be very helpful.   You might want to incorporate some or all of these ideas (or none, possibly) into a homebuilt spectrometer or you may wish to specify these features in commercial spectrometers.

The first feature is a display of the current number of scans in real time.   We find it very convenient to have the current number of scans showing at all times.   This is especially true for a series of fairly short accumulations of the order of half an hour because misestimating the duration of the run by a few minutes wastes a significant fraction of the total time. Without a counter readout, you either have to interrogate the computer (thus interrupting the data taking process) or just wait around for the accumulation to finish.

This is an easy feature to implement.   You need a display device (like LED's with decoders included) and a latch which will remember what the number is.   The computer will output the updated count each cycle to the latch.

A second useful feature is a set of analog inputs into the computer from knobs.   The knob readings are digitized by an ADC and put into specific locations in computer memory where they can be assigned to any parameters. For example, two of the knobs can be assigned to the phase correction parameters corresponding to the phase independent and the dependent corrections, respectively. A knob can also be used to specify a point in the data (by brightening that point in the visual display, for example). It can be used to continuously and quickly change the rf pulse length during the set up procedure, too. A good number for these knobs is three or four. Each knob can be assigned to a particular parameter by software.

The construction of such circuits is also quite easy. The knobs operate variable resistors which feed variable voltages into a multiplexed ADC. The latter should have a latched output, i.e., it should put out the last value converted until the new conversion. The speed of the ADC is not critical and you can use an inexpensive one.

It is also very convenient to have the capability to assign numbers, say from the terminal keyboard, to the same parameters mentioned above because this can be done more reproducibly than by the use of analog knobs. For example, this feature could be used to set the scale factor to a predetermined value for displaying a spectrum.

The third hardware suggestion is not a feature to be built but rather to be bought. It used to be customary to output the spectra, after all the data massaging was done, to an analog recorder through a digital-to-analog converter. With the advent of good moderately priced digital recorders, we think the days of analog recorders in NMR are numbered. In particular, there are many digital recorders designed to work with desktop computers which are ideal for NMR. They cost no more than good analog recorders and can draw axes, label them, title the graph, write the date, etc. There is even a model on the market (in 1981) which will make any number of these graphs unattended and stack them for you.

Look for such recorders as accessories to desktop calculators, although others exist as well. They are "smart" plotters in that many of the operations can be input as macro commands, thus vastly simplifying the programming. In particular, all alphanumeric characters (i.e., letters and numbers) need only be specified as such rather than having to construct each character, segment by segment. Commonly, they are available with IEEE-488 interface (also known as GPIB or HPIB, standing for general purpose interface bus or Hewlett-Packard interface bus, respectively) which can be interfaced to an NMR spectrometer via, for example, a CAMAC GPIB interface module, discussed in V.C.3.

Another device which may be very useful is called by various names such as a reflectance bridge, a standing wave meter, or a directional wattmeter. It can be used to measure the impedance match between a coil and its voltage source, for example, the transmitter or the decoupler, by monitoring the power reflected at an impedance mismatch.

Directional wattmeters are not very expensive and can be used to monitor the state of the impedance match under actual operating conditions. Thus, they are extremely useful when the impedance match is susceptible to changes, as when a sample absorbs so much power from the decoupler coil that the coil becomes detuned from the heat.

CHAPTER VI

## PRACTICAL TECHNIQUES

## VI.A.    TURNING IT ON AND TUNING IT

In these sections, we discuss the actual steps you need
to take to get and optimize an NMR signal.

## VI.A.1   A PROCEDURE FOR TUNING UP A PULSE NMR
APPARATUS

While pulse and FT spectrometers of different manufacturers will vary in their design details, the following should
be a good general guide to tuning up.  If the spectrometer
has recently been tuned there may be no need to adjust it
before using it, but if the spectrometer is being changed to a
new frequency or if it has not been tuned up for some time,

Eiichi Fukushima and Stephen B. W. Roeder, Experimental Pulse NMR: A Nuts and Bolts Approach

some adjustments are probably necessary.  The need for tuning will be indicated by an increased 90° pulse time or a less than normal S/N on a known sample.  An essential thing to do is to write down all the operating parameters and signal strengths with standard samples <u>while</u> <u>the</u> <u>machine</u> <u>is</u> <u>working</u> <u>well</u> so that you will know when something is even slightly wrong.

For a pulse NMR experiment, the following components are required as a minimum and most of them must be adjusted in some suitable way before an NMR signal can be observed.

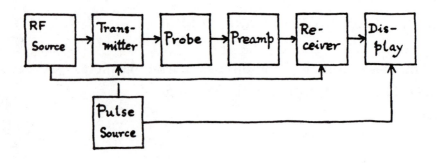

### RF signal source

A suitable Larmor frequency is chosen for the experiment.  The output amplitude must be sufficient to drive the transmitter.  In addition, this source provides the reference signal for phase sensitive detection and this signal level must be adjusted, independently of the level to the transmitter, for optimal S/N.

### Pulse source

For tuning purposes, a simple repetition of pulses can be used with sufficient repetition rate so that the signal will appear continuously on an oscilloscope.

## Probe tuning

As mentioned elsewhere, the probe has to be tuned to resonance and to a certain impedance value at resonance. You could use a vector impedance meter or even a magic-tee (see V.C.11.) to tune the tank circuit to the desired parameters but, in the final analysis, some adjustments may be necessary after the probe is hooked up in the final configuration. A valuable aid in tuning up an NMR system is a dip meter which provides a dummy NMR signal. The dip meter is a solid state analog of the old grid dip meter and is inferior to the grid dip meter in its output level. So, if you have an old grid dip meter, keep it.

Aside from its canonical use as a device for finding the resonance frequency of a parallel tank circuit, the dip meter can be used as an rf source to mimic an NMR signal. Because of the high sensitivity of NMR receivers, the dip meter need only be close to the NMR apparatus. But remember the output level comment in the last paragraph. An old grid dip meter can be put, say, one meter away from the receiver system and tuned close to the phase detector reference frequency to produce an observable signal. A dip meter usually has to be closer.

Crossed coil probes are easy to tune because the transmitter and the receiver functions are independent. Tune the receiver tank circuit for a maximum signal through the receiver from the dip meter. The transmitter coil can be optimized by maximizing the rf field amplitude at the coil as described in the section on transmitter tuning below.

For a single coil probe, the receiving function and the transmitting function should be optimized simultaneously if the rf circuits were designed correctly. However, for a surprising number of spectrometers, the adjustment which yields the

highest signal S/N does not correspond to the largest $H_1$ amplitude.  A better impedance match can reduce this discrepancy.
If you want to use the set-up as is and ignore the problem,
tune the probe to maximize the received signal by looking, for
example, at the detected signal from a dip meter.  The loss in
S/N of the received signal due to an improperly tuned probe
is irrecoverable while that is not so for the transmitter pulse.
In principle, an inadequate $H_1$ amplitude due to inefficient
coupling between the transmitter and the coil can be overcome
by simply having a more powerful transmitter.

Finally, the decoupler circuit can be tuned with a directional wattmeter, after the coil temperature has stabilized under
power.

Transmitter tuning

Many  transmitters  have  output  tuning  and  impedance
matching  adjustments.    It  is  important  to  make  the  final
adjustment with the apparatus in the final (working) configuration.   This requires some means of measuring and maximizing
the rf current in the coil.  (The quantity which directly affects the magnitude of $H_1$ is the current through the coil.
For a fixed impedance, this is proportional to the voltage
across the coil.)  There are two different approaches to monitoring the rf pulse.  Many transmitters have a pulse monitor
port for a transmitter output suitably divided, say by 1000,
to be viewed on an oscilloscope.  This, however, is not the
most direct measurement since it is only the input to the tank
circuit.  A more direct method would involve an antenna (pickup coil) or a current probe close to the sample coil.  Such a
pickup coil can be a permanent loop of wire to intersect the rf
magnetic field lines in the probe, or a temporary loop of wire
to lower into the sample coil, say, inside an empty sample

tube. If the probe is for a large sample, it is even possible to insert an oscilloscope probe into the sample coil. All adjustments in the transmitting stage, in the vast majority of cases, should maximize the rf current in the coil. Once an approximate maximum has been attained, a final adjustment is called for with an actual NMR signal as described later under "final tuning."

## Preamp/receiver tuning

The receiver and preamp can be tuned with the dip meter signal at the same time the probe is tuned by adjusting the preamp/receiver for a maximum output. If the receiver has a built-in filter, it must be adjusted to the lowest cutoff (or threshold) frequency so that it will not distort the signal. Another common practice is to maximize the noise by the tuning. We find that the dip-meter signal is easy to generate and makes the tuning more accurate.

## Duplexer tuning

The duplexer, if tunable, must minimize the transmitter pulse feed-through to the receiver while maintaining good continuity between the probe and the receiver for the signal. If the transmitter pulse itself is used for this adjustment, turn the amplitude way down (by several orders of magnitude) so the receiver chain will not be saturated. Monitor the pulse through at least one preamplifier stage so the monitoring process will not disturb the circuit. The pulses should also be extra long so that the frequency spectrum of the pulse is fairly narrow, i.e., there should be a long constant amplitude center section to the pulse compared to the rapidly changing end sections as shown in the sketch on p. 401. This is because most duplexers are tuned devices which work best at

a specific frequency.   Alternatively, a cw signal can be used
to tune a duplexer and this is much easier than using rf
pulses.   With the usual spectrometer, this involves some cable
switching which may be a nuisance but probably no more so
than the changes in the pulse parameters described above
needed to do the tuning with the pulses.

## Final tuning with an NMR signal

After finding an FID as per the prescriptions given in a
later section, a final set of adjustments can be made.   First
the preamp/receiver can be tuned for a maximum FID amplitude. Then the transmitter pulse length should be adjusted so
that it is definitely shorter than a $\pi/2$ pulse.   Now adjust the
transmitter tuning and the probe tuning for the maximum FID
amplitude.   The transmitter pulse length may have to be shortened again during this adjustment.   The manner in which the
transmitter pulse decays upon being turned off usually depends
on the tuning and impedance adjustments so that extra care in
making this adjustment can reap valuable dividends in decreased deadtime and, therefore, in S/N.

All these final adjustments are easy if a sample is
available with a large signal.   For a low sensitivity nucleus
with no large S/N standard sample, you simply have to do the
best you can without the sample and then adjust the parameters according to how the signal looks when it emerges from
the noise upon averaging.   It will be a slow process.

## Miscellaneous

In general, the tuning becomes easier as the tuning
becomes better.   The most difficult situation is when the
apparatus is being assembled from scratch and nothing is
tuned with anything.   The transmitter can be approximately

tuned by observing the output at the desired frequency. So
can the receiver, but not so directly. The problem, if any,
arises because the receiver is not only a series of amplifiers
but includes a detector. Thus, a constant amplitude signal
at the detector reference frequency shows up as no signal
after detection. In order to check out the entire device, a
modulated rf signal which results in a detected signal corresponding to this modulation is needed. If a modulated rf source
is available, the operation is simple. Maximize the detected
output at the desired carrier frequency. What if a modulated
rf source is not available? Then use our favorite tuning aid,
the dip meter. Its output, when it is close to the carrier frequency, will beat with the carrier to produce a modulated signal.

This all sounds simple but what do we do when nothing
is in tune and we can't even trust the dip meter dial? We
have two choices: One is to look at the rf signal in the receiver before the detector and tune the receiver to maximize
it; and the other is to have faith in the dip meter dial and
hope that it is accurate enough to give rise to detectable
beats, at least if the coupling between the dip meter and the
receiving circuit is good. The dip meter can be moved closer
to the probe or the receiver. It is also possible to use the
aforementioned pickup coil to inject the rf signal close to the
receiver coil. Finally, it may be fairly easy to remove some
shielding either on the probe box or on the front end of the
preamplifier. Once the receiver is even crudely tuned to the
desired frequency, the dip meter can be tuned to give appropriate beats and it can also be moved away to a more convenient location. When the system is close to being tuned, we
can have our grid dip meter (which has a higher output than
the modern dip meter) one or two meters away from the probe

and its associated circuitry.

Sometimes the operator's proximity to the adjustment knobs will affect the tuning condition. This condition should have been avoided in the design of the apparatus. An unsightly but quick-and-sure fix is to extend the adjustment shaft with an insulator. Standard diameter shafts, e.g., 0.125" and 0.250" diameter, can be extended with plastic shafts of the same diameter using commercial shaft couplers. Insulated shafts can also be adapted for various screwdrivers and similar tuning tools, either commercial or homemade. We have seen insulated shaft extensions as long as one meter, which attests to the practicality of some NMR people.

## VI.A.2.   HOW TO SET THE PHASE OF AN FID

Phase considerations intrude even in the simplest experiments of observing an FID or an echo. Accurately adjusting the phases of rf pulses can be very important, particularly in experiments involving trains of pulses such as the Carr-Purcell Meiboom-Gill train or the multiple pulse line narrowing sequences. In other sections we have considered how phase shifts originate and how to cope with them.

In section V.A.1., we discussed the advantage of phase sensitive detectors (PSD) over diode detectors. The presence of a PSD raises the question of how to set the reference phase to observe the absorption mode or the dispersion mode signal.

In order to get an absorption signal, the obvious thing to do is to adjust the reference phase to maximize the FID on resonance. Unfortunately, the signal is quite insensitive to the phase when it is close to the absorption mode. This is because the signal amplitude is proportional to the projection of the magnetization along some axis and so it is proportional to the cosine of the angle between the magnetization and the axis. For an angle close to zero, the cosine function is essentially flat; i.e., it is difficult to tell the difference between exactly on phase and, say, 5 degrees off because the two projections, then, will have the ratio 1000:996. By definition, this is good enough for measuring amplitudes of FID's but not good enough for many experiments requiring careful phase adjustments like multiple pulse sequences.

The method described here is quite simple. There are many more complicated and sensitive methods involving multiple pulse sequences which we will not consider here. Choose a sample which has an isolated symmetric line. Examine the resulting FID as a function of the resonance condition and alter the phase so that the FID shape is symmetric or anti-symmetric about resonance. In the former case, the area under the FID is a maximum at resonance and we have the absorption mode. The latter corresponds to the dispersion signal and ideally is an absolute null at resonance. This follows because the integral of the FID is proportional to the value of the spectrum at that field or frequency and, conversely, the t=0 value of the FID is proportional to the area under the spectrum.

The easiest method for going off resonance by a certain amount in both directions is to change the field so that there is a measurable frequency to the beats, say on an oscilloscope, and then reverse the field past resonance and go off on the other side, so the beats are in the same places as before. This method may not work as advertised if the lineshape is determined by the field inhomogeneity rather than the sample.

It almost always works for solid samples with symmetric lines which have short enough $T_2$'s so that the field inhomogeneity can be ignored.

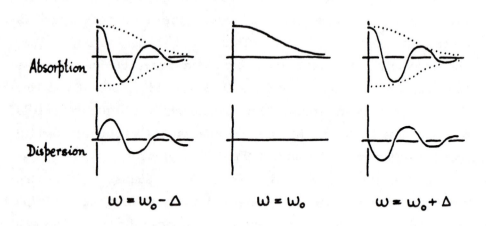

$$\omega = \omega_o - \Delta \qquad \omega = \omega_o \qquad \omega = \omega_o + \Delta$$

## VI.A.3   HOW TO SET PULSE LENGTHS

As with the phase adjustment, the most sensitive adjust-ment of the pulse length can usually be made during an experiment which requires such an adjustment. For example, a straight Carr-Purcell sequence with a long $T_2$ sample is ideal for adjusting the pulse length to be exactly 180 degrees because small errors in the pulse length will accumulate and become evident during the sequence. A less accurate, but simpler, 180 degree pulse adjustment is to simply null the FID on resonance after the pulse. This is a fairly accurate adjust-

ment, compared to that for a
90 degree pulse by optimizing
the FID amplitude because a
null of a sinusoidal function is
much better defined than its
maximum.

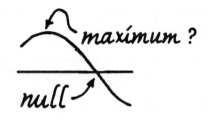

So, what is a quick way to adjust the 90 degree pulse
besides shooting for a maximum of the FID? A common method
is to first obtain a good 180 degree pulse by one of the above
mentioned methods and then cut the pulse length in half.
This is good in principle but there are two problems with it,
one more serious than the other. The basic problem is that
the actual pulse is usually not rectangular or otherwise symmetric. So, to go from a $\pi$ pulse to a $\pi/2$ pulse, the area of
the pulse has to be halved, not necessarily the length. The
other problem is that the actual transmitter pulse is usually
not the same length as the gating pulse which was sent to the
transmitter. This is because the resonant circuit containing
the sample coil distorts the pulse. Therefore, any pulse
length adjustment of the sort we are considering must be made
on the actual rf pulse and not on the gating pulse.

As you can imagine, the above technique of choosing a
90° pulse to be one-half of a 180° pulse works least well for
experiments with short rf pulses because the relative distortion
of the pulse is the largest then. With a good transmitter
which can maintain a constant rf amplitude during the pulse,
most high resolution NMR experiments do not suffer too much
from this problem because the pulses are likely to be quite
long, say, many tens of microseconds. Furthermore, most
high resolution work at the present time is performed at quite
high frequencies where the ringing effects of the resonant

circuit are not so bad.   The problem should be most acute with short $T_2$ and small gyromagnetic ratio samples.

There is a simple method for getting a $\pi/2$ pulse which is easy and works quite well.   Instead of creating a $\pi$ pulse, two identical pulses are formed closely together so that the combined effect of the two equals a $\pi$ pulse.   Provided that they are really identical, i.e., the two pulse lengths can be made identical and the power supply voltage does not "droop" from one pulse to the next, each of the pulses will be a $\pi/2$ pulse. In practice, there are a couple of problems which can occur. You may be bothered by echoes so that it is hard to tell when a null occurs; or, you may simply not be able to get a good null.   This will happen if there is non-zero magnetization at the beginning of the second pulse due to $T_1$ relaxation.   Solid samples with $T_1 \gg T_2$ usually do not exhibit these problems.

What if you can't guarantee making the two pulses identical?   For samples with $T_1 \gg T_2$ a variant of the two pulse method just described works just as well.   Two pulses are applied with a spacing longer than $T_2$ and much shorter than $T_1$.   Set each pulse independently so that each is nominally a $\pi/2$ pulse by maximizing the FID after each.   Next, adjust the length of the first so that there is no FID following the second.   The first pulse, then, is an exact $\pi/2$ pulse (within the limitation of the method) regardless of the setting of the second pulse because the magnetization after a $\pi/2$ pulse is zero and, provided that you wait a period $\tau$ such that $T_1 \gg \tau \gg T_2$, there should be no FID after the second pulse.   If the two pulses must be set independently, only the first pulse can be set to be exactly a $\pi/2$ pulse unless the positions of the two pulses can be exchanged or a third pulse can be used with the second.

A variation of this method will work even if, for some

reason, you cannot set $\tau \ll T_1$ so that there is a finite FID after the second pulse. If the first pulse is a true $\pi/2$ pulse, the FID after the second pulse will be independent of the repetition rate of the pair for a fixed $\tau$ because the first pulse will have knocked down whatever value of magnetization there might have been to zero.

Pulses which are multiples of $\pi/2$ pulses are the easiest to set, but other pulse lengths are occasionally needed, like a 45 degree pulse for Jeener echoes (IV.B.4.), or arbitrary length pulses for variable nutation $T_1$ experiments (III.D.5.) and the DANTE sequence (II.D.2.). For any submultiple of $\pi/2$ pulses, the trick is to repeat the pulse an appropriate number of times so that the combined effect is that of a $\pi/2$ pulse or multiples thereof. For example, you could go for 30 degree pulses by having three identical pulses act like a 90 degree pulse. Any odd length pulses must be interpolated after that.

## VI.A.4.   ESTIMATING S/N

When studying a new sample, a sample containing a nucleus with which you are not familiar, or a sample about which you have some suspicion, you should try to estimate the S/N so that you will not be taken by total surprise when you get no signal at all. This task is considerably easier for solution samples than solids because the line broadening mechanisms are more predictable and you do not have to worry about losing intensities to satellite transitions as you might for quadrupolar nuclei in solids. The most likely circumstance under

which an NMR line in solution is broadened beyond detectabil-
ity is in the presence of paramagnetic ions and even then the
line is not likely to be so broad that it will escape detection
on a spectrometer for broadline solids.

The main source of information for S/N is an NMR
table such as the one by Lee and Anderson (Appendix A)
which gives the calculated relative sensitivity per nucleus.
These values are calculated with several assumptions (such as
$T_1$=$T_2$) and represent the maximum signal possible so that the
actual signal may not scale according to these values. One of
the effects not taken into account in the table is the degradation of the usual NMR coil quality factor as the resonance
frequency is increased and coil inductance reduced if the
same probe cannot be used at the higher frequency. The net
effect of this is that the actual signal has something close to
a $w^{3/2}$ dependence rather than the predicted $w^2$ dependence
(which is used in calculating the NMR tables). Thus, at
least in this one respect, the signal from a small $\gamma$ nucleus
will not be as bad as predicted at constant field.

In a constant frequency experiment the NMR tables
should be quite reliable for giving the maximum possible
signal compared to a known signal from some standard nucleus.
Just adjust the calculated sensitivity according to the relative
number of nuclei in the sample which is easy to do if you
know the isotopic abundances, the actual densities of the
samples, and their molecular weights. Because the predicted
signal size is proportional to the integrated intensity, the
signal heights in the frequency domain will depend not only
on the splittings (such as the spin-spin and the chemical
shift in solution and quadrupolar in solids) but also on the
line widths, of course.

For the much more common constant field case, if you

estimate the maximum signal in the same way for a small $\gamma$ or a large I nucleus, then it may turn out to be an underestimate. Alternatively, you can use the $\omega^{3/2}$ dependence rather than the $\omega^2$ one. The NMR tables are calculated assuming that the signal is proportional to $[(I+1)/I^2]\mu^3H^2$ or $(I+1)\mu\omega^2$. Then the constant field expression is $[(I+1)/I^2]\mu^3$, whereas it will become $[(I+1)/I^{3/2}]\mu^{5/2}$ if we assume a $\omega^{3/2}$ dependence. The sensitivity for nitrogen-15, for example, is listed as $1.04 \times 10^{-3}$ at constant field whereas it is three times better if an $\omega^{3/2}$ dependence is assumed, a significant effect.

For polycrystalline quadrupolar solids, estimating S/N is difficult because the range of possible quadrupole splitting and broadening is so wide. As mentioned in II.D.1., the quadrupole splitting depends on the square of the quantum number m. As is made clear by the energy diagram in that section, half integer spins will always have an unperturbed $\Delta m=1$ transition between the m=½ and m=-½ levels because the two levels shift by the same amount regardless of the angular dependence of the shift which is identical for the two levels (at least for quadrupole interactions small compared to the Zeeman). For integer spins, this is not the case. The m=0 level shifts independently of the other levels so that even for a single crystal the m=1 to m=0 transition will be different from the m=-1 to m=0 transition and, furthermore, this difference will depend on the angular orientation of the crystallite axis with the magnetic field. Thus, there is virtually no chance in detecting an integer spin nucleus in a powder, unless the spin is in a cubic environment, while the ½↔-½ transition in a polycrystalline sample containing half integer spins is quite often detectable. The ½↔-½ transition is very easy to detect in spin-3/2 boron-11 and sodium-23, for example. This transition has also been observed in spin-5/2 manganese-

55 and antimony-121, although in these cases the line was no longer symmetric due to the second order quadrupole shift which modifies the $\frac{1}{2} \leftrightarrow -\frac{1}{2}$ transition which was not affected to first order in the perturbation.

As mentioned in II.D.1., the central $\frac{1}{2} \leftrightarrow -\frac{1}{2}$ transition becomes weaker with increasing I and, in addition, this is more severe in pulse NMR than in cw.  It is possible, in some cases, to detect the satellites in a powder (Creyghton, et al., 1973) but usually by cw NMR.  The only satellite detection by pulse NMR that the authors are aware of involved a sophisticated double resonance quantum coherence effect (Polak and Vaughan, 1978).

REFERENCES

J. H. N. Creyghton, P. R. Locher, and K. H. J. Buschow. "Nuclear magnetic resonance of $^{11}B$ at the three boron sites in rare-earth tetraborides," Phys. Rev. B7, 4829-4843 (1973).

M. Polak and R. W. Vaughan, "Nuclear double resonance interferometric spectroscopy," J. Chem. Phys. 69, 3232-3241 (1978).

VI.A.5     SEARCHING FOR RESONANCE IN HIGH
              RESOLUTION NMR

Searching for a resonance is usually no big deal in high resolution NMR.  Because the required parameters for various samples with a given nucleus are not so different from each other compared to solids, the spectrometer can be used over and over with virtually the same settings.  The only time to worry about finding a resonance, then, is when the apparatus is being put together for the first time or when it is reassembled after an overhaul or a major repair. If major problems are encountered, it may be necessary to read the following section on searching for resonances in general, since the considerations described there also apply to high resolution solution NMR. Normally, there should be a standard operating conditions list which can be followed. What if you never made such a list? Then the problem is more significant. The spectrometer must be tuned as described in the previous section and then a suitable standard sample should be used to search for a resonance. Use a wider than normal "sweep width," i.e., faster digitizing rate, to locate signals for the first time.

Consider a carbon-13 spectrometer. If the probe will take large sample tubes, say, 10 mm diameter, pure benzene will be a good tuning sample with FID's usually detectable by eye without signal averaging. If you have some confidence in the decoupler frequency, use it to enhance the S/N by NOE. Otherwise, consider perdeuterobenzene which, in the absence of decoupling, will give a triplet in the transformed spectrum due to the coupling with the spin-1 deuterons but unlike the doublet from the undecoupled benzene is a very narrow multiplet. Of course, samples with carbon-13 enriched would have even greater S/N. These days, such compounds are quite reasonably priced compared to what they were a few years ago so if the spectrometer is rearranged often, for example, for different nuclei or different magnet configurations, the purchase of a standard sample with carbon-13 enrichment will be worthwhile.

It is important to write down as many of the parameters as possible such as the observe, lock, and decoupling frequencies and various gains, levels, and standard sample signal sizes and shapes when the spectrometer is working well. This

list can help in the process of reactivating the spectrometer
after some change of state.  A separate notebook strictly on
spectrometer behavior, settings, repairs and modifications,
reference numbers, service facility telephone number, etc.
should be kept somewhere close to the machine.

## VI.A.6.   SEARCHING FOR RESONANCE:
### GENERAL CONSIDERATIONS

It is important to have a good tuning sample with short
$T_1$, good S/N, and a simple spectrum.  The importance of
short $T_1$ sample cannot be overemphasized.  Early attempts at
NMR detection failed because long $T_1$ samples were used.
[This is a rare situation, at least in science, where distilled
water is much less preferable to dirty water.]  Good S/N and
a simple spectrum usually go hand-in-hand since the intensities are widely distributed in a complex spectrum.  Use the
largest sample tube the probe is designed for and have an
adequate amount of the sample so its length exceeds the length
of the sample coil.

A commonly used tuning sample for protons is water
doped with a paramagnetic salt to reduce the $T_1$.  Don't
forget that the paramagnetic ions will shift the water resonance slightly to higher frequency (or lower field).  Vacuum
pump oil and glycerol are two other often used proton tuning
samples, especially in spectrometers for broad lines.  A
convenient solid sample is a piece of plastic:  polyethylene for
protons and KEL-F or Teflon for $^{19}F$.  Actually, one of the
nicest solid sample containing protons is rubber.  An appro-

priate size rubber stopper can be used without modification but stoppers and erasers can easily be formed into the necessary sizes and shapes. Alternatively, rubber bands can be handy; either cram them into a sample tube or lower a few into the probe.

For nuclei with less inherent NMR sensitivity than protons and fluorine, tuning samples are even more important. It is imperative that the field/frequency relation for the particular nuclei be established to within the possible chemical or other shifts and splittings before the "unknown" sample is inserted into the probe. It is quite often necessary to see a resonance of a different nucleus first and use it as a magnetometer to establish the magnetic field strength. Then the frequency can be changed to the appropriate value for the desired nucleus. The converse procedure is much less precise because the frequency can be directly set much more accurately than the field. Of course, whenever the frequency is changed, the spectrometer must be retuned as described in the earlier sections of this chapter.

Let us consider an example. Suppose the desired nucleus is $^{23}$Na in a non-cubic powder where the quadrupolar nucleus will interact with the electric field gradient in the crystal to give rise to a broad and weak spectrum. If an accurate magnetic field calibration exists so the settability is much better than about 1%, it is possible to go directly to the desired field/frequency and try a tuning sample. Otherwise, a different nucleus can be observed first. A common "general" tuning sample is $D_2O$ with a $\gamma/2\pi$ of 653.566 Hz/G. Since $^{23}$Na has $\gamma/2\pi$ = 1126.2 Hz/G, the Larmor frequencies of the two nuclei will be in the ratio 1126.2/653.57 = 1.7232 in the same field. If the desired Larmor frequency for $^{23}$Na is 20 MHz, the field must be set so the $D_2O$ resonance will occur at

20/1.7232 or 11.607 MHz.   So the apparatus should be tuned for 11.607 MHz (as described elsewhere) and the field adjusted to see a $D_2O$ signal.   (The field strength will, of course, be 11.607 MHz divided by 653.566 Hz/G or 17.759 kG.)   Then, the field should be left alone and the spectrometer retuned to 20.0 MHz before a signal from a sodium tuning sample is sought.

There is another complication to this procedure.   If two samples with different nuclei are to be used, the transimitter pulse lengths probably need to be different for the two nuclei. There are three reasons for these effects.   One is that $H_1$ usually depends on the frequency because the transmitter efficiency is frequency dependent.   It is easy enough to calibrate $H_1$ as a function of the frequency at various power settings to cope with this problem.   The second is that the tip angle $\theta$ through which the magnetization is rotated during the pulse duration t depends on $\gamma$ as $\gamma H_1 t = \omega_1 t$.   The third is that the relation $\gamma H_1 = \omega_1$ given in Section I.B. for the pre- cession rate $\omega_1$ around the field $H_1$ in the rotating frame does not hold for magnetization associated with quadrupole split Zeeman levels.   Therefore, even if $H_1$ were the same at two different frequencies, a $\pi/2$ pulse for $^{23}Na$ central ($\frac{1}{2} \leftrightarrow -\frac{1}{2}$) transition must be shorter than that for $D_2O$ by a factor of two in addition to the 1.7 from the ratio of $\gamma$'s.   Therefore, the pulse length should be cut down by about a factor of 3.4 from the optimum pulse length for $D_2O$ when the frequency is changed to 20 MHz for sodium.   In Section II.D.1. we dis- cussed these relations between the experimental parameters for quadrupole split levels.

Now that we have come this far with this example, we will tell you that $^{23}Na$ is a relatively easy nucleus to see compared with most others and it is possible to bypass many of the previously described steps and straight away look for a signal from a standard Na sample. (At the same Larmor frequency $^{23}Na$ has 1.3 times the sensitivity of an equal number of protons and 3.3 times that of deuterons but only in the absence of quadrupole splitting, e.g., in solution.) The deuterium-sodium pair is a good one to practice changing between because of the moderately good S/N which allows direct observation of the signals without signal averaging and because of the factor 1.7 between the $\gamma$'s which is small enough to usually permit the use of the same sample coil but is large enough for the operation not to be too trivial. If your spectrometer uses different coils for $^2H$ and $^{23}Na$, you should choose a more appropriate pair from the NMR tables (Lee and Anderson, Appendix A).

An obvious candidate for a standard Na sample is NaCl dissolved in water. A better solid standard at room temperature is NaCN (preferably sealed in a tube -- it is poisonous). Even though its crystal structure is the same as that of NaCl, i.e., the sodium nuclei reside at sites of cubic symmetry, the sodium nuclei have much shorter $T_1$'s due to the hindered rotation of the $CN^-$ units which modulate the EFG (about the average value of zero) at the sodium sites. The cubic site is important because it allows the full intensity to be observed. In compounds with distorted Na sites, the electric field gradient will interact with the quadrupole moment to split the line so that only the central ($\frac{1}{2}\leftrightarrow-\frac{1}{2}$) transition with only 20% of the unsplit intensity will be detectable, although 40% of the total is contained in the central transition in a cw experiment. Again see Section II.D.1.

For a sample expected to have very small S/N for reasons of small $\gamma$ or small number of nuclei in the sample, the adjustment of the field/frequency condition described in this section is only the beginning, though necessary, step. Before the spectrometer is tuned for such a sample, the expected sensitivity should be estimated, as described in Section VI.A.4., and a decision reached that this is not a hopeless experiment. Now try to make an intelligent guess about the sample $T_1$ by comparison with known $T_1$'s of similar compounds. Then choose a convenient pulse interval $\tau$ and a tip angle $\alpha$, called Ernst angle, so that

$$\cos\alpha = \exp(-\tau/T_1),$$

as discussed in the following section. Now signal average and hope for the best. If no signal is detectable in the allotted time, change the field (or frequency) by about one-half of the spectral extent of the rf pulse; i.e., if the pulse length is $\tau$, the field should be changed by an amount corresponding to a frequency change of $(2\tau)^{-1}$; and try again. Then try the other side by an equal displacement.

If everything fails, we recommend one more thing, change the field, if possible, and examine a proton or fluorine sample at the same frequency and spectrometer settings to make sure that the settings are, indeed, optimal. If they are not, write down the optimal parameters (again) and repeat the whole procedure. Good luck.

## VI.A.7   THE CHOICE OF TIP ANGLE FOR OPTIMUM S/N

For many of the experiments already discussed, the tip angle $\gamma H_1 \tau$ (for spin-$\frac{1}{2}$) where $\gamma$ is the gyromagnetic ratio of the nucleus, $H_1$ is the rotating rf magnetic field amplitude, and $\tau$ is the duration of the pulse (see section I.C.1.), must be well defined. For example, the Carr-Purcell sequence and the inversion recovery $T_1$ experiment require $\pi/2$ and $\pi$ pulses as already described. Some of the other experiments such as the saturating comb technique for $T_1$ are deliberately designed to be relatively insensitive to the tip angle (see section III.D.2.). What should the tip angle be for accumulating the best FID as far as S/N is concerned?

For a one-shot experiment with the magnetization in thermal equilibrium to begin with, the optimum tip angle for the best FID is obviously $\pi/2$. However, for accumulating FID's over some finite time, the relaxation time $T_1$ affects the ultimate S/N for any given tip angle. This is because most of the time required for data acquisition by accumulation of FID's is spent waiting for the magnetization to relax towards the static field direction so that the system can be pulsed for another FID. If a combination of $T_1$, tip angle, and repetition rate can be chosen so that a moderately good FID can be obtained moderately often, the data acquisition might be quite efficient compared to the case, for example, where each FID is very strong but it is not possible to generate such an FID very frequently.

Consider the $\pi/2$ pulse. After the pulse, there is no magnetization left along the static field so the recovery must begin from zero. Now consider an 80 degree pulse instead. An FID after such a pulse would have $\sin(80°)$ or about 98% of the amplitude of an FID after a $\pi/2$ pulse. At the same time, the component along the static field is $\cos(80°)$ or about 17% of full magnetization after the pulse. The time required after a $\pi/2$ pulse for the magnetization to recover 17% is $0.19T_1$, as you may easily verify. Thus we have the potential for time-saving between pulses of 19% of $T_1$ for a loss of 2% of signal if we always choose the repetition time to be the same in units of the relaxation time. In the same way, a $\pi/4$ pulse will yield 0.707 of the full FID while keeping 70.7% of the magnetization along the static field.

The previous paragraph discussed the relation between tip angle and the net S/N for a given repetition rate. A similar relation exists between the repetition rate and the net S/N for a given tip angle. Taking both of these effects into account, Ernst and Anderson (1966) pointed out that the optimum tip angle $\alpha$ and the pulse spacing $\tau$ are related to $T_1$ by

$$\cos\alpha = \exp(-\tau/T_1),$$

where $\alpha$ is called the Ernst angle. Waugh (1970) has pointed out that the optimum S/N is a slowly varying function of the $(\tau, \alpha)$ combination for $\alpha$ less than about 45 degrees. This is fortunate since otherwise both $\tau$ and $\alpha$ would have to be adjusted for each different $T_1$. As it is, a $T_1$ estimate would allow you to choose an optimum $\tau$ for a fixed $\alpha$, the latter being more inconvenient to adjust than the former, usually. For $\alpha$ of 30 and 45 degrees, the optimum pulse spacing $\tau$ should be 0.14 and 0.35 times $T_1$. An important side benefit to these shorter pulses is the increased spectral coverage as already discussed. For a related discussion, see Ernst (1966).

REFERENCES

R. R. Ernst, "Sensitivity Enhancement in Magnetic Resonance," Adv. Magn. Resonance 2, 1-135 (1966).

R. R. Ernst and W. A. Anderson, "Application of Fourier

transform spectroscopy to magnetic resonance," Rev. Sci. Instrum. <u>37</u>, 93-102 (1966).

J. S. Waugh, "Sensitivity in Fourier transform NMR spectroscopy of slowly relaxing systems," J. Mol. Spectry. <u>35</u>, 298-305 (1970).

## VI.A.8.   WHEN YOU CANNOT USE LOCK SOLVENTS

The most common internal locking scheme in high resolution solution NMR is to lock on the deuterium in deuterated solvents and many spectrometer locks are set up only for deuterium. What if the sample does not contain any deuterium? (The following argument is true for any nuclei chosen for locking.) An instance of this would be with an air sensitive sample well sealed in a tube and not containing any deuterium.

An obvious and well used solution to this problem is to use an appropriate deuterium containing solution in the same NMR coil but physically separated from the sample under study. There are two ways to do this. If you have a sealed sample tube which is not to be contaminated, the lock sample must go on the outside of the sample tube. For example, an 8 mm o.d. sample tube fits nicely in a 10 mm o.d. tube with just enough room for an annulus of lock solution.

A better scheme, if the sample can be opened to air, is to use a capillary containing the lock material in the center of the sample tube. This scheme is better in the sense that you do not need as much locking material so that the signal S/N can be better than in the annulus scheme. In both schemes, it is important to engineer precision spacers so the cylindrical symmetry of all components is maintained in order to avoid spinning sidebands. The problem is especially acute for paramagnetic samples because of the significant demagnetization fields induced in them by the static field.

We mention one more difficulty with using a locking solution which is physically not coincident with the sample of interest and that has to do with tuning for the best homogeneity of the static field. With many spectrometers, the most convenient way to maximize the homogeneity is to maximize the lock signal level. This is convenient and works well for samples dissolved in deuterated solvents. In schemes like those discussed here, this procedure does not work because maximizing the field homogeneity over the lock solution volume usually does not maximize it over the sample of interest. Thus, in cases like these, it is better to optimize the field with the signal from the actual sample of interest.

## VI.A.9    HELPFUL HINTS

#1. It is very useful to have a table of a few characteristic compounds, listing the lock frequency, the observe frequency, decoupler frequencies, and other relevant settings that would enable one to pick good starting parameters. The S/N should also be recorded for these spectra. Periodically running a standard sample and comparing the S/N and instrument settings to those of the standard can be helpful in spotting poor tuning or other instrument problems. When looking for a difficult resonance, it is a great assurance that the instrument is working right.

#2.   Get in the habit of writing down all relevant information on each spectrum that you take. How many times have you forgotten to record some observation that you needed to remember later?

#3.   When first looking for a spectrum which is likely to be quite wide or where you do not know exactly the correct carrier frequency, use a higher than normal digitization rate to create a larger than normal spectral window. Then move the spectrum to a convenient position and, in steps, close down the window. This way you can progressively adjust the spectral position without the difficulties of aliasing and folding.

#4.   When observing a line very close to the Nyquist frequency, you may find that the S/N is worse there. This is due to the noise being aliased into the spectrum through an imperfect filter. You can improve the S/N in this region by increasing the digitizing rate so that any noise components that might alias back are thoroughly attenuated. See V.A.3. for more details.

#5.   One should do as much spectrometer tuning as possible with real time signals as opposed to stored signals. For example, when setting up the FT NMR spectrometer, it is useful to tune the receiver coil with a real time signal. If it is difficult to have a signal that you can see in one pulse, you can do several things to see a real time signal. Use an isotopically enriched sample or a sample with a single strong line and then optimize the spectrometer settings to make it observable in one pulse. In the case of $^{13}$C, use a benzene sample, set the line close to the carrier, and then reduce the filter bandwidth to increase S/N. Display the FID on an oscilloscope, and while pulsing at a moderate rate, adjust the probe tuning for maximum signal. Be sure to have the tip angle set to less than 90 degrees for this operation. Otherwise, you may be optimizing the receiver match while inducing smaller FID's from tipping the magnetization too far by optimizing the transmitter match.   This approach can be applied to any nucleus for which you have a strong, single-line signal.

#6.   Be sure to use the maximum number of nuclei of interest possible.   After all, the total number of nuclei being observed is crucial even if the nuclei have inherently high sensitivity according to the NMR tables.   This means that solids should have the maximum density.   For powder samples, it pays to pack the sample tube by tapping it against a hard surface or even pressing it into a cylindrical pellet in a hydraulic press, if possible.   For solutions, it goes without saying that the more sample dissolved in the solvent the bigger the signal.

## VI.B.     WHEN THINGS GO WRONG

In these sections, we discuss some of the actual problems you may encounter in an NMR experiment.   In addition, Section VI.B.2. deals with suggested equipment to have on hand for calibration and maintenance.

## VI.B.1.   TROUBLESHOOTING SUGGESTIONS

No signal?   Don't panic -- this happens to everyone. Do you see any noise on the output of the receiver?   Turn

up the scope and receiver gain and set the scope trigger on
AUTO to get a continuous display. Make sure that the filter
is cut back enough, i.e., the cutoff high enough, to allow
noise to come through. If you do see noise, go on to the
next paragraph. (The usual noise amplitude should have
been recorded with the apparatus in working condition for
comparison.) If you do not, this means that something is not
turned on, something is disconnected, or that a failure has
occurred in the receiver. A not unheard of problem is an
electrically bad solder joint in the probe or a bad cable connector. Be sure to check the reference rf level to the phase
detector, too.

You see noise but do not see an NMR signal. Is there a
sample in the probe? Is it the right sample? Is the sample
positioned correctly? Is the magnet on? Is the field set at
the right value? You should use a tuning sample with a
strong signal and short relaxation time, to enhance your
search for a signal.

Perhaps the transmitter is not putting out pulses. Can
you see pulse feedthrough on the start of the scope trace?
Can you use the scope to check the output of the transmitter
to make sure that it is pulsing? (Use an indirect method.
See Section VI.A.1.)

If you have checked everything indicated above, you
must have another major problem: Do you have the right
nucleus conversion module in a multinuclear instrument? Does
the sample you are tuning on contain nuclei of the kind the
spectrometer is set up for? Are there any disconnected
cables? Look around for something obvious.

If there is a signal but it is very weak, there are several possible causes. 1) The transmitter pulse length might be
off. 2) The pulse repetition rate might be too fast for the

relaxation time of the sample.   3) The field/frequency condition might be off, i.e., it is off resonance.   4) The receiver
or the tank circuit might need to be tuned.   5) The sample
was hung up so that it is not centered in the coil in the
probe.   The reader is referred to the write-ups in sections
VI.A. for some of the procedures required to correct these
problems.

If you think you see a weak signal but aren't sure, try
removing the sample to see if it goes away.   Or change the
field, if possible, by an amount large enough to go off resonance but not so large that it might change the magnetic field
induced effects on the probe such as the spurious ringing
discussed in VI.B.5.   If the signal does not go away, it is
not an NMR signal!

## VI.B.2.   MAINTENANCE EQUIPMENT SUGGESTIONS

What follows is a list of useful equipment for maintenance,
troubleshooting, and modest instrument modification.

A good oscilloscope is indispensible.   It should have
sufficient bandwidth for observing the highest frequency
used.   The bandwidth can be obtained from the manufacturer's
instruction book or catalog.   The shortest timebase does not
represent the risetime of the scope because its bandwidth is
limited by the vertical amplifier and not the timebase.   Thus,
the bandwidth will usually be less than that suggested from
the shortest timebase.   Note that the bandwidth normally
refers to the 3 dB point.   Thus, for some applications, the
bandwidth must exceed the highest carrier frequency and a

good rule of thumb is to obtain a scope with the next higher bandwidth available than the highest carrier frequency.

In using the scope, bear in mind that the probe capacitance (printed on the probe) is applied across the circuit under investigation. While usually negligible, it may not always be, especially at high frequency. Note also that some probes are ×10 probes (also indicated on the probe) meaning that the voltage is attenuated by an order of magnitude by the probe. On many modern scopes, the gain indicators for the input amplifiers automatically change for such probes, but on many older scopes it is up to the operator to remember that factor of 10.

The vertical amplifiers used in oscilloscopes can be of very high quality. On many scopes, there is a connector in the back which gives access to the output of the amplifier, a very useful feature if you find yourself in need of an amplifier having a modest gain. In a dual channel amplifier, one output can be fed into the other channel for further amplification.

While scopes are usually used as voltage measuring devices and therefore have high input impedances, there are probes available which, by clamping around a wire, measure the ac current. This can be useful in monitoring the current flowing in NMR transmitter coils.

An oscilloscope camera is a useful but not an essential accessory. When the instrument is operating normally, a photograph of the waveform characteristics after each stage in the receiver and transmitter should be taken, to be used later in troubleshooting. In the absence of a scope camera, sketches of the waveform characteristics will have to do.

A VOM (Volt-Ohm Meter) is necessary to check power supply voltages, circuit continuity, resistances, etc. Convenience is more important than accuracy for most applications. Thus, there is definitely a place in the NMR lab for a small battery-operated portable VOM which can be either analog or digital. In this day and age, an input impedance of $10^7$ ohms is commonplace in solid state meters but even that is not necessary for many of the routine uses. If a small and convenient high impedance ($\sim 10^7$ ohms) meter is not readily available, acquire any small VOM that is easy to use and keep it in a drawer near the machine. When a high impedance meter is needed, you can borrow one or use an oscilloscope.

A dummy load with sufficient power dissipation to be attached to the output of the transmitter will help in troubleshooting and tuning the transmitter. This can be made out of a network of resistors assembled to match the cable impedance from the probe with sufficient dissipation to exceed the average output power. (In a train of pulses, what is significant is not the peak power but the rms power. In many cases this average power is not very much; for example, a 1 kW pulse 1 µs long occurring every second leads to an average power of 1 mW.) Hefty commercial dummy loads are also available with the resistors immersed in oil for better heat dissipation.

A set of non-magnetic screwdrivers and/or tuning wands is very useful for adjusting the probe while it is in a magnetic field. Those available from electronic supply sources and manufacturers of components requiring such tools, e.g., piston capacitors, usually can be made to work well with probes in iron electromagnets. Plastic wands are particularly preferable for this purpose because they do not impart a significant capacitance to the circuit. Unfortunately, they are weak if they are all plastic. Thus a hybrid of a stainless tip and a plastic handle and possibly a plastic shaft is a good compromise. Similar tuning tools for SC magnet probes must be made up specially.

A frequency meter is used to measure frequencies but the modern frequency source usually contains either a frequency meter or some other indication of the frequency such as the button or knob settings of a synthesizer. In the absence of a meter, the frequency source can be calibrated with a second source whose frequency is known by beating the two signals together.

A signal generator is very useful for trouble-shooting the receiver, adjusting the probe, etc. Many modern instruments use a frequency synthesizer as a master oscillator and this can be used in lieu of purchasing an additional signal generator.

For building probes, making tank circuits, and as an rf source which acts as a artificial NMR signal without any electrical connections to the circuit, a dip meter is nearly essential. In addition, it is useful in its intended role to passively measure the parallel resonant tank circuit's resonant frequency.

Another useful gadget is a Q-meter for measuring coil parameters if you are going into probe building. Q-meters can also measure inductances. At the present time in the U.S., the World War II vintage Q-meter is still available in the used market. There are, of course, solid state Q-meters as well.

To go first class in building probes, you should have a vector impedance meter. As pointed out in Section V.C.11., the magic-tee may be substituted at two orders of magnitude reduction in cost but with signficantly more fiddling.

Modern NMR relies so much on digital electronics that some of the standard tools for digital work should be in the lab. These include a good logic probe (a small one with a logic 0 or 1 and pulse indications near the probe tip), integrated circuit insertion and removal tools (so you don't mistreat those IC's which is an easy thing to do), and an audible continuity checker.

An <u>audio oscillator</u> capable of sinusoidal, triangular, and square outputs with a choice of offsets so that the zero level can be at the top, middle, or the bottom of the waveform is often useful. It should have a maximum frequency of 1 MHz and variable output amplitude. One possible use is as a dummy detected NMR signal for checking out the digitizing system although the pulse sequence generator output is better because it can by synchronized to an external signal like the digitizer trigger.

<u>Variable voltage power supplies</u> are nice to have around either to power some trial circuits, to substitute for a malfunctioning power supply, or to supply a variable voltage for testing, for example, an ADC.

## VI.B.3.   PROBE ARCING

One possible difficulty with probes is high voltage arcing.

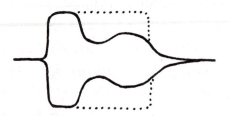

It can usually be detected by observing the rf pulse which may partially collapse during arcing as shown. Sometimes the arcing can be heard as a snapping or popping sound and it may be observable in a dark room. A more subtle form of arcing can occur when the breakdown occurs only at the top of each cycle of the rf. The result is an rf pulse which looks feathery and irregular at the edges. Usually this does not produce the distinctive snap of large, isolated arcing. A common symptom of probe arcing is irreproducible FID amplitudes with reproducible FID shapes (signifying that the magnetic field is stable).

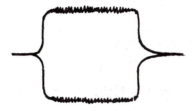

The probe design can be modified to suppress arcing by moving rf components away from grounds. Arcing is more likely to occur at sharp corners so they should be rounded off as much as possible. With a probe that you do not wish to rebuild (perhaps because it is someone else's probe) the problem is to suppress arcing as best as you can. High breakdown voltage dielectric like Teflon tape may work. Coating the suspected parts with a corona dope, glyptal varnish, or RTV silastic can often cure the problem. Even though such materials may give rise to unwanted NMR signals, they do not contribute noticeably to high resolution spectra because of their short $T_2$'s. In a probe used to observe broad lines, greater care must be exercised. Another potential problem is that the insulating material may have a dielectric dispersion at the spectrometer frequency. Only materials having low dielectric losses should be used. As a last resort, you may have to manage with less transmitter power.

There are some special situations in which the arcing problem is worse than normal. The presence of helium gas, which is common in cryogenic experiments, is one such case. Being at high altitude is another. The laboratory in Los Alamos is 2200 meters above sea level and we are very careful in this regard.

## VI.B.4.  H$_1$ INHOMOGENEITY

The rotating magnetic field H$_1$ can be quite uniform for a crossed coil system with the transmitter coil in a Helmholz configuration or a single coil system with the coil in the solenoid configuration if the sample is small compared with the coil. Unfortunately, this is not the normal state of affairs in a single coil experiment. Usually, the sample is much longer than the dimension of the coil so that the part of the sample outside the coil sees a smaller rotating field than the part inside. There are two reasons for having the sample much longer than the coil: to maximize the signal strength and to minimize susceptibility variations as seen from the receiver coil in order to maximize resolution.

Another source of field inhomogeneity in a coil is due to the discrete nature of the coil wire. Consider a solenoidal coil with a tightly fitting cylindrical sample inside it. The field due to the coil on the surface of the cylinder depends on whether the point in question is adjacent to a wire or between two consecutive turns of wire. One way to reduce this $H_1$ variation is to wind the turns closer together, if possible. Another is to shape the wire cross section so that the side facing the inside of the solenoid is flat. This can be accomplished by squashing the wire between two rollers so that its cross section is approximately rectangular (Rhim, 1980). A third cure, usually applied incidentally, is some form of an electrostatic shield like a piezoelectric resonance shield or a Faraday shield used for some reason other than for $H_1$ homogeneity (see VI.B.5.).

Another kind of spatial inhomogeneity of $H_1$ cannot be blamed on coil imperfections but rather is intrinsic to metallic samples. The skin effect determines how large an $H_1$ field will be at a given depth below the sample surface at a given frequency.

The above problems are problems relative to the spatial homogeneity of $H_1$. Another kind of problem is encountered when the resonance offset is a significant fraction of the $H_1$ field used. This kind of problem is encountered in FT NMR of broad spectra in which the effective $H_1$ is not uniform across the spectrum when the offset frequency is a measurable fraction of $\gamma H_1/2\pi$. A typically good $H_1$ is on the order of 60 G which is about 1000 ppm in a superconducting magnet and the effects of such offsets may be noticeable in a 500 ppm spectrum. The consequences of having such an $H_1$ field which does not cover the entire spectrum uniformly is discussed in section II.A.2.

For a simple lineshape analysis (by, for example, a Fourier transform of the FID) this $H_1$ inhomogeneity, as long as it is not excessive, is of no consequence because it is software correctable (II.B.4.) but there are many experiments for which either the precise knowledge of $H_1$ or a good spatial homogeneity of $H_1$ is necessary. One example is the inversion recovery experiment in which $T_1$ is determined from the time required to reach a null, as the incomplete inversion would put this time at a value different from $T_1 \ln 2$. Correcting for the error introduced into the null method by $H_1$ inhomogeneity is discussed in an article by Van Putte (1970). Note that if $T_1$ is extracted from the slope of the log of the relaxation process vs. time, rather than from the null time, this error does not occur. Another example occurs in some multiple pulse experiments which cannot tolerate cumulative errors in the forced nutation of the magnetization.

A simple test to determine how severe $H_1$ inhomogeneity problems (of either source) are is to compare the FID amplitude following a 90° pulse and a 90°+360°=450° pulse. The discrepancy is a measure of the inhomogeneity. An obvious extention is to keep repeating the 360° pulses with the pulse intervals just long enough to monitor the initial heights of the FID's. The decay of the magnetization is the Fourier transform of the $H_1$ inhomogeneity provided that $T_2^* \ll T_2$ in complete analogy with the relation between $T_2^*$ and the $H_0$ inhomogeneity (Lowe and Whitson, 1977).

A novel suggestion to alleviate both the problem of moderate missetting of the 180° pulse condition plus the problem of moderate resonance offsets is due to Levitt and Freeman (1979), and involves forming composite pulse sandwiches to do the work of individual pulses. For example, a single 180° pulse where the magnetization is rotated about the x' axis, which we denote as $180_{x'}$, can be replaced with a composite pulse composed of the pulses $90_{x'}$-$180_{y'}$-$90_{x'}$ with the shortest possible delay between them.

Levitt and Freeman performed some trajectory calculations of the magnetization under several conditions to show the effect of the composite pulse in refocussing the magnetization. They also performed an experiment in which they measured $T_1$ by an inversion recovery zero-crossing method and showed that even at an offset of 36% of $H_1$, the error in the measured $T_1$ using the composite pulse was only 12% compared with 66% for the same determination with the standard 180° pulse. For more details, see a later paper by the same group (Freeman, et al., 1980).

## REFERENCES

R. Freeman, S. P. Kempsell, and M. H. Levitt, "Radiofrequency pulse sequences which compensate their own imperfections," J. Magn. Resonance 38, 453-479 (1980).

M. H. Levitt and R. Freeman, "NMR population inversion using a composite pulse," J. Magn. Resonance 33, 473-476 (1979).

I. J. Lowe and D. W. Whitson, "Homogeneous rf field delay line probe for pulsed nuclear magnetic resonance," Rev. Sci. Instrum. 48, 268-274 (1977).

W.-K. Rhim, private communication, 1980.

K. Van Putte, "Elimination of $H_1$ inhomogeneity and spin-spin relaxation in the determination of spin-lattice relaxation times," J. Magn. Resonance 2, 174-180 (1970).

## VI.B.5.  SPURIOUS RINGING

Often a large spurious ringing signal, which can be strong enough to obscure the FID, is caused by the transmitter pulse. Such a ringing is usually coherent with the rf pulse so that signal averaging will not attenuate it nor can it be distinguished from an NMR signal by reversing the rf phase.

One possible cause of such a signal is an acoustic ringing of some probe parts. An acoustic standing wave can be set up by an rf pulse in a conductor placed in a magnetic field (Buess and Peterson, 1978; Fukushima and Roeder, 1979). This acoustical wave is converted back into an rf radiation by a reciprocal mechanism and is then picked up by the receiver coil. It is most commonly observed under high magnetic field, high pulse power, and low frequency. The ringing detected is directly proportional to the $H_1$ field and also to the square of the static field intensity $H_0$ through the conversion efficiency E where

$$E = kH_0^2/[mv_s(1 + \beta^2)] \tag{1}$$

with k a proportionality factor, $H_0$ the static magnetic field, m the mass density, $v_s$ the acoustic shear velocity, and

$$\beta^2 = 2.5 \times 10^{13} \, (\rho^2/v_s^4)\nu^2$$

in which $\rho$ is the resistivity and $\nu$ is the frequency.

Of the three common probe parts which can ring, the body is the easiest to deal with. A good choice of the probe body material usually will take care of the problem. The following table of room temperature parameters of metals (Fukushima and Roeder, 1979) shows that aluminum is a poor choice because of its small $\beta^2$ and m. Calculation of the denominator of eq. (1) shows that brass is about 2.4 times better than aluminum and stainless steel about 14 times better. Furthermore, $\beta^2$ decreases rapidly with decreasing temperature for pure metals making aluminum even less attractive at low

| | m ($10^3$ kg/m$^3$) | v$_s$ (m/sec) | $\rho$ ($10^{-8}$ $\Omega$-m) | $\dfrac{mv_s}{mv_s}$ (A$\ell$) | $\beta^2$ (at 5 MHz) |
|---|---|---|---|---|---|
| ALUMINUM | 2.7 | 3040 | 2.7 | 1.0 | .005 |
| BRASS (70% Cu) | 8.5 | 2110 | 6.2 | 2.2 | .121 |
| COPPER | 8.9 | 2270 | 1.7 | 2.5 | .007 |
| GOLD | 19.7 | 1200 | 2.4 | 2.9 | .173 |
| LEAD | 11.4 | 690 | 20.7 | 1.0 | 118 |
| PLATINUM | 21.4 | 1730 | 10.6 | 4.5 | .783 |
| SILVER | 10.4 | 1610 | 1.6 | 2.0 | .024 |
| STAINLESS (347) | 8.0 | 3100 | 73 | 3.0 | 3.61 |
| TUNGSTEN | 19.3 | 2640 | 5.7 | 6.2 | .042 |
| ZINC | 7.1 | 2440 | 5.9 | 2.1 | .061 |

The m, v$_s$, and $\rho$ values are taken from Handbook of Chemistry and Physics, 50th ed., pp. E-41, F-140, F-141, The Chemical Rubber Co., Cleveland, OH, 1969; Metals Handbook, 8th ed., vol. 1, T. Lyman, ed., pp. 52, 56, Am. Soc. for Metals, Metals Park, OH, 1961.

temperature.  From these considerations, non-magnetic stainless steel is an ideal material for NMR probes, especially at
low temperature.  For a given probe material, the effects of
acoustic ringing can be reduced by maximizing the distance
between the coil and the probe body or alternatively by putting a shield between them.  (Speight, et al., 1974).

Besides the probe body, the coil itself can ring but since
copper is such a desirable material for winding coils because
of its favorable electrical and mechanical properties, it is difficult to find a substitute.  The spurious ringing in the coil
may be reduced by changing the wire size, usually to a
smaller size, in order to change the standing wave pattern in
the material.  Also constraining the wire against a coil form
may attenuate the acoustical wave.  A useful combination of
the above is a stranded insulated wire.  Finally, if the sample
is metallic and the ringing originates in the sample itself,
powder the sample, arrange it as a stack of thin slices, or
soak it in some damping fluid like mineral oil.

A second possible cause of spurious ringing which is
limited only to some solids is piezoelectric resonance of the
sample.  (Of course the cause need not be limited to a sample.
More on this later.)  A piezoelectric crystal has the property
that a mechanical deformation of the material takes place in the
presence of an electric field.  Therefore, an acoustic ringing
can be induced in a piezoelectric material located within an
NMR probe by the rf electric field associated with the rf magnetic field.  Such a ringing due to piezoelectricity has the
distinguishing property that it is independent of the applied
magnetic field intensity in contrast with the magnetic field
induced effect described in the first part of this section.

Gibson and Raab (1972) found that piezoelectric resonances of the NMR sample can be attenuated in one of two

ways. The first is an acoustic damping scheme which in their case was to immerse the sample in silicon oil. The second, and a far more practical and less sticky method is to use an electrostatic shield between the coil and the sample. They found that a suitable shield at frequencies of 25-30 MHz can be made from 30 μm thick aluminized Mylar whereas 120 μm thick copper foil was necessary at 3.3 MHz. The screen should have an insulated overlap of at least one third of a turn, should extend beyond the sample and the coil, assuming that they are the same length, by one coil diameter, and be grounded. They found that these screens degraded the Q of the coil by a factor of two, a price well worth paying to be able to see the signal at all.

As hinted above, piezoelectric resonance can occur in some probe parts besides the sample so that it could even affect probes for solution NMR. Therefore, if there seems to be a problem with a magnetic field independent spurious ringing, check for possible piezoelectricity of a component in the probe. Remember that quartz is very piezoelectric. In addition to the obvious uses as tubes and whatnot, quartz is also used sometimes as a filler in epoxy.

## REFERENCES

M. L. Buess and G. L. Peterson, "Acoustic ringing effects in pulsed magnetic resonance probes," Rev. Sci. Instrum. 49, 1151-1155 (1978).

E. Fukushima and S. B. W. Roeder, "Spurious ringing in pulse NMR," J. Magn. Resonance 33, 199-203 (1979).

A. A. V. Gibson and R. E. Raab, "Proton NMR and piezoelectricity in tetramethylammonium chloride," J. Chem. Phys. 57, 4688-4693 (1972).

P. A. Speight, K. R. Jeffrey, and J. A. Courtney, "A probe modification for pulsed nuclear magnetic resonance to eliminate spurious ringing," J. Phys. E 7, 801-802 (1974).

## VI.C.    NOISE

These sections deal with the very important issue of noise. We discuss some theory as well as give practical examples of how to cope with the noise problems.

## VI.C.1.  REMOVAL OF COHERENT NOISE

In pulse NMR, there is always noise coherent with the pulse sequence because of the transient nature of the experiment. This is in addition to the usual random noise as well as other systematic noises such as the 50 or 60 Hz "hum" from the ac source which are not synchronized with the pulses. These non-synchronous components of noise can be reduced either by appropriate filtering or by multiscan averaging or both. In this section we will deal with noise coherent with the pulse sequence. Even if such noises were very small, they will add upon signal averaging (just like the desired signal) and may, therefore, be troublesome in experiments where very many FID's are accumulated due to inherently poor S/N.

The best solution to any problem, of course, is to treat it at the source. Much of the incoherent as well as some of the coherent noise are due to "cross-talk" between different electronic components. Grounding the different components to a common point (as opposed to grounding everything to each other) with husky copper ground straps goes a long way in reducing much of this kind of noise, especially the 60 or 120 Hz hum. Some magnets, especially those iron-core ones with field regulation, can cause problems if they are a part of this ground loop. Some people have found that electrically isolating the magnet from the rest of the electronics, for example, by insulating the probe fom the probe holder, helped reduce low frequency noise.

Another source of noise may be the computer. Monitor the receiver output on an oscilloscope as a function of computer operation. The noise pattern will change upon changes in the program being run (or even more clearly on stopping the execution) if computer noise is getting into the receiver. Try an isolation transformer or even a good line regulator in series with the primary power cable of the computer to reduce the problem. Much of the computer noise will be at or below tens of MHz to hundreds of kHz so choosing a high IF in an up-conversion scheme will also help (see under Receiver in V.A.1.).

Since the noise coherent with the FID is usually a ringing of some kind generated by the high intensity rf pulse, such ringing must be suppressed as much as possible. Some of the major sources of ringing were discussed in section VI.B.5.

The most obvious way to cancel unwanted signal coherent with the FID is to accumulate the total signal and the signal without a contribution from the nuclei in the sample and subtract them. The sample contribution may be removed in several ways. The sample may be physically removed; the magnetic field may be changed to go off resonance; or the magnetization may be saturated by some pulse sequence. The removal of the sample is obviously the easiest to contemplate and, often, to implement.  Its implementation is not at all trivial, however, for sideways sample experiments in super-conducting magnets and even more difficult for cryogenic experiments in which the sample coil is deep inside a dewar. Finally, the removal of the sample will not accomplish its intended aim for some samples, e.g., metals, because its removal may alter the characteristic of the spurious signal we are trying to cancel.

The second method for removing the signal contribution from the sample is to go off resonance.  Here we are talking about changing the field rather than the frequency because the latter usually involves having to retune resonance circuits after which the experimental conditions will have changed. Going off resonance to remove the sample contribution of the signal works as long as the amount of field shift required to obliterate the signal is not enough to change the character of the undesirable signal.  Another possible problem is the ability of the magnet to come back to exact resonance afterwards.  It turns out that it is not necessary to go so far off resonance that the signal is absolutely zero.  Later in this section, we discuss a digital lock-in method wherein the field is changed by an amount sufficient only to change the character of the desired signal rather than to eliminate it.

The problems with the above methods of generating a "background" signal to be subtracted from the total signal are largely eliminated with the saturation of the NMR signal by some means where the sample and the resonance condition remain untouched.  As is clear from our previous discussions, there are many ways to saturate the magnetization so that a subsequent pulse will yield little or no NMR signal.  The de-sirable methods should not require a separate sequence to accumulate the "background" signal.

One of the ways to accomplish this objective is to pulse the system when there is no net magnetization. Such is the case after a $\pi/2$ pulse if $T_1$ and $T_2$ are sufficiently different so that a delay time $\tau$ can be chosen where $T_2 < \tau < T_1$. If it is not possible to satisfy this condition because the two relaxation times are too similar, we must be more clever. Cueman and Soest suggested making the first pulse a little longer than 90 degrees, say 95 degrees, which hardly affects the FID. After a suitable delay, the z component of the magnetization will be exactly zero because it is crossing over from being negative to positive. A 90 degree (or more conveniently a 95 degree) pulse at that instant will result in no NMR signal at all; just the noise. This method works only for single lines and requires more adjustments than the long comb but the pulse sequence is easier to generate and the power requirements on the transmitter are much less (Cueman, 1975).

Another way to record the background signal in the same sequence as the total signal is to do a saturating comb sequence discussed in section III.D.2. and record the signal after the last pulse in the sequence as well as the desired FID after the first pulse. Because of the special properties of such a comb, this works especially well for a spectrum which is quite spread out, for example, for metallic samples in which there is a large distribution of $H_1$ amplitudes, or for quadrupolar nuclei in a non-cubic solid with an electric field gradient. In addition, the pulse length adjustment is not critical since it is the cumulative effect of the comb which makes the magnetization vanish. The method does require two digitizers working at different times because there is not enough time, usually between the desired FID and the "background" signal at the end of the sequence, to gather both

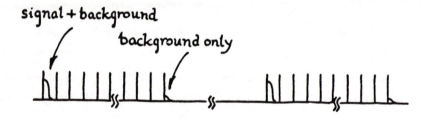

with the same digitizer.   See the example in V.C.3.

As already mentioned, it is undesirable to have to make a separate run to get the background signal because of the extra time required.   Therefore, any method which utilizes the same sequence for obtaining both the "background" and the "total" signals is preferable to any which does not. There are several methods other than the two just discussed which are good in this respect.   The methods to be discussed differ from the two previous in that only one FID is recorded per sequence, thus eliminating the need for a second digitizer. Instead, the NMR signal character is changed a part of the time and the difference is recorded.   The hope is that the unwanted noise components will cancel and the desirable signal will still be manageable.   Above all, every sequence will have a contribution towards the desirable FID so that the method will be efficient.

The more common of the two methods reverses the phase of the rf carrier for exactly one-half of the sequences.   Then the signals from the two halves are subtracted.   Reversing phase will turn the desirable FID upside down so the subtraction will simply accumulate the two halves together.   If the spurious transients are not coherent with the rf, they will not respond to the phase reversal and will cancel in the process. The phase reversal can be performed on successive sequences with concomitant juggling of the data, or N signals can be accumulated with one phase and N more with the other. The latter is easier to implement but is more susceptible to instrumental drifts. In addition, for high resolution NMR in liquids, reversing the phase on successive pulses has the added advantage that spin echoes are suppressed in cases where the data acquisition is performed in a time less than $T_2$.

The phase reversal method mentioned above does not work in cancelling the unwanted signal if the unwanted signal is phase coherent with the rf so that it is indistinguishable from the desired signal as far as the phase is concerned. This is the case, for example, for electromagnetically generated acoustic standing waves (see VI.B.5.). Duncan, et al. (1979) have proposed a solution whereby the rf detection pulses remain the same while the magnetization is reversed by 180 degree pulses in front of every other 90 degree detection pulse after which the alternate FID's are subtracted.

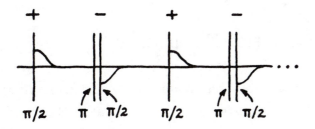

We end this section with another method which is not widely used but deserves mention if for no other reason than its originality. Suppose we accumulate N FID's and then subtract the sum from an accumulation of N other FID's which were taken as a slightly different magnetic field. The coherent noise is cancelled and we only need to worry about what is left which turns out to be quite manageable and yields

essential spectral information which comes from both signals.
Intuitively, it is not surprising that this results in a derivative spectrum so what we have here is a digital form of a
lock-in detector commonly used in cw NMR of broad lines
(Hatch, et al., 1974).

REFERENCES

M. K. Cueman, "Quadrupole modulation of NMR free induction
decays," PhD. thesis, College of William and Mary, 1975;
listed in Diss. Abstr. <u>36</u>, 5666B (1976).

T. M. Duncan, J. T. Yates, and R. W. Vaughan, "$^{13}$C NMR
of CO chemisorbed on Rh dispersed on $Al_2O_3$," J. Chem. Phys.
<u>71</u>, 3129-3130 (1979).

G. F. Hatch, J. W. Neeley, and R. W. Kreilick, "Wideline
pulsed-Fourier transformation NMR," J. Magn. Resonance <u>16</u>,
408-416 (1974).

VI.C.2  THE ANALOG FILTER AND THE FID

More often than not, we have to battle poor signal to
noise ratios in NMR experiments. The greatest boon to this
end is signal averaging, as mentioned in the introduction but
the practically available S/N enhancement in time t is limited
because it is proportional only to $t^{\frac{1}{2}}$. Therefore, it clearly
pays to first maximize S/N by whatever means available, other
than signal averaging, such as 1) using a low noise preamplifier, 2) choosing the optimum combination of the tip angle and
the repetition rate, and 3) filtering parameters. After all, if
twofold S/N improvements were possible in each of these three
areas, the total S/N improvement is 8 and the time required to

achieve the equivalent improvement by signal averaging is cut down by a factor of 64! This means that an overnight (16 hr) run can be compressed into a 15 minute run or alternatively, a weekend run, say, of 60 hours can be performed routinely whereas with the inferior S/N an equivalent run would require over 5 months. The moral of the story is to minimize noise in the NMR signal as much as possible by any means possible because the resulting savings in time will be well worth the effort. In this section, we discuss the third means listed above, and in particular the filtering of random high frequency noise. The low frequency random noise is more likely due to poor grounding and should be attacked by improving the grounding system since it is more difficult to filter.

The optimum low pass filter, i.e., a filter which cuts out all frequencies above a certain critical frequency, has its cut-off frequency just above the highest frequency we want to keep. We can illustrate this by a diagram in the frequency domain. Here the signal is drawn separately from the noise but we measure the sum of the two. In the diagram, S/N is improved optimally (within our limitation) because all noise above the signal of interest is cut out with a sharp filter, i.e., the filter transmits fully below the cutoff frequency and

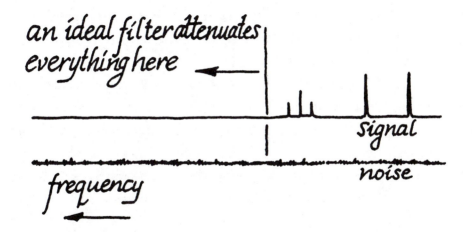

attenuates completely above it.   Why do we want to attenuate
noise components which are at frequencies different from the
signal?   Because the sampling theorem (section II.B.1.) tells
us that all signal and noise are reflected across the Nyquist
frequency.

Thus, the filter is needed to keep the higher frequency noise
from folding into the spectrum.

In actual practice, all filters have a distributed cutoff
frequency so that none are infinitely sharp, and the way
in which the attenuation "rolls off" with frequency affects
the attainable S/N.   The world of electrical engineering
knows of many different filters (such as the Bessel and the
Butterworth) which are characterized by different amplitude
rolloff and phase characteristics near the cutoff frequency.
A commonly used filter is the RC filter because of its ease of
implementation.   It consists simply of a capacitor C and a
resistor R.   It has the time
constant RC (check it; it has
the unit of time) and this
simply means that it will not
respond to signals that change
appreciably in times shorter
than RC so it is a low pass

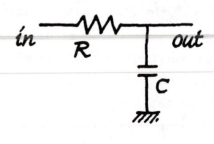

filter.   Its response to a step function in time is exponential
so that the rolloff in the frequency domain, i.e., its Fourier
transform, is a Lorentzian and the cutoff is very broad.

Now, if the rolloff is very gentle as is the case with an
RC filter, there is an obvious problem.   Since we do not want
the spectral features to be distorted, the cutoff has to be set
far enough from the highest frequency of interest which will
be quite far away.   If the Nyquist frequency is set fairly
close to the spectrum of interest, the noise from that part of

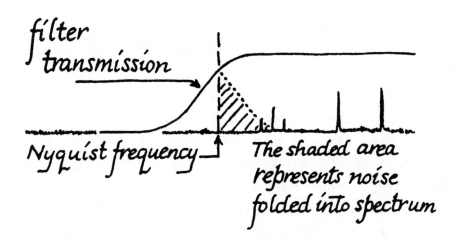

*filter transmission*

*Nyquist frequency*

*The shaded area represents noise folded into spectrum*

the spectral region above the Nyquist frequency, but not cut off by the RC filter, will be folded back into the spectrum. One obvious way to solve this problem is to use a filter with a sharper cutoff close to Nyquist frequency. We discuss other filters a bit later. Since the culprit in the above scheme, if we ignore the RC filter, is the Nyquist frequency about which the noise gets folded into the spectrum, another solution is to simply move it out by sampling at a higher rate. It costs you a faster digitizer and possibly a larger memory, since the resolution depends on the total time of data acquisition which is the total number of points divided by the digitizing rate. Thus, in order to maintain the original resolution while over-sampling, the total number of points must go up in proportion to the digitizing rate. This scheme is further discussed in the following section.

Of the more sophisticated filters, the two most commonly used are the Butterworth and the Bessel. They can provide much sharper cutoff than the RC filter. The difference between the Butterworth and the Bessel are subtle and are usually described in terms of the rolloff characteristics, i.e., the frequency dependence of the filter attenuation. If the rolloff characteristics are Fourier transformed, they become the time response of the filter following an impulse. (Note the analogy with FT NMR.) The Butterworth and the Bessel differ in the shape of the rolloff. For a similar average rolloff, the Butterworth has a sharper rate of rolloff than the comparable Bessel but suffers from an overshoot of the time response. thus it has better S/N but also more distortion (which mostly affect the baseline). Within each type, various rolloff rates may be chosen. See a discussion by Schaefer and Stejskal (1974) for more details.

Consider the frequency domain again. The requirement for a sharp cutoff arises because we want to cut off as much noise as possible and the sharper the filter cutoff the closer it can be set to the spectral feature to be saved. It is clear that the filter cutoff needs to be only as sharp as the spectral feature. Therefore, a careful consideration of filter characteristics is most important for high resolution NMR, whereas the RC filter is tolerably good for very broad lines such as those in solids observed by normal NMR. But since the opportunity for some oversampling is much greater for high resolution NMR than in wideline NMR, the absolute necessity for an "ultrasharp" filter is not as great as it is supposed in either case.

Analog filters cannot be fully effective in cleaning up many free induction decays because of the sudden initiation of the FID. Because the frequency spectrum of this step contains frequencies higher than those of interest, it is impossible to filter out enough noise without distorting the beginning of the FID. One possible solution to this problem is to notice that the dispersion mode signal contains more information than the absorption FID because we know that it

is constrained to be zero at t=0, thus allowing us to interpolate the missing section of the FID. But most people want
the absorption component. We can get rid of the discontinu-
ous jump at the beginning by creating a function earlier in
time which smoothly joins onto the beginning of the FID. An
empirical method for doing this is offered by Karlicek and
Lowe (1978) who provide an artificial signal immediately preceding the FID and joining it smoothly so that the discontinu-
ous jump affects only the artificial portion of the signal.
Other solutions to this problem are discussed in the following
section on digital filtering.

REFERENCES

R. F. Karlicek, Jr. and I. J. Lowe, "A pulsed, broadband
NMR spectrometer," J. Magn. Resonance 32, 199-225 (1978).

J. Schaefer and E. O. Stejskal, "Baseline artifacts in high-
resolution Fourier transform NMR spectra," J. Magn. Resonance 15, 173-176 (1974).

VI.C.3.   DIGITAL FILTERING

We have already discussed some aspects of filtering in
the previous section. In this section, we will consider digital
filtering, i.e., filtering of the NMR signal after the FID has
been digitized and (usually) stored. Much of what we discussed under analog filters applies here as well because it is
possible to duplicate the operation of any analog filter digitally.
Furthermore, it is possible to perform many operations digitally,
which are impossible by analog methods. We devote most of

this section to those operations which are possible only with digital methods or nearly so.

Some of the advantages of digital over analog signal processing are accuracy, stability, ease of changing parameters during the data taking operation, e.g., changing the filter characteristics for different parts of an FID, and being able to try different filtering (and other) operations on the same set of data.

First, consider a simple example which can illustrate a filtering operation which is not realizable with an analog filter. Suppose we want to create a lowpass filter to smooth the high frequency noise while preserving the low frequency information. One obvious way to do this is to form a moving average, i.e., for the n-th point of the data array, we average a certain number of points about the n-th point. For example, a three-point running average can be formed by summing the n-th point with the point before and the point after and dividing by three and repeating the operation for all other points in the data set. As simple as it is, this is clearly a filtering operation which is not realizable with an analog filter because a given filtered point utilizes not only the points earlier in time but those which came later.

Moving averages as described above are special cases of convolution as defined in the section (II.A.1.) on Fourier transform theorems. The digital equivalent of the definition stated therein for the convolution of f(t) with g(t) is

$$h_i = \sum_{j=i-n}^{i+n} f_j g_{ij} \qquad (1)$$

where 2n+1 is the number of points summed around the i-th point. In our example of a 3-point running sum, $f_{i\pm1}=f_i=1/3$

and all other $f_j$'s were zero. Since, by the convolution theorem, the Fourier transform of a convolution is the product of the Fourier transforms, the FID smoothed by a running sum transforms into a spectrum multiplied by a Fourier transform of a square pedestal which is $\sin\pi x/\pi x$ where x is the frequency parameter. One can consider the $\sin\pi x/\pi x$ function to be the frequency profile (the transfer function) of this digital filter. As long as the running sum is taken over an interval short compared to the duration of the highest frequency information desired, there will not be much distortion, and the $\sin\pi x/\pi x$ function being broad cuts out only rather high frequencies.

The above discussion is somewhat academic because there are other convolutions which are much more useful or meaningful than the running sum. One such improvement is to modify the three point running sum so that $f_{i\pm1}=\frac{1}{4}$ and $f_i=\frac{1}{2}$. Now the frequency profile of the filter is much narrower and more closely matched to the shape of an NMR line and its implementation is still trivial.

Now that we have used the convolution formalism, let us consider, as an example, the digital implementation of a physically realizable analog filter, namely the RC filter as described in the last section. In this case, the $g_{ij}$'s would represent an exponential function like the one shown at right with the time constant set to RC. It is intuitively clear that the convoluted, i.e., the filtered, signal will be skewed towards the

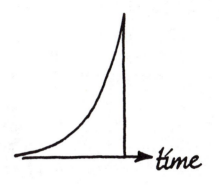

right or later time, as is true for the analog filter.

With this background, consider how we might achieve the best fit to a set of data points. The usual criterion for a best fit curve is the least squares fit in which the rms deviation between the data points and the fitted curve is minimized. A small enough section of the data must be considered for each segment so that it can be fitted by a polynomial. This will be a tedious process for a complex function like the FID from a multiline spectrum.

Savitzky and Golay (1964) pointed out that a least squares fit can be implemented with a convolution which avoids the laborious process of evaluating the coefficients of the polynomial. Their procedure yields only the filtered point which is all we want from a filtering routine anyway. They give the coefficients $g_{ij}$ of eq (1), for example, for n up to 12 and for up to a quintic polynomial fitting function. [They also give the coefficients for fitting derivatives up to the 5th which may not be as useful for NMR.] Thus, their work allows a rapid least squares fit to the digitized data, be they in the time or frequency domain.

A very clever digital data manipulation already alluded to in the last section has as its aim coping with the rapid initial transient while filtering the remaining FID. The problem comes about because the pulse feedthrough occurs much quicker, i.e., with higher frequency components than the FID and optimum filter parameters for filtering the FID would lead to distortions of the portion of the FID immediately following the pulse feedthrough. Smith and Cohn-Sfetcu (1975) proposed that the FID be reversed in time before being filtered. In that way the distortion due to the initial pulse feedthrough would occur earlier in real time and would not bother the FID itself so that the filtering parameter can be optimized for the FID.

Another way in which digital data manipulation cannot be matched by analog methods easily is in sampling. It is straightforward to pick off the tops of CPMG echoes, for example, (see section I.C.2.) or the magnetization to be monitored during a WAHUHA sequence (section IV.E.4.).

REFERENCES

A. Savitzky and M. J. E. Golay, "Smoothing and differentiation of data by simplified least squares procedures," Anal. Chem. 36, 1627-1639 (1964).

M. R. Smith and S. Cohn-Sfetcu, "On the design and use of digital filters in certain experimental situations," J. Phys. E: Sci. Instrum. 8, 515-522 (1975).

## VI.D.    SPECIAL TECHNIQUES

We now have several unrelated sections dealing with specialized techniques which may not apply to your needs but, then again, may be essential to your particular experiment.

## VI.D.1.  FIELD CYCLING EXPERIMENTS

Often, it is necessary to obtain the spin-lattice relaxation rates at different frequencies in order to study the spectral density or to differentiate different relaxation mechanisms (Noack, 1971). With a tunable spectrometer, it is possible to cover a frequency range of one or two orders of magnitude before you run out of sensitivity at the low field end. (Remember that S/N goes roughly as $H_0^{3/2}$.) Besides, it is difficult to change the magnetic field intensity and still keep the same field homogeneity with iron-core electromagnets whereas it is awkward to change the field, period, for persistent mode SC magnets. Therefore, variable field spectrometers are quite uncommon, most of them being homebuilt.

We have discussed two ways in which spin-lattice relaxation rates can be measured at fields much smaller than the usual laboratory fields. In section IV.B.1. we introduced the concept of dipolar order in fields of neighboring nuclei and in IV.C.1. we talked about the spin-lattice relaxation in the rotating frame. In this section, we suggest a third method of measuring relaxation at low fields, which works only if the $T_1$ to be measured is moderately long.

This method, field cycling, involves having the sample in two different fields at different times. To measure the $T_1$ at a low field, the nuclei are saturated at high field before the sample is moved to a much smaller field. After a certain amount of relaxation in the reduced field, the sample is brought back into the large field and the magnetization is sampled. Clearly, this technique is limited to systems with $T_1$'s long enough so that the magnetization recovery is negligible during the field changing operations. The nuclei will continue to precess around the changing magnetic field with the same entropy, i.e., without flipping the spin, provided that the quantity $(dH/dt)(1/H)$ is small compared to the Larmor precession frequency $\gamma H$, exactly as in adiabatic fast passage discussed in I.B.

Another way to perform field cycling is to actually change the magnetic field without moving the sample. Most magnets are very sluggish because of the large inductance of the coil so a special low inductance coil and a power supply must be used for field cycling. The advantage of this method is the ease in doing variable temperature work because the sample need not be moved. In an experiment of this sort, a special SC magnet was used in which the field could be switched by 7 kG in 20 ms and remained stable thereafter (Hallenga and Koenig, 1976, and references therein.)

For any sizeable field, the field cycling experiment can be performed by pneumatically moving the sample between the two magnets. As mentioned above, variable temperature work is significantly more difficult. A typical transit time is of the order of 500 ms. See, for example, the papers by Jones, et al. (1969) and Edmonds (1977).

As long as we are changing fields, there is no reason to use just two. We could, for example, have a very high field with only a minimal homogeneity in which the initial polarization takes place. Then that large polarization is inverted in a homogenous intermediate field and then allowed to relax

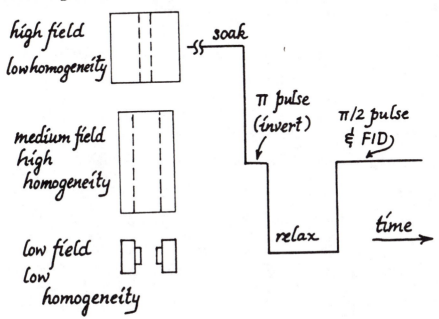

in a low field magnet, again with no stringent homogeneity requirements. Finally, the magnetization is monitored in the high homogeneity intermediate field. Such a sequence is sketched above. The function of the high field magnet is simply to induce a large initial magnetization $M_o$ for better S/N. Remember, though, that the fully recovered value will not be $M_o$ but some value scaled down by the ratio of the two higher fields.

Usually, low field $T_1$ measurements by field cycling are not as time consuming as the corresponding measurements at the high field because $T_1$ is often correlated with the applied magnetic field. However, if it is still very long even at the low field, consider some of the one shot methods for $T_1$ determination, as discussed in III:D.4. If the field is changed rather than the sample moved, the magnetization can be monitored by adiabatic fast passage.

## REFERENCES

D. T. Edmonds, "Nuclear quadrupole double resonance," Phys. Rep. 29C, 233-290 (1977).

K. Hallenga and S. H. Koenig, "Protein rotational relaxation as studied by solvent $^1H$ and $^2H$ magnetic relaxation," Biochemistry 15, 4255-4264 (1976).

G. P. Jones, J. T. Daycock, and P. T. Roberts, "A sample moving system for nuclear magnetic resonance adiabatic demagnetization experiments," J. Phys. E: Sci. Instrum. 2, 630-631 (1969).

F. Noack, "Nuclear magnetic relaxation spectroscopy" in NMR, Basic Principles and Progress, vol. 3, edited by P. Diehl, E. Fluck, and R. Kosfeld (Springer-Verlag, Berlin, 1971), pp. 83-144.

## VI.D.2 HIGH SPEED SPINNING OF SAMPLES

Slow sample spinning (like few tens of Hz) has been used for decades to average out small magnetic field inhomogeneities seen by different parts of the sample. It has also been known for a long time that spinning the solid sample rapidly about an axis tilted with respect to the laboratory magnetic field by the magic angle (angle $\theta_m$ where $3\cos^2\theta_m - 1 = 0$) eliminates those parts of interactions like the dipolar and the chemical shift anisotropy which are secular (Andrew, et al., 1958; Lowe, 1959).

[We digress to point out that secular usually means worldly or of things which exist through the ages and in the specialized usage here describes those terms in the Hamiltonian operator (which describe the interactions of the spins) which commute with the Zeeman term. A simply stated consequence is that secular terms do not cause nuclear transitions but do contribute to line positions and widths.]

Since the spinning rate required to average an interaction is approximately equal to the interaction itself, a spinning rate of a few to tens of kHz is required for the usual solid with an abundant NMR nucleus having a dipolar interaction of several kHz. As Zilm, et al. (1978) have pointed out, the centripetal acceleration at the rim of a 1 cm diameter sample spinning at 10 kHz is two million g's which gives an indication of why this technique has remained a difficult one for so long.

At the time of this writing, a rapidly growing area of research within NMR is the high resolution study of solids brought about by the development of certain multiple pulse sequences (such as the WAHUHA and MREV-8) on the one hand and an adaptation of the Hartmann and Hahn cross polarization experiment to obtain spectra of dilute spins on

the other (see IV.E.).   Schaefer, Stejskal, and co-workers
recognized that magic angle spinning can be used in conjunction with cross polarization to yield spectra for polycrystalline solids in which the powder pattern due to the chemical
shift anisotropy is collapsed (Schaefer, et al., 1977, and
references therein).   Because of the reduction in the dipolar
interaction inherent in these experiments even without the
spinning, the requirements for spinning rates are much more
lenient than they were in the absence of the concomitant line
narrowing mechanism.   So, the use of magic angle spinning is
on the rise, predominantly to remove chemical shift anisotropy
broadening in high resolution NMR of rare spin-1/2 isotopes
(such as $^{13}C$ and $^{15}N$) in the presence of a group of abundant spins (most commonly protons) in polycrystalline solids.

We conjecture that much new technology will be forthcoming, both in the spinning methods and in rotor materials.
Because the technology of fast spinning is advancing a spinning rate of 10 kHz, although still difficult or impossible for
many samples, especially powders, does not seem so awesome
as it did a few years ago as described in an early review by
Andrew et al. (1969).   The important things to remember are
to keep sample sizes small and to optimize the air cushion between the rotor and the stator by some means.   The reader
is referred to the article by Zilm, et al. (1978) for one such
scheme.

A novel suggestion by these authors is to make the
spin-axis be 45 degrees with respect to the support stem.   In
that way, the adjustment of the magic angle is easier in an
iron magnet because of the relatively small dependence of the
angle on the orientation of the support stem about its symmetry axis.

The Andrew-type rotor has a conical air bearing surface. The other common styles have a bearing on each end of a cylindrical rotor (Lowe, 1959; van Dijk, et al., 1980). Some modern cylindrical rotors use cylindrical air bearings at each end. Because of the geometry, the air bearing thickness is fixed so that it must be optimized for a particular spinning rate. A geometry with two conical ends may be a good compromise. See, for example, Eckman, et al. (1980) and references therein. The air bearing thickness of such rotors can be adjusted by changing the separation of the stators.

The critical parts of the spinner can be made from well chosen materials to operate the spinner at other than room temperature. Fyfe, et al. (1979) describe spinners which work at temperatures down to 77K by using the spinner driving gas to control the sample temperature.

The most common method of measuring the spinning rate is by some optical means, for example, by a photo-sensitive transistor and a light source. The spinning can be monitored on an oscilloscope in which case the quality of spinning can be kept track of as well as the speed. Alternatively, a tachometer can be constructed in a fashion analogous to counting rate meters used in nuclear physics (Elmore and Sands, 1949) and appropriately calibrated.

Kendrick, et al. (1980) have suggested a novel method of monitoring the spinning rate. Instead of the usual optical method or the not so usual acoustic method, both mentioned in their article, these authors suggest a method whereby an induced emf from an insulating rotor spinning in an insulating stator can be picked up in a nearby wire due to the separation of the static electrical charges.

## REFERENCES

E. R. Andrew, A. Bradbury, and R. G. Eades, "Nuclear magnetic resonance spectra in solids: Invariance of the second moment under molecular reorientation," Arch. Sci. 11, 223-226 (1958).

E. R. Andrew, L. F. Farnell, M. Firth, T. D. Gledhill, and I. Roberts, "High-speed rotors for nuclear magnetic resonance studies on solids," J. Magn. Resonance 1, 27-34 (1969).

R. Eckman, M. Alla, and A. Pines, "Deuterium NMR in solids with a cylindrical magic angle sample spinner," J. Magn. Resonance 41, 440-446 (1980).

W. C. Elmore and M. Sands, Electronics, Experimental Techniques (McGraw-Hill, New York, 1949), pp. 249-256.

C. A. Fyfe, H. Mossbruger, and C. S. Yannoni, "A simple 'magic-angle' spinning apparatus for routine use in NMR spectroscopy of solids," J. Magn. Resonance 36, 61-68 (1979).

R. D. Kendrick, R. A. Wind, and C. S. Yannoni, "A simple method for measuring the spinning frequency in magic angle spinning experiments," J. Magn. Resonance 40, 585 (1980).

I. J. Lowe, "Free induction decays of rotating solids," Phys. Rev. Letters 2, 285-287 (1959).

J. Schaefer, E. O. Stejskal, and R. Buchdahl, "Magic-angle $^{13}$C NMR analysis of motion in solid glassy polymers," Macromolecules 10, 384-405 (1977); and E. O. Stejskal, J. Schaefer and R. A. McKay, "High-resolution, slow-spinning magic-angle carbon-13 NMR," J. Magn. Resonance 25, 569-573 (1977).

P. A. S. van Dijk, W. Schut, J. W. M. van Os, E. M. Menger, and W. W. Veeman, "A high speed spinner for magic angle spinning NMR," J. Phys. E: Sci. Instrum. 13, 1309-1310 (1980).

K. W. Zilm, D. W. Alderman, and D. M. Grant, "A high-speed magic angle spinner," J. Magn. Resonance 30, 563-570 (1978).

## VI.D.3.  THE CHEMICAL SHIFT CONCERTINA

Ellett and Waugh (1969) proposed the chemical shift concertina which is a method for scaling down the chemical shifts by arbitrary amounts at a fixed magnetic field while leaving spin-spin splittings unchanged in liquid samples. This may be a very useful technique in some circumstances.

First, spectral analysis can be simplified in some cases by varying the ratio of chemical shifts to spin-spin coupling constants. While in principle the same result can be obtained by using a spectrometer operating at a lower field, the S/N will be smaller. Also, in practice the number of variable field/frequency spectrometers is limited. This proposed multiple pulse sequence, in principle, could give you a continuous range of chemical shifts without loss of sensitivity. Secondly, two spectral lines whose chemical shift difference is comparable to the frequency of exchange are required for studying chemical exchange processes and being able to adjust the chemical shift difference to this condition might be useful.

The correct explanation for how the chemical shift concertina works is complex but the following discussion provides a plausibility for the experiment. Suppose there is a train of equally spaced $\pi/2$ pulses whose phase alternates between $0°$ and $180°$.

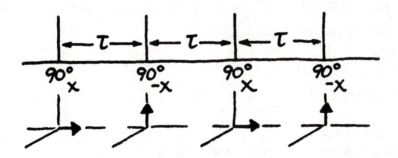

Each pair of pulses may be viewed as a cycle with the first $\pi/2$ pulse rotating the magnetization into the x-y plane where it evolves in time for a short period $\tau$ before the second pulse rotates the magnetization back into the $H_0$ direction and leaves it there for an equal time $\tau$.

As the magnetization spends an equal amount of time in the x-y plane and along the z direction, the time average properties of the magnetization resembles those of a magnetization which is tilted 45° from the static field. As the projection of $M_0$ along $H_0$ is $M_0 \cos 45°$, it seems reasonable that the chemical shifts be scaled down by $\cos 45°$. In fact, it is aproximately correct to assume that for any such series of cycles in which the pulse used to rotate the magnetization $\theta$ degrees will scale the chemical shifts by $\cos(\theta/2)$. So, as the pulse length is increased from 0 to 180°, the chemical shifts are scaled down from their full value as normally observed to zero. As far as homonulcear spin-spin couplings are concerned, each pulse treats both spins of a coupled pair alike, resulting in no change in spin-spin coupling.

How is the signal observed? The FID can be sampled for that period $\tau$ following the first pulse in each cycle. If, in this fashion, one point is sampled and stored per cycle, the resulting FID, obtained as a sequence of these points taken one per time window, can be Fourier transformed to give the spectrum. For a detailed description, the reader is referred to the original paper.


REFERENCE

J. D. Ellett, Jr. and J. S. Waugh, "Chemical-shift concertina" J. Chem. Phys. 51, 2851-2858 (1969).

## VI.D.4.  ZERO TIME RESOLUTION OF FID

The early part of the FID is strongly influenced by the second and higher moments.  Thus, it is essential to be able to record the FID with the smallest possible delay after the rf pulse and the associated deadtime.  However, there are three difficulties in doing this.  (a)  The FID actually starts in the middle of the 90° pulse (Barnaal and Lowe, 1963); (b) there may be a significant deadtime after the pulse because of ringing in the probe and instrumental saturation of the receiver (See also V.A.4.); and (c)  the S/N may be poor, particularly in samples with short $T_2$ so that the signal is significantly reduced before you can observe it.

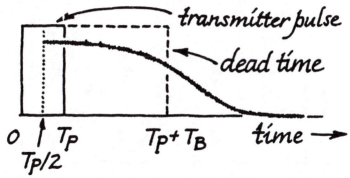

As seen in the diagram, we cannot begin to observe the FID until a time $t_p + t_b$ from the start of the pulse even though we would like to observe from the time $t_p/2$ onward.  There are several ways to minimize this problem:  The two covered already are the solid echo and the Jeener echo techniques (IV.B.3. and IV.B.4.).  Another, which we only mention, is a double resonance echo method which is applicable to solids containing two spin systems (Terao and Matsui, 1980).  The method we now describe not only enables the measurement of the FID at discrete points right up to the start of the FID within the initial 90° pulse, but it also gives an improved

S/N as well.  This method, called the zero time resolution (ZTR) method, was originated by Lowe and his co-workers and the following discussion will closely parallel that presented by these authors (Lowe, et al., 1973; Lowe, 1977; Vollmers, et al., 1978).

An application of a spin-locking pulse at time $\delta$ after the FID has started results in an FID following the spin-locking pulse which depends on magnetization $M_o$, the unperturbed FID, and $T_{1\rho}$.  In particular, the magnetization at a time $\tau$ after the spin locking pulse has been turned off is

$$M(t=\delta+\Delta+\tau) = M_o f(\delta)g(\Delta)f(\tau) = Cf(\delta) \tag{1}$$

where $\Delta$ is the duration of the spin-locking pulse.

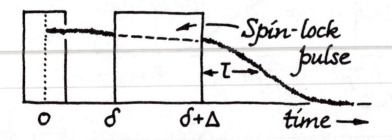

Even though it is not possible to measure the signal until a time $\tau$ after the spin-locking pulse is over, where $\tau > t_b$, such a value measured is proportional to $f(\delta)$ and $\delta$ can be made as small as $t_p/2$ at which point the spin locking pulse is contiguous with the 90° pulse.  Therefore, one can measure the shape of the FID, point by point, at times down to $t_p/2$ by observing an arbitrary point on the final FID and varying $\delta$.  This shape will differ from the original FID only by a scaling factor.  This method, however, does not enable the measurement at times between 0 to $t_p/2$ and the S/N is no better than if the FID were measured directly. Both of these problems can be circumvented in the following manner.

Consider a sequence of N+1 spin-locking pulses as indicated in the diagram below. The magnetization at a time $\tau$ after the final locking pulse, where N+1 locking pulses were used is

$$M = M_o f(\delta)[g(\Delta)]^{N+1}[f(t)]^N f(\tau)$$

$$= M_o K[f(t)]^N \quad . \tag{2}$$

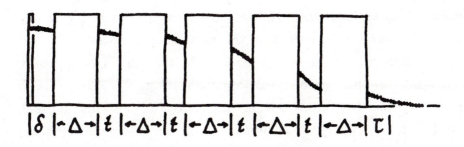

Note that t, the interval between spin locking pulses, can be made small without limit, and that $f(t)$ goes as the N-th root of $M(\tau)$. By measuring $M(\tau)$ for various values of t down to zero, one can extract the FID shape right down to t=0. The signal to noise ratio is also improved by a factor of N. This can be seen by considering incremental changes dM and df.

$$dM = M_o KN[f(t)]^{N-1} df(t)$$

$$= M_o KN[f(t)]^N df(t) f(t)$$

$$= MN df(t)/f(t), \tag{3}$$

where the last equality holds true because of eq. (2). We can rearrange eq. (3) as

$$f(t)/df(t) = NM/dM \qquad (4)$$

which completes the proof.

It is interesting to note, and possibly surprising to some, that the signal-to-noise ratio in a derived quantity can be better than it is in the original. It clearly depends on the functional relation between the original and the derived quantities and we get the improvement in this case because $M \propto f^N$. Had the derived quantity been the N-th power of the measured quantity, the S/N of the derived quantity would have been much worse than of the measured quantity.

As always, there is a practical limit to the S/N improvement and in this case it is approximately N. First, it is limited by the fact that $\Delta \cdot N$ is limited by $T_{1\rho}$. This in itself would not be a serious limitation if $\Delta$ can be made as short as we wished but $\Delta \gg T_2$ in order for the signal after the spin-locking pulse to be the same as the true FID. Roughly speaking, $N < T_{1\rho}/T_2$. See the discussion in Vollmers, et al., (1978).

Another caveat is that the ZTR method only works for single crystals, if at all, because of the nondistributive character of the N-th root operation. Again, see the discussion by Vollmers, et al. (1978).

Despite these limitations, ZTR is the best method to date for obtaining the shape of the FID at the time origin accurately. Engelsberg and Lowe (1974) used this technique to obtain the even moments for $CaF_2$ up to the 14-th!

## REFERENCES

D. Barnaal and I. J. Lowe, "Effects of rotating magnetic fields on free-induction decay shapes," Phys. Rev. Letters 11, 258-260 (1963).

M. Engelsberg and I. J. Lowe, "Free-induction-decay measurements and determination of moments in calcium fluoride," Phys. Rev. B 10, 822-832 (1974).

I. J. Lowe, "Motionally narrowed NMR line shapes in solids" in Magnetic Resonance in Condensed Matter -- Recent Developments, Proceedings of the IVth Ampere International Summer School, Pula, Yugoslavia, Sept. 13-23, 1976, edited by R. Blinc and G. Lahajnar (Univ. of Ljubljana, Ljubljana, 1977).

I. J. Lowe, K. W. Vollmers, and M. Punkkinen, "New method for measuring the short time behavior of the free induction decay with applications to calcium fluoride and ammonium chloride" in Pulsed Nucl. Magn. Resonance Spin Dyn. Solids, Proceedings of the First Specialized Colloque Ampere, edited by J. W. Hennel (Inst. Nucl. Phys., Krakow, Poland, 1973), pp. 70-79.

T. Terao and S. Matusi, "Indirectly induced NMR spin echoes in solids," Phys. Rev. B 21, 3781-3784 (1980).

K. W. Vollmers, I. J. Lowe, and M. Punkkinen, "A method of measuring the initial behavior of the free induction decay," J. Magn. Resonance 30, 33-50 (1978).

## VI.E.    NMR MATH

Here, we have three elementary sections on how to deal with "NMR math."

VI.E.1.   CALCULATIONAL AIDES IN NMR

Calculations involving formulas like

$$1/T_1 = (2/5)[\gamma^4 \hbar^2 I(I+1)/r^6]\{[\tau/(1+\omega^2\tau^2)]+[4\tau/1+4\omega^2\tau^2)]\} \quad (1)$$

are easy to do unless you are totally unfamiliar with physics, mathematics, or formulas like these.  Since this book is written for those people who may be totally unfamiliar, we will now discuss simple rules and procedures to help in such calculations and others.  In particular, we will consider some rules relating various quantities in NMR which are very useful to know.  In discussing these rules, it is mandatory to understand units, so we will also talk about how to check units.  [We discuss units further in the next section.]  The material in this section should be enough to get you to understand most practical formulas in NMR books.

Everybody knows that "=" means the quantities related by the sign, however complicated, are identical.  Problems arise when a calculation is performed with quantities having various units and one must choose the appropriate units to be used in the first place so that the answer has the desired unit.

For instance, consider the equation $E=mc^2$.  If m is in grams and c in cm/sec, E will have the unit of $gm \cdot cm^2/sec^2$.  Now further suppose that you need E in joules.  Then, you have to decide what a joule is in terms of $gm \cdot cm^2/sec^2$.  It turns out to be $10^7$ $gm \cdot cm^2/sec^2$ so that we can write equations for these units like

$$1 \text{ erg} = 1 \text{ } gm \cdot cm^2/sec^2$$

and

$$1 \text{ joule} = 10^7 \text{erg} = 10^7 \text{gm} \cdot \text{cm}^2/\text{sec}^2,$$

and the energy in units of $\text{gm} \cdot \text{cm}^2/\text{sec}^2$ should be multiplied by $10^{-7}$ to change it to joules. All relations like this are straightforward, only some are much messier. In this section, we will develop relations like these which will be handy for NMR.

NMR concerns nuclear moments, gyromagnetic ratios, fields, and other things like that. The nuclear moment $\bar{\mu}$ is a vector whose magnitude $\mu$ is a measure of how strong a magnet the nucleus is. The moment arises from the protons and neutrons making up the nucleus which also have angular momenta associated with them.

How much magnetic moment there is on a nucleus depends not only on the angular momentum but also on how the protons and neutrons are distributed in the nucleus which are different for different nuclei. Therefore, the proportionality factor $\gamma$ between the angular momentum and the magnetic moment is a property of the nucleus and is called the gyromagnetic or magnetogyric ratio. [The distinction, to us, is akin to that between an ant-eater and an eat-anter according to the cartoon character B. C. (Hart, 1960).] This simple fact, namely that there is a property (measurable with great precision) which is unique to each kind of nucleus, is a basic reason for the usefulness of NMR. More precisely, the foregoing may be written

$$\mu = \gamma I \hbar \qquad \text{(Rule \#1)}$$

where $\mu$ is the magnetic (dipole) moment of the nucleus, I is the spin of the nucleus, $\hbar = h/2\pi$ is the unit of angular momentum so that the angular momentum associated with a spin I is $I\hbar$, and $\gamma$ is the gyromagnetic ratio. h is Planck's constant and has a value $6.624 \times 10^{-27}$ erg·sec. For the time being, we will leave the units for $\mu$ and $\gamma$ open. (The spin I is just a number. It is 1/2 for nuclei like the proton, carbon-13, fluorine-19, and phosphorous-31, while it is 1 for nitrogen-14 and the deuteron, 3/2 for boron-11, sodium-23, and aluminum-27, and so on.)

When a magnetic moment (any magnetic dipole moment -- even a compass needle) is in a static magnetic field H it feels a torque, which tries to align it parallel to the field. But the nucleus with an angular momentum will, due to this torque, precess about the static field with a frequency proportional to both the field strength and the moment. Let us rephrase it. A nucleus with a magnetic dipole moment $\mu$ (and an associated gyromagnetic ratio $\gamma$) in a static uniform magnetic field H will precess about the field direction with angular velocity

$$\omega = \gamma H \qquad \text{(Rule \#2).}$$

You can see this by multiplying both sides by $\hbar$, and using the definition of $\gamma$.

The only other rule that we will give you now is for finding the magnitude of the field H due to a magnetic moment $\mu$ at a distance r much greater than the physical extent of the magnetic dipole.

$$H \sim \mu/r^3. \qquad \text{(Rule \#3).}$$

This comes from eq.(1) in IV.E.4. and you may try to check it as a simple exercise.

We talk about units now for all three rules. If we take H in G (for gauss) and $\omega$ in rad/sec (for radians per second), rule #2 gives us a working unit for $\gamma$ of [rad/(sec·G)] where the brackets indicate the unit for the quantity under consideration.

Now what about $\mu$? Rule #3 gives us one working unit for $\mu$, namely $\mu$ = [G·cm$^3$]. On the other hand, rule #1 together with the unit for $\gamma$ gives us another: $\mu$ = [erg/G]. This means that anytime you see G·cm$^3$ you can replace it with erg/G. In fact, it's better than that. Let's write out what we know: [G·cm$^3$] = [erg/G]. You can treat this as a real equation and get, for example, [erg = G$^2$·cm$^3$]. This is so handy so let us write it on a separate line:

$$[\text{erg} = G^2 \cdot \text{cm}^3].\tag{2}$$

This means, for example, that the unit for $\hbar$ is not only erg·sec/rad but is also G$^2$·cm$^3$·sec/rad. These manipulations are very useful as we will show in some examples.

The first rule, defining the gyromagnetic ratio $\gamma$, is useful for dimensionality checks and for derivation of relationships between units as we showed previously. The second rule is used to calculate the NMR frequency from only the knowledge of $\gamma$ and H. Since the calculation of NMR frequencies is such a common activity among people looking at a variety of nuclei, we will go through the steps once.

Suppose you want to know where to set the frequency to observe boron-11 in a field of 21.2 kG. You go to an NMR table (an excellent one is by Lee and Anderson, Appendix A) and look for $\gamma$. There is no such listing? Then look at the

heading marked "NMR frequency in a field of 10 kG" or some
such.   That almost sounds like the gyromagnetic ratio $\gamma = \omega/H$
(rule #2) except that $\omega$ is an angular velocity (radians/sec)
instead of frequency (hertz).   Since $\omega = 2\pi\upsilon$, the numbers in
the NMR table are $\upsilon/H = \omega/(2\pi H) = \gamma/2\pi$.   In analogy with
$\hbar = h/2\pi$, some people write $\gamma/2\pi = \not\gamma$.

So the table gives $\gamma/2\pi$ for boron-11 as 13.66 MHz/10kG
or 1.366 kHz/G.   Clearly, the Larmor frequency at 21.2 kG
is 13.66 × (21.2/10.0) MHz = 28.96 MHz.   If the value of $\gamma$
itself is desired, it is $2\pi$ times the $\gamma/2\pi$ in the NMR table --
for boron-11 it is $8.583 \times 10^3$ rad·$G^{-1}$·$sec^{-1}$.

For another example in the use of rule #2, consider a
Gaussian carbon-13 line whose second moment $\langle\Delta H^2\rangle$ is known
to be $10G^2$.   How long might its FID last?   First figure out
what that second moment is in angular velocity.   Rule #2
leads to $\langle\Delta\omega^2\rangle = \gamma^2\langle\Delta H^2\rangle = 4.52 \times 10^8$ $rad^2/sec^2$ and since the
second moment is related to $T_2$ by $\sqrt{\langle\Delta\omega^2\rangle} = \sqrt{2}/T_2$, the characteristic relaxation time of the FID is $T_2 = \sqrt{2}/\sqrt{\langle\Delta\omega^2\rangle} =$
$\sqrt{2}/(\gamma\cdot\sqrt{\langle\Delta H^2\rangle}) = 66.5$ μsec/rad.   (Remember that $T_2$ has a
different meaning for a Gaussian as it does for an exponential
decay.   See III.B.)

As a third example, consider a proton chemical shift of
10 Hz at a field of 25 kG.   First, the Larmor frequency is
106.2 MHz because $\gamma/2\pi = 42.5759$ MHz/10kG.   Then the fractional shift is $10/(106.2 \times 10^6) = 0.1$ part per million (ppm).
The 10 Hz, by rule #2 corresponds to an extra field at the
nucleus of (10/4257.59) G = $2.4 \times 10^{-3}$ G which is certainly
miniscule compared with the applied field of 25 kG since this
ratio is the same 0.1 ppm.   Furthermore, we can show that it
is miniscule compared even with the field of another proton a
couple of angstroms away like another proton on the same
molecule.   To do this, we turn now to rule #3.

The NMR table tells us that the proton moment is 2.79 in units of nuclear magneton which, in turn, is $5.05 \times 10^{-24}$ erg/G so that $\mu = 14.1 \times 10^{-24}$ erg/G. The field due to this moment at a point $2\overset{\circ}{A}$ away is, then, roughly 2G by rule #3. So a shift such as the $2.4 \times 10^{-3}$G corresponding to the 10 Hz proton shift would be swamped by a nearby proton, or for that matter, any nucleus with a sizeable magnetic moment. So, how can such shifts be observed in a real molecule? In a liquid sample, the tumbling motion of the molecule averages out the fields due to the neighboring nuclei on the time scale of NMR. In nonviscous liquids the NMR lines can be very narrow, like 1/10 Hz, and the study of narrow line spectra is called high resolution NMR. To many chemists, the term NMR means high resolution NMR in solution, and that is somewhat unfortunate. There are many experiments outside the narrow category of "high resolution solution NMR" which can yield much chemical and physical information. In fact, one purpose of this book is to make these other techniques better known. They are so much fun and often yield information unavailable from high resolution solution NMR experiments.

We will do one more example, this time using the relation $[erg = G^2 \cdot cm^3]$, eq. (2). Suppose you wish to calculate the relaxation rate $1/T_1$ given by the first equation of this section. We will give you the parameters and go through the dimensions and suggest the appropriate checks but you figure it out.

We repeat eq. (1) for convenience.

$$1/T_1 = (2/5)[\gamma^4 \hbar^2 I(I+1)/r^6]\{[\tau/(1+w_0^2\tau^2)]+[4\tau/(1+4w^2\tau^2)]\}$$

This is an expression for a spin-lattice relaxation rate $1/T_1$ of a nucleus defined by $\gamma$ and $I$, due to randomly varying magnetic fields of a like neighboring nucleus. The random motion is an isotropic rotational diffusion with a characteristic time $\tau$. The distance between the two nuclei is r while $w_0$ is the Larmor velocity in the applied magnetic field. If we use the three rules given above, we may transform the relaxation rate in the following ways. Note that the relationships are not rigorously true but are approximate. By rule #1,

$$1/T_1 \cong (2/5)(\gamma^2\mu^2/r^6)[\tau/(1+w_0^2\tau^2)]. \tag{3a}$$

By rule #3

$$1/T_1 \cong (2/5)\gamma^2 H_n^2[\tau/(1+w_0^2\tau_n^2)]. \tag{3b}$$

By rule #2

$$1/T_1 \cong (2/5)w_n^2[\tau/(1+w_0^2\tau_n^2)]. \tag{3c}$$

In eq. (3b), $H_n$ is the field at a nucleus due to the other nucleus located at a distance r, so that $w_n$ is the angular velocity of a nucleus if it just experienced the field from the other nucleus. The $\tau/(1+w_0^2\tau^2)$ on the right side is a factor accounting for the effect of the randomly varying motion and was explained in III.A. and III.C.1.

Just from these simple considerations, we now know that the relaxation rate due to another nucleus interacting with the first is proportional to the square of the magnetic interaction between them regardless of the rate of the rotational tumbling (assuming, of course, that this is the only relaxation mechanism.)

In talking about rule #3, we estimated a field due to a proton at a distance of 2Å to be 2G which is our $H_n$ in eq. (3b). With this value for $H_n$, $w_n \ll w_0$ since in a normal NMR experiment $H_0$ is in the kG range.

We can further rewrite eq. (3c) according to whether or not $w_0\tau$ is much greater or much less than 1. (The former corresponds to a much lower temperature than the latter.) Therefore,

$$
1/T_1 \cong
\begin{cases}
w^2\tau, \quad w_0\tau \ll 1 \quad \text{(high temperature)} \\[4em]
(w/w_0)^2(1/\tau), \quad w_0\tau \gg 1 \quad \text{(low temperature).}
\end{cases}
\tag{4, 5}
$$

Consider eq. (1) again. We know from the units of $\gamma$, $\hbar$, and I that $\gamma^4\hbar^2 I(I+1)$ has units of

$$(Hz/G)^4 \cdot rad^4 \cdot erg^2 \cdot sec^2 \cdot rad^{-2}.$$

If interatomic distance has the unit of cm and the correlation time factor $\tau/(1+w_0^2\tau^2)$ has the units of sec/rad, the entire expression has the units $(erg^2 \cdot rad)/(G^4 \cdot sec \cdot cm^6)$. This may be momentarily baffling but we can simplify it with eq. (2) which says $[erg = G^2 \cdot cm^3]$. Cancelling the $erg^2$ against $G^4 \cdot cm^6$ we end up with rad/sec for $1/T_1$. Remember that $1/T_1$ and $1/T_2$ both have units of rad/sec and not Hz.

REFERENCE

Johnny Hart, Hey! B.C. (Funk and Wagnalls, New York, 1960).

VI.E.2.          B VS. H

There has been an age-old argument about the relation of $\overline{B}$, commonly called the magnetic induction or flux density, and $\overline{H}$, the magnetic intensity. There is a documented case where the question of whether $\overline{B}$ and $\overline{H}$ were the same kind of quantity or not was even put to a vote (Griffiths, 1932)! The low temperature physicist Casimir (1968) has written a fable about this question.

It has been pointed out periodically that since $\overline{B}$ contains the induced magnetization as well as the direct effects of $\overline{H}$, it is the appropriate quantity to use in NMR of finite permability materials. We have taken the easy way out and used $\overline{H}$ throughout this book, however, simply because its use is nearly canonical in NMR and, despite the philosophical problems, therefore is unambiguous by usage. This is akin to a newly coined word in a language becoming an accepted word by usage, even though it never should have.

There is a slightly more sticky question of units. Traditionally, a gauss is a unit (in cgs) of the magnetic induction $\overline{B}$. The magnetic field intensity $\overline{H}$ (in cgs) has the unit oersted, abbreviated Oe. Thus, to use $\overline{H}$ in units of gauss as we have done in this book is inconsistent. Again we invoke the overwhelmingly common usage of this combination as a justification.

With the increasing use of SI units, we should point out that the appropriate unit is the tesla (T) which equals $10^4$ gauss. (The appropriate SI unit for H is the ampere/meter which equals $4\pi/1000$ Oe.) The tesla is a very convenient unit for describing magnets but is awkward for expressing local field or the rotating rf field. The typical local field is of the order of 100's of microteslas rather than of a few gauss.

REFERENCES

H. B. G. Casimir, "On electromagnetic units," Helv. Phys. Acta 41, 741-742 (1968).

E. Griffiths, "Electrical and magnetic units," Nature 130, 987-989 (1932).

VI.E.3.   WHAT IS A dB?

In Section V.B.1. we said that an 80 dB attenuation meant a factor of $10^4$ difference in voltage. There are other situations in NMR when you will come across something in dB (decibel). One is in the specification of noise figure for an amplifier. Another might be as an indication of power output of something like a signal generator in which case the common unit is dBm. Often, filter characteristics are given in dB per octave. What does dB measure?

A dB is basically a relative measure of power. As already stated in V.B.1.,

$$\text{relative power (dB)} = 10\log_{10}(P/P_o)$$

where we want to characterize the power $P$ compared with $P_o$. Thus, if $P=P_o$ then the relative power is 0 dB and if $P=100P_o$ then the relative power is 20 dB.

Now, remember from electricity that the power is proportional to the square of the voltage $V$ for a fixed impedance $R$, that is $P=V^2/R$. Therefore, we can also express relative voltages in dB's as relative voltage (dB) $= 20\log_{10}(V/V_o)$

where we are expressing the voltage V in terms of $V_o$. Thus, a voltage attenuation of $10^4$ is, indeed, 80 dB.

What about dBm? It is a power ratio as already discussed with $P_o$ fixed at 1 mW of power. So, if a signal generator puts out 10 mW, its output is said to be 10 dBm. Usually, we want to know what the output voltage is rather than power. That is easily calculated with a knowledge of the impedance by the above relation $P=V^2/R$. For a 50 Ω system, 1 dBm=0.22 v, 7 dBm=0.5 v, and 13 dBm=1.0 v. For a 100 Ω system, 1 dBm=0.32 v, 4 dBm=0.5 v, and 13 dBm=1.4 v.

Finally, the noise figure F of a signal processing device such as an amplifier or a mixer is also given in dB's and is a measure of how much total noise power there is coming out of the device in question compared to how much noise power is going into it. If $N_I$ is the noise voltage impinging on the device and $N_D$ is that generated by the device,

$$F \text{ (dB)} = 10\log_{10}[(N_I^2 + N_D^2)/N_I^2].$$

We began this book by noting the equivalence between a representation in the time domain and its conjugate representation in the frequency domain. The Fourier transform enabled us to transform the information, without loss, from one domain to the other. It was not only more efficient to obtain the information in the time domain and then convert but, also, going back and forth between the two domains permitted many informational manipulations such as trading S/N against resolution or tailored excitation.

We also described elementary ideas in pulse and multiple pulse NMR. Many of the pulse trains manipulate the Hamiltonian, that is, the nuclear interactions, rather than just having the operator passively observe its time evolution. This manipulation of the Hamiltonian is used to select information: In the Carr-Purcell train it suppresses information about the external field inhomogeneity; in the multiple-pulse line-narrowing sequence, it suppresses information about the dipole-dipole interactions; in the broad-band decoupling experiments it suppresses information about spin-spin couplings; and in selective excitation, it suppresses information about certain lines in the spectrum. We note that information suppression plays an important role because often some of the information present (e.g., the inhomogeneity of the magnetic field or information about a solvent peak) obscures the information we want and, at other times, the information present (e.g., heteronuclear spin-spin couplings) may prevent us from determining any useful information because it is too complex and overwhelming. We try to get information that is simple enough to digest by suppressing parts of it.  Sometimes, we can use techniques to admit a bit of the previously suppressed information in an amount that we can use (e.g., the off-resonance decoupling suppresses most heteronuclear spin-spin couplings except one).

Pulse sequences can be used to manipulate information in other ways, as well.  The chemical shift concertina uses it to manipulate the relative importance of two terms in the Hamil-tonian, the chemical shift term and the spin-spin coupling term.  Echo sequences can be viewed as causing the system to evolve back in time to enable us to view information that existed earlier but was obscured by instrumental problems. The same is true of the ZTR experiment to observe the FID at zero time.  The cross polarization experiments enable us to exploit the thermodynamic properties of a spin system to enhance the spectrum S/N of a second spin system.

The revolution in magnetic resonance spectroscopy which began in the 1960's was grounded in the understanding of where the information was contained and how it might be manipulated.  The first stage was the implementation of the fast Fourier transform which permitted us much better access to that information.  The second stage was the recognition that suppression of undesired information was possible by manipulating the Hamiltonian while performing the experiment. The result of all of this was to change the scientist from a passive observer to a manipulator of nature (or at least of spins) to extract the desired information.  These developments have provided not only new techniques but new ways of thinking about NMR.

Many scientists feel that the only way to work is to pur-chase instruments as turnkey systems and use them without modification. Such distinguished scientists as P. B. Medawar (1980) gives the advice that one should not construct equipment if it can be bought. We view the situation as more complex.

Basically, there are three approaches: (a) Only work with turnkey systems purchased from a manufacturer; (b) construct your own electronics from scratch; (c) construct your own system out of readily available commercial components including parts of NMR spectrometers. Rather than siding with any one of these approaches, we believe that each choice has merit under appropriate circumstances. Surely much fine scientific work is done on commercial instruments without modification. However, the history of NMR is full of powerful experiments which at first could not be done on commercially available instruments without significant modification. Of course, most of the pioneering work in developing a new NMR capability is complete before such a capability becomes available commercially, and some of us find the development of techniques and their early applications to be as exciting as the employment of well-known techniques to study matter. Much of this book is aimed at increasing the options open to the experimenter beyond that given in (a) above. In particular, this book provides option (c) as an alternative which greatly increases the scope of work available to the experimentalist. Often, to the novice there is something intimidating about making changes in or constructing an instrument and we hope that this book reduces the barrier to doing so. But in the end, one simply must get in, try, and learn by one's self. [Some well-known philosophical advice on dealing with machines are due to Robert M. Pirsig (1974).]

The experimenter needs to know the different techniques available to him and what the details of their implementation are. An understanding of these methods and the practical details in employing them has formed the backbone of the book. Knowledge of techniques and knowledge of instrumentation are synergistic; we have attempted to discuss both, each as a counterpart to the other.

Finally, pulse NMR is experiencing an explosion, not only in the number of techniques being developed but also in the heterogeneity of the applications and backgrounds of the practitioners. We have tried to keep the discussion simple in order to be useful to as large an audience as possible, particularly to newcomers to the field. Beginners without affiliation with an active NMR research group need to get plugged into the practical information that abounds on how to do this or that. We have tried to assemble and write down some of this NMR folklore. We hope readers will write to share their favorite technique or procedure with us.

## REFERENCES

P. B. Medawar, Advice to A Young Scientist (Harper and Row; New York and London, 1979).

R. M. Pirsig, Zen and the Art of Motorcycle Maintenance (Wm. Morrow, 1974).

# APPENDIX A. REFERENCES, GENERAL AND REVIEW

General References in Magnetic Resonance:

A. Abragam, The Principles of Nuclear Magnetism (Oxford University Press, Oxford, 1961). Although 20 years old, this is the premier reference work in nuclear magnetic resonance.

A. Carrington and A. D. McLachlan, Introduction to Magnetic Resonance: With Applications to Chemistry and Chemical Physics (Harper and Row, New York, 1967). A readable introduction to NMR and ESR written for the chemist.

C. P. Slichter, Principles of Magnetic Resonance, (Springer-Verlag, Berlin, 1978). The emphasis in this basic work is on the physics of magnetic resonance with examples from solid state physics.


Experimental Pulse and FT NMR References:

N. Boden "Pulsed NMR Methods", in Determination of Organic Structures by Physical Methods, vol. 4, edited by F. C. Nachod and J. J. Zuckerman (Academic Press, New York, 1971), Chap. 2, pp. 51-137. A review of broadline solid state pulse NMR methods.

T. C. Farrar and E. D. Becker, Pulse and Fourier Transform NMR (Academic Press, New York, 1971). This was the first book on time domain NMR. It is elementary and readable.

M. L. Martin, J.-J. Delpuech, and G. J. Martin, Practical NMR, Spectroscopy (Heyden and Sons, London, 1980). A detailed and wide ranging book on modern experimental methods useful in high resolution FT NMR.

K. Müllen and P. S. Pregosin, Fourier Transform NMR Techniques: A Practical Approach (Academic Press, London, 1976). An experimental FT NMR book which is not as detailed as the book by Shaw or Martin, et al.

D. Shaw, Fourier Transform NMR Spectroscopy (Elsevier, Amsterdam, 1976). A book on high resolution solution FT NMR methods with a narrower emphasis than Martin et al.

I. D. Weisman, L. J. Swartzendruber, and L. H. Bennett, "Nuclear Magnetic Resonance and Mössbauer Effect" in Techniques of Metals Research vol. VI, pt. 2 (Measurement of Physical Properties), edited by E. Passaglia, Wiley-Interscience, New York 1973), pp. 165-504. A rather long survey of various methods used through the 1960's with emphasis on special needs for NMR in metals.


Particularly Useful Specialized References:

Ron Bracewell, The Fourier Transform and Its Applications, 2nd edition (McGraw Hill, New York, 1978). This is a very useful and readable book on Fourier transforms.

J. Cooper, The Minicomputer in the Laboratory (John Wiley and Sons, New York, 1977). Even though this book is written in terms of the PDP-11 computer, it is an excellent treatment of software used in FT NMR, as well as mini-computer programming in general.

K. Lee and W. A. Anderson, "Nuclear Spins, Moments, and Magnetic Resonance Frequencies" in The Handbook of Chemistry and Physics, edited by Robert C. Weast (The Chemical Rubber Company, Cleveland). This yearly publication has one of the best NMR tables in it.

# APPENDIX B. REVIEW JOURNALS IN MAGNETIC RESONANCE

Annual Reports on NMR Spectroscopy (Academic Press, London), edited by G. A. Webb.

Advances in Magnetic Resonance (Academic Press, New York), edited by J. S. Waugh.

Advances in Nuclear Quadrupole Resonance (Hayden and Son, London), edited by J. A. S. Smith.

Bulletin of Magnetic Resonance, the quarterly review journal of the International Society of Magnetic Resonance (The Franklin Institute, Philadelphia), edited by D. Fiat.

Magnetic Resonance in Biology (John Wiley & Sons, New York), edited by Jack S. Cohen.

Magnetic Resonance Review, a quarterly literature review journal (Gordon and Breach, New York), edited by Charles P. Poole, Jr.

NMR: Basic Principles and Progress (Springer-Verlag, Berlin), edited by P. Diehl, E. Fluck, and R. Kosfeld.

Nuclear Magnetic Resonance, A Specialist Periodical Report (The Chemical Society, London).

Progress in Nuclear Magnetic Resonance Spectroscopy (Pergamon Press, Oxford), edited by J. W. Emsley, J. Feeney, and L. H. Sutcliff.

Topics in Carbon-13 NMR Spectroscopy (John Wiley & Sons, New York), edited by George C. Levy.

# APPENDIX C.  CONFERENCES IN MAGNETIC RESONANCE

Those interested in attending conferences on various aspects of NMR should examine current issues of the Journal of Magnetic Resonance, Physics Today, Chemical and Engineering News, Science, and Nature for up-to-date information. Some of the important conferences are listed here.

## The Ampere Congress or Colloque Ampere

Meets in Europe in even years, usually in late summer or early fall. Oriented towards physicists and physical chemists in that high resolution solution NMR is considered to be outside its domain. Proceedings are published and prove useful in providing up-to-date reviews of physical aspects of magnetic resonance. In addition, Specialized Colloque Ampere meetings are held in odd years in Europe.

## The Ampere Summer School

Held in Europe usually just prior to the Ampere Congress. It consists of several tutorial lectures on current topics in NMR.

## The Experimental NMR Conference (ENC)

Annual meetings alternate between California and eastern locations in the U. S. usually in late winter or early spring. This conference deals with experimental techniques and problems with emphasis on chemistry.

## The Gordon Conferences

Each year a variety of Gordon Conferences are held on diverse topics at various locations in the U. S. Conferences on magnetic resonance have been held in odd years in New Hampshire, U. S. A. usually in June. There are no proceedings. Announcements are made in Science.

## The International Conference on Magnetic Resonance in
### Biological Systems

Concerned with those aspects of NMR and ESR relevant to biological systems, this conference meets in different countries in different years.

## The International Symposium on Magnetic Resonance

This conference, meeting every three years in various countries, deals with all aspects of magnetic resonance in physics, chemistry, and biosciences. The 1980 meeting was in combination with the Ampere Congress.

## The International Symposium on Nuclear
### Quadrupole Resonance Spectroscopy

The scope of the conference is less restrictive than the title implies; the conference deals with all aspects of quadrupole interactions, including NMR, Mössbauer, and other spectroscopies. Meetings occur in odd years world wide.

## NATO Advanced Study Institutes

These occur at irregular intervals on various topics, with an occasional one on an aspect of magnetic resonance.

## The NMR Summerschool

Invited speakers present overviews of areas of NMR to participants. Emphasis is on the basic physics and physical chemistry of magnetic resonance rather than applications. It is held in odd years at the University of Waterloo, Waterloo, Ontario, Canada, usually during the week preceding or following the NMR Gordon Conference. Interested individuals should contact

Director, NMR Summerschool
Department of Physics
University of Waterloo
Waterloo, Ontario, Canada   N2L   3G1

## RAMIS

Biennial conference on Radio Spectroscopy sponsored by the Institute of Molecular Physics of the Polish Academy of Sciences. Occurs in odd years in Poznań, Poland. For further information, write:

> Institute of Molecular Physics
> Polish Academy of Sciences
> Ulica Mariana Smoluchowskiego 17/19
> 60-179 Poznań
> Poland

## Rocky Mountain Conference on Analytical Chemistry

Annual (mid-summer) meeting in Denver, USA, usually with a prominent NMR sub-session. Sponsored by Rocky Mountain Section of Society for Applied Spectroscopy and Rocky Mountain Chromatography Discussion Group.

# APPENDIX D.   HARDWARE SOURCES (USA & CANADA)

We list some representative sources of NMR hardware in North America.   The list is by no means complete nor is there any endorsement implied in any way by either authors or institutions they are affiliated with.   Furthermore, there is no guarantee for the accuracy of the information given here.


## Amplifiers, Receiver Preamp (Broadband)

Radiation Devices Co. Inc., Box 8450, Baltimore MD   21234.


## Amplifiers, Receiver Preamp (tuned)

JANEL, 33890 Eastgate Circle, Corvallis, OR   97330.

Lunar Electronics, 2785 Kurtz St., Suite 10, San Diego, CA
    92110. (GaAs)


## Amplifiers, RF Power (broadband)

Amplifier Research, 160 School House Rd., Souderton, PA
    18964.

ENI, 3000 Winton Rd. South, Rochester, NY   14623.

Instruments for Industry, 151 Toledo St., Farmingdale, NY
    11735.


## Amplifiers, RF Power (tuned, for amateur radio use)

ARCOS, P.O. Box 546, East Greenbush, NY   12061.   (high
    frequency)

Heath Co., see under Dip Meters.

Henry Radio, Inc., 11240 W. Olympic Blvd., Los Angeles, CA
    90064.

## Antenna Tuners

Wm. M. Nye Co., Inc., 1614 130th Ave., Bellevue, WA 98005.


## CAMAC Modules and Supplies

Bi Ra Systems, 3520-D Pan American NE, Albuquerque, NM
87107.

Jeorger Enterprises, 166 Laurel Rd., East Northport, NY
11731.

Kinetic Systems, 11 Maryknoll Dr., Lockport, IL 60441.

Le Croy Research Systems, 700 S. Main St., Spring Valley,
NY 10977

Nuclear Enterprises, 935 Terminal Way, San Carlos, CA 94070.


## Capacitors, Piston (non-magnetic)

Johanson Manufacturing, 400 Rockaway Valley Rd., Boonton,
NJ 07005.

Voltronics, West St., East Hanover, NJ 07936.


## Capacitors, Vacuum Variable

Jennings Division, ITT; 970 McLaughlin Ave., San Jose, CA
95122.


## Digitizers (high speed)

Bi Ra, see under CAMAC.

Bruker Instruments, Manning Park, Billerica, MA 01821.

Gould, Inc. (Biomation), Santa Clara Operation, 4600 Old
Ironside Dr., Santa Clara, CA 95050.

Le Croy, See under CAMAC.

Nicolet Instrument Corp., 5225 Verona Rd., Madison, WI 53711.

Physical Data, 8220 S. W. Nimbus, Beaverton, OR   97005.

EG&G Princeton Applied Research, P.O. Box 2565, Princeton, NJ 08540.

Tektronix, P.O. Box 1700, Beaverton, OR   97075.  (ultra high speed)

## Dip Meters

Heath Co., Benton Harbor, MI   49022.

## Directional Watt Meters

Bird Electronics Corp. 30303 Aurora Rd., Cleveland, OH 44139.

Heath Co., see under Dip Meters.

## Filters, Analog

CIRQTEL, 10504 Wheatley St., Kensington, MD 20795.

Rockland Systems Corporation, Rockleigh Industrial Park, Rockleigh, N.J.   07647.

## Filters, Crystal

Tyco Crystal Products Div., 3940 W. Montecido, Phoenix, AZ 85019.

## Frequency Synthesizers (high frequency)

GenRad, 300 Baker Ave., Concord, MA   01742.

Hewlett-Packard, 1820 Embarcadero Rd., Palo Alto, CA 94303.

Programmed Test Sources, 194 Old Pickard Rd., Concord, MA 01742.

Rockland Systems Corp., see under Filters, Analog.

## Inductors, Variable

Mallory Timers Co., 3029 E. Washington St., Indianapolis, IN
    46206.

Multronics, 12307 Washington Ave., Rockville, MD   20852.

## Molecular Sieves (Oxysorb)

MG Scientific, 1100 Harrison Ave., Kearny, NJ   07029

## NMR Components

Arenberg Sage, Inc., P. O. Box 250P, 57 Cornwall St.,
    Jamaica Plain, MA   02130.

Chemagnetics, Inc., 208 Commerce·Dr., Ft. Collins, CO 80524.
    (including magic angle spinning apparatus)

F&H, 10535 Cambridge Ct., Gaithersberg, MD   20760.

MATEC, 60 Montebello Rd., Warwick, RI   02886.

Novex, Inc., P.O. Box 3006, Gaithersburg, MD 20760.

SEIMCO, Box 51, Parnassus Station, New Kensington, PA
    15068.

Spin-Lock Electronics, 403-28 Helene St. N., Port Credit,
    Ontario, Canada L5G 3B7.

Spin-Tech Electronics, 2951 Eddystone Crescent, North
    Vancouver, British Columbia, Canada V7H 1B8.

## Passive RF Components

Anzac Division, Adams-Russell, 80 Cambridge St., Burlington,
    MA   01803.

Merrimac, 41 Fairfield Pl., West Caldwell, NJ   07006.

Mini-Circuits Laboratory, 837-843 Utica Ave., Brooklyn, NY
    11203.

Vari-L Co., 3883 Monaco Parkway, Denver, CO   80207.

## Q-Meters

Hewlett Packard, see under Frequency Synthesizers.

## Spectrometers (analytical)

IBM Instrument Systems, 1000 Westchester Ave. White Plains, NY 10604.

Perkin-Elmer Corp., Main Ave., Norwalk, CN 06856.

Praxis Corp., 5420 Jackwood, San Antonio, TX 78238.

## Spectrometers (research and analytical)

Bruker Instruments, see under Digitizers.

JEOL USA, 235 Birchwood Ave., Cranford, NJ 07016.

Nicolet Technology, 145 East Dana St., Mountain View, CA 94041.

Varian, Instrument Division, 611 Hansen Way, Palo Alto, CA 94303

## Superconducting Magnets for NMR

Bruker Instruments, see under Digitizers.

Cryomagnet Systems, 5508 Elmwood Ave. (#410), Indianapolis, IN 46203.

Nicolet Magnetics, 2300B Stanwell Dr., Concord, CA 94520.

Oxford Instruments, 9130-H Red Branch Rd., Columbia, MD 21045.

## Timing Simulator

Interface Technology, 150 East Arrow Highway, San Dimas, CA 91773.

## Variable Temperature Accessories

Air Products and Chemicals, Inc., APD-Cryogenics, P. O. Box
    2802, Allentown, PA  18105.  (cryogenic temperature
    apparatus)

Janis Research Co., Inc., 22 Spencer St., Stoneham, MA
    02180.  (variable temperature cryogenic dewars)

Lake Shore Cryotronics, Inc., 64 E. Walnut St., Westerville,
    OH  43081.  (temperature sensors and controllers)

Linear Research, Inc., 5231 Cushman Pl., Suite 21, San
    Diego, CA  92110.  (temperature controllers and liquid
    nitrogen level controllers)

Pope Scientific, Inc., N90 W14337 Commerce Dr., Menomonee
    Falls, WI  53051.  (glass and quartz dewars)

## Vector Impedance Meters (Vector Volt Meters, Network Analyzers)

Hewlett-Packard, see under Frequency Synthesizers.

# APPENDIX E.   ABBREVIATIONS

| | |
|---|---|
| ADC | analog to digital converter |
| A/D | see ADC |
| ADRF | adiabatic demagnetization in the rotating frame |
| CAT | computer of average transients |
| CP/MAS | cross polarization magic angle spinning |
| CPMG | Carr-Purcell Meiboom-Gill sequence |
| cw | continuous wave |
| DAC | digital to analog converter |
| D/A | see DAC |
| DBM | double balanced mixer |
| DEFT | driven equilibrium Fourier transform |
| EFG | electric field gradient |
| EPR | see ESR |
| ENDOR | electron nucleus double resonance |
| ESR | electron spin resonance |
| FFT | fast Fourier transform |
| FID | free induction decay |
| FT | Fourier transform |
| IF | intermediate frequency |
| IRSE | inversion recovery spin echo |
| $\ell$-He | liquid helium |
| $\ell$-N$_2$ | liquid nitrogen |
| MAS | magic angle spinning |
| NAR | nuclear acoustic resonance |
| NMR | nuclear magnetic resonance |
| NOE | nuclear Overhauser enhancement |
| NQR | nuclear quadrupole resonance |

| | |
|---|---|
| PRFT | partially relaxed Fourier transform |
| PSD | phase-sensitive detector |
| QD | quadrature detection |
| rf | radio frequency |
| SC | superconducting |
| SEFT | spin echo Fourier transform |
| S/N | signal-to-noise ratio |
| $T_1$ | spin-lattice relaxation time |
| $T_{1D}$ | dipolar spin-lattice relaxation time |
| $T_{1\rho}$ | spin-lattice relaxation time in the rotating frame |
| $T_2$ | spin-spin relaxation time |
| VOM | volt-ohm meter |
| VT | variable temperature |
| WEFT | water elimination Fourier transform |

Numbers underlined designate those page numbers on which the complete literature citations are given.